Prospects for Antisense Nucleic Acid Therapy of Cancer and AIDS

Prospects for Antisense Nucleic Acid Therapy of Cancer and AIDS

Editor

Eric Wickstrom

*Departments of Chemistry,
Biochemistry and Molecular Biology,
and Surgery
University of South Florida
Tampa*

A JOHN WILEY & SONS, INC., PUBLICATION

New York • Chichester • Brisbane • Toronto • Singapore

Address All Inquiries to the Publisher
Wiley-Liss, Inc., 605 Third Avenue, New York, NY 10158-0012

Copyright © 1991 Wiley-Liss, Inc.

Printed in the United States of America.

While the authors, editors, and publisher believe that drug selection and dosage and the specifications and usage of equipment and devices, as set forth in this book, are in accord with current recommendations and practice at the time of publication, they accept no legal responsibility for any errors or omissions, and make no warranty, express or implied, with respect to material contained herein. In view of ongoing research, equipment modifications, changes in governmental regulations and the constant flow of information relating to drug therapy, drug reactions and the use of equipment and devices, the reader is urged to review and evaluate the information provided in the package insert or instructions for each drug, piece of equipment or device for, among other things, any changes in the instructions or indications of dosage or usage and for added warnings and precautions.

Recognizing the importance of preserving what has been written, it is a policy of John Wiley & Sons, Inc., to have books of enduring value published in the United States printed on acid-free paper, and we exert our best efforts to that end.

Library of Congress Cataloging-in-Publication Data

Prospects for antisense nucleic acid therapy of cancer and AIDS /
    editor, Eric Wickstrom.
        p.   cm.
    Includes bibliographical references and index.
    ISBN 0-471-56880-5
    1. Antisense nucleic acids—Therapeutic use—Testing. 2. Cancer—
Genetic aspects. 3. AIDS (Disease)—Genetic aspects.
4. Oncogenes. 5. Gene expression. I. Wickstrom, Eric.
    [DNLM: 1. Acquired Immunodeficiency Syndrome—drug therapy.
2. Neoplasms—drug therapy. 3. Oligonucleotides, Antisense—
therapeutic use. QZ 267 P966]
RC271.A69P76 1991
616.99'4061—dc20
DNLM/DLC
for Library of Congress   91-22880
                                                                                            CIP

Contents

Contributors

Sudhir Agrawal, Worcester Foundation for Experimental Biology, Shrewsbury, MA 01545 **[143]**

Adrienne L. Block, Department of Biochemistry and Molecular Biology, Louisiana State University Medical Center, New Orleans, LA 70119; present address: Laboratory of Molecular Ophthalmology, Departments of Biochemistry and Molecular Biophysics and Ophthalmology, College of Physicians and Surgeons, Columbia University, New York, NY 10032 **[71]**

Samuel Broder, Clinical Oncology Program, National Cancer Institute, National Institutes of Health, Bethesda, MD 20892 **[159]**

Bruno Calabretta, Department of Pathology and Fels Cancer Research Institute, Temple University School of Medicine, Philadelphia, PA 19140; present address: Department of Microbiology, Jefferson Cancer Institute, Thomas Jefferson University School of Medicine, Philadelphia, PA 19107 **[53]**

Esther H. Chang, Departments of Pathology and Surgery, Uniformed Services University of the Health Sciences, Bethesda, MD 20814-4799 **[115]**

Frederick O. Cope, Ross Laboratories, Columbus, OH 43215 **[125]**

Geneviève Degols, UA CNRS 1191, Université Montpellier II, Sciences et Techniques du Languedoc, 34095 Montpellier Cedex 5, France **[25]**

Peter B. Dervan, Division of Chemistry and Chemical Engineering, California Institute of Technology, Pasadena, CA 91125 **[227]**

John P. Doucet, Department of Biochemistry and Molecular Biology, Louisiana State University Medical Center, New Orleans, LA 70119 **[71]**

Ross H. Durland, Center for Biotechnology, Baylor College of Medicine, The Woodlands, TX 77381 **[219]**

A. Felder, Ciba-Geigy AG, Division Pharma, CH-4002 Basel, Switzerland **[63]**

Alan M. Gewirtz, Hematology Sections, Departments of Pathology and Internal Medicine, University of Pennsylvania School of Medicine, Philadelphia, PA 19104 **[53]**

P. Herrlich, Kernforschungszentrum Karlsruhe, Institut für Genetik und für Toxikologie, and Institute of Genetics, University of Karlsruhe, D-7500 Karlsruhe 1, Germany **[63]**

Michael Hogan, Center for Biotechnology, Baylor College of Medicine, The Woodlands, TX 77381 **[219]**

B. Kaina, Institute of Genetics, University of Karlsruhe, D-7500 Karlsruhe 1, Germany **[63]**

Donald J. Kessler, Center for Biotechnology, Baylor College of Medicine, The Woodlands, TX 77381 **[219]**

Dmitri G. Knorre, Novosibirsk Institute of Bioorganic Chemistry, Siberian Division of the USSR Academy of Sciences, Novosibirsk 90, 630090, USSR **[195]**

Bernard Lebleu, UA CNRS 1191, Université Montpellier, Sciences et Techniques du Languedoc, 34095 Montpellier Cedex 5, France **[25]**

The numbers in brackets are the opening page numbers of the contributors' articles.

Jean-Paul Leonetti, UA CNRS 1191, Université Montpellier II, Sciences et Techniques du Languedoc, 34095 Montpellier Cedex 5, France **[25]**

Lee Leserman, Centre d'Immunologie, INSERM-CNRS de Marseille-Luminy, 13288 Marseille Cedex 9, France **[25]**

Patrick Machy, Centre d'Immunologie, INSERM-CNRS de Marseille-Luminy, 13288 Marseille Cedex 9, France **[25]**

L. James Maher III, Division of Chemistry and Chemical Engineering, California Institute of Technology, Pasadena, CA 91125; present address: Eppley Institute for Research in Cancer and Allied Diseases, University of Nebraska Medical Center, Omaha, NE 68198-6805 **[227]**

Makoto Matsukura, Department of Pediatrics, Kumamoto University Medical School, Kumamoto 860, and Department of Pediatrics, Ashikita Institute for Handicapped Children, Ashikita 869-54, Kumamoto Prefecture, Japan; present address: Department of Child Development, Kumamoto University Medical School, Kumamoto City 860, Japan **[159]**

Nadir Mechti, UA CNRS 1191, Université Montpellier II, Sciences et Techniques du Languedoc, 34095 Montpellier Cedex 5, France **[25]**

Dan Mercola, Department of Pathology, University of California at San Diego, La Jolla, CA 92093, and the Veterans Administration Medical Center, San Diego, CA 92161 **[83]**

Paul S. Miller, Department of Biochemistry, School of Hygiene and Public Health, The Johns Hopkins University, Baltimore, MD 21205 **[115]**

Hiroaki Mitsuya, Clinical Oncology Program, National Cancer Institute, National Institutes of Health, Bethseda, MD 20892 **[159]**

Susan L. Morrow, Department of Biochemisty and Molecular Biology, Louisiana State University Medical Center, New Orleans, LA 70119 **[71]**

Michael Pawlita, Institut für Virusforschung, Deutsches Krebsforschungszentrum, Im Neuenheimer Feld 280, D-6900 Heidelberg, Germany **[179]**

H. Ponta, Kernforschungszentrum Karlsruhe, Institut für Genetik und für Toxikologie, D-7500 Karlsruhe 1, Germany **[63]**

Karola Rittner, Institut für Virusforschung, Deutsches Krebsforschungszentrum, Im Neuenheimer Feld 280, D-6900 Heidelberg, Germany **[179]**

A. Schönthal, Institute of Genetics, University of Karlsruhe, D-7500 Karlsruhe 1, Germany **[63]**

Georg Sczakiel, Institut für Virusforschung, Deutsches Krebsforschungszentrum, Im Neuenheimer Feld 280, D-6900 Heidelberg, Germany **[179]**

Stephen P. Squinto, Department of Biochemisty and Molecular Biology, Louisiana State University Medical Center, New Orleans, LA 70119; present address: Regeneron Pharmaceuticals, Inc., Tarrytown, NY 10591 **[71]**

L. David Tomei, Comprehensive Cancer Center, Ohio State University, Columbus, OH 43210 **[125]**

S. van den Berg, Institute of Genetics, University of Karlsruhe, D-7500 Karlsruhe 1, Germany **[63]**

Valentin V. Vlassov, Institute of Bioorganic Chemistry, Siberian Division of the Academy of Sciences of the USSR, Novosibirsk 630090, USSR **[243]**

Eric Wickstrom, Departments of Chemistry, Biochemistry and Molecular Biology, and Surgery, University of South Florida, Tampa, FL 33620 **[7]**

John J. Wille, Convatec-Squibb, Skillman, NJ 08558 **[125]**

Barbara Wold, Division of Biology, California Institute of Technology, Pasadena, CA 91125 **[227]**

Leonid A. Yakubov, Institute of Bioorganic Chemistry, Siberian Division of the Academy of Sciences of the USSR, Novosibirsk 630090, USSR **[243]**

Kazushige Yokoyama, Gene Bank and Frontier Research Program, Tsukuba Life Science Center, RIKEN (The Institute of Physical and Chemical Research), Tsukuba, Ibaraki 305, Japan **[35]**

Paul C. Zamecnik, Worcester Foundation for Experimental Biology, Shrewsbury, MA 01545 **[1]**

Valentina F. Zarytova, Novosibirsk Institute of Bioorganic Chemistry, Siberian Division of the USSR Academy of Sciences, Novosibirsk 90, 630090, USSR **[195]**

Foreword

Since the earliest attempts by Zamecnik and Stephenson more than ten years ago to inhibit gene expression using antisense oligodeoxynucleotides, it has become clear that the introduction of an antisense oligodeoxynucleotide could be used to investigate normal and pathological gene function and possibly to inhibit the expression of genes involved in pathologic processes as different as viral infection or malignant transformation. Wickstrom et al. have used a 15-mer antisense oligodeoxynucleotide complementary to the c-*myc* initiation codon and downstream sequences to demonstrate that c-*myc* expression is required throughout the cell cycle and that triggering of myeloid differentiation of HL-60 leukemic cells is the result of down regulation of c-*myc* expression. Zamecnik and colleagues have synthesized oligodeoxynucleotides targeted against RNA encoded by HIV, the etiologic agent of acquired immunodeficiency syndrome (AIDS), to obtain substantial inhibition of viral replication and expression, thus demonstrating the usefulness of synthetic oligodeoxynucleotides as inhibitors of viral gene expression. Recently synthetic oligodeoxynucleotides have also been utilized to inhibit the expression of many cellular genes, especially those encoding proteins involved in the regulation of cell proliferation. For example, Heikkila et al. reported that a 15-mer oligodeoxynucleotide complementary to the c-*myc* initiation codon and downstream region specifically inhibited mitogen-induced c-*myc* protein expression in T-lymphocytes and prevented entry into S phase. A similar effect on T-lymphocyte proliferation was obtained by Gewirtz et al., who have shown inhibition of entry into S phase of mitogen- or antigen-stimulated purified lymphocytes in the presence of synthetic oligodeoxynucleotides complementary to c-*myb* mRNA from codons 2–7. Inhibition of proliferating cell nuclear antigen (PCNA) expression was associated with cessation of the proliferation of BALB/c 3T3 mouse fibroblasts, further demonstrating that antisense oligodeoxynucleotides are very useful to assess the functional role of gene products implicated in the regulation of cellular proliferation. Another important application of this approach could be the study of the functional role of cancer suppressor genes (antioncogenes). This could provide some significant new information on the mode of action of these genes.

Thus, the antisense oligodeoxynucleotide approach shows remarkable applications in studies of gene function and of the role of cellular and viral genes in normal and pathologic processes. I strongly believe that this approach could also be exploited very effectively to investigate the functional role of genes involved in mammalian development and in cell differentiation.

I would like to stress the important implications of the use of antisense oligonucleotides in human therapy. As I mentioned above, it is possible by using antisense oligodeoxynucleotides to inhibit the expression and replication of viruses such as HIV. It becomes feasible, therefore, to attempt to eliminate virus replication and spreading by this approach. Even more far-reaching are the implications of the use of synthetic oligodeoxynucleotides in cancer therapy, in particular in the case of malignancies of the hematopoietic system. An obvious problem with the use of antisense oligodeoxynucleotides in human therapy is the necessity to deliver the synthetic oligos to all target cells of a given target tissue. For example, it is difficult to imagine at the present time being able to target all the cells of a solid human tumor with synthetic antisense oligodeoxynucleotides. It is, however, feasible to transfect human bone marrow cells in vitro with the synthetic oligodeoxynucleotides. Thus, this approach could be used to shut off selectively the expression of activated oncogenes in bone marrow, making it suitable for autologous bone marrow transplantation in the treatment of leukemias and lymphomas. The highly specific gene rearrangements observed in most human hematopoietic tumors could be exploited to produce tumor-specific antisense oligodeoxynucleotides to be used in cancer treatment. For example, the use of antisense oligodeoxynucleotides for the BCR–ABL junctions should result in the inhibition of the expression of the BCR–ABL chimeric proteins responsible for the expression of malignancies. Thus the molecular definition of the genetic changes occurring in human cancer and the production of tumor-specific antisense oligodeoxynucleotides should provide very powerful tools in the treatment of human neoplasms.

The recent and exciting progress on the use of synthetic oligodeoxynucleotides in cell biology and virology makes this volume particularly useful for investigators interested in most fields of biology and medicine.

Carlo M. Croce

Preface

Traditional chemotherapy agents for cancer or AIDS are designed to inhibit the enzymes necessary for cell growth or production of new viruses. Chemotherapeutics have strong side effects, which often make patients very ill. The diseased cells often become resistant to existing drugs, and so new approaches are needed. Cloning and sequencing of pathogenic genes over the last decade have made possible a direct genetic approach to the treatment of disease, using nucleic acid therapeutics. The ability to turn off individual disease-causing genes at will in growing cells provides a powerful tool for elucidating the role of a particular gene and for therapeutic intervention when that gene is pathogenically activated.

Gene-specific nucleic acid therapy has gone from theory to practical possibility in a very short time. These new DNA and RNA agents are intended to stop the growth of cancerous cells, or the production of HIV from infected cells, with fewer side effects than before, and to make the transformed cells more easily recognized by the immune system. The purpose of this book is to inform clinical and biomedical investigators of the applications that appear to be plausible at this time, by presenting a few examples of the wide variety of current studies in this field.

In principle, one needs to identify a unique target sequence in the gene of interest and to prepare a complementary oligonucleotide against the target sequence (Belikova et al., 1967). Antisense DNAs were first successfully utilized against Rous sarcoma virus (Zamecnik and Stephenson, 1978). Antisense DNA and RNA have been applied since then to inhibit the expression of many different target genes in viral, bacterial, plant, and animal systems, both in cell-free extracts and in whole cells (Cohen, 1989; Murray, in preparation).

In this volume, a brief history of the field is provided by Dr. Paul Zamecnik, followed by a section on applications against oncogenes, a section on applications against human immunodeficiency virus, and a section on the chemistry, structure, and pharmacology of oligonucleotide derivatives.

In the oncogene section, both DNA and RNA approaches to control of activated oncogenes are presented. Inhibition of c-*myc* is discussed by Dr. Wickstrom, Drs. Leserman, Degols, Machy, Leonetti, Mechti, and Lebleu, and Dr. Yokoyama, while c-*myb* is analyzed by Drs. Gewirtz and Calabretta. The impact of down-regulation of c-*fos* is presented from several points of view by Drs. van den Berg, Schönthal, Felder, Kaina, Herrlich, and Ponta, Drs. Block, Doucet, Morrow, and Squinto, and Dr. Mercola. The frequently activated oncogene c-Ha-*ras* is distinguished from its normal allele with a series of sequence-specific derivatives by Drs. Chang and Miller, and the regulation of retinoid receptor genes is described by Drs. Cope, Wille, and Tomei.

In the HIV section, successful inhibition of viral expression and replication are demonstrated with antisense DNA derivatives by Dr. Agrawal and Drs. Matsukura, Mitsuya, and Broder. Similar efficacy with antisense RNA is reported by Drs. Sczakiel, Rittner, and Pawlita.

In the chemistry section, Drs. Knorre and Zarytova discuss the synthesis of an elegant array of antisense DNA derivatives, including alkylators, intercalators, and cleavers. In view of the opportunity to preclude transcription of a pathogenic gene by formation of a triplex DNA structure, two approaches to triplex formation are presented by Drs. Durland, Kessler, and Hogan, and by Drs. Maher, Dervan, and Wold. Looking ahead to practical problems of therapy, Drs. Vlassov and Yakubov then describe modes of uptake, excretion, and immunological response of some oligonucleotide derivatives.

Investigators in nucleic acid therapeutics are amazed themselves at the pace of progress over the last ten years. At present, several groups are carrying out animal trials that are encouraging but too premature for inclusion in this volume. It is now plausible to predict human trials by the middle of the decade and regular application by the turn of the century.

REFERENCES

Belikova AM, Zarytova VF, Grineva NI (1967): Synthesis of ribonucleosides and diribonucleoside phosphates containing 2-chloroethylamine and nitrogen mustard residues. Tet. Lett. 7:3557–3562.

Cohen J (ed) (1989): Oligodeoxynucleotides. Antisense Inhibitors of Gene Expression. Boca Raton, FL: CRC Press.

Murray JAH (ed) (in preparation): Antisense RNA and DNA. New York: Wiley-Liss.

Zamecnik PC, Stephenson ML (1978): Inhibition of Rous sarcoma virus replication and cell transformation by a specific oligodeoxynucleotide. Proc. Natl. Acad. Sci. U.S.A. 75:280–285.

Eric Wickstrom
University of South Florida

Introduction: Oligonucleotide Base Hybridization as a Modulator of Genetic Message Readout

Paul C. Zamecnik

Worcester Foundation for Experimental Biology, Shrewsbury, Massachusetts 01545

HISTORICAL ASPECTS

With the discovery of the double helical structure of DNA, and of its role in replication (Watson and Crick, 1953), the basis for a new form of chemotherapeutics was set in place. A few years thereafter, transfer RNA was found (Hoagland et al., 1957). Hybridization of a trinucleotide portion of a negatively stranded RNA (the anticodon segment of tRNA) with a complementary trinucleotide of a positively stranded messenger RNA provided the crucial feature of the translation of the genetic code (Hoagland et al., 1959). Puromycin, an amino acid-nucleoside analogue, was then found to cause premature termination of protein synthesis by occupying the A (amino acid) site on the ribosome, by accepting the peptidyl chain of the growing protein, and then aborting the continuation of peptide synthesis by an inability to hybridize with the 50 S ribosomal attachment site for tRNA (Yarmolinsky and de la Haba, 1959; Allen and Zamecnik, 1962). Deciphering of the genetic code (Nirenberg and Matthaei, 1961; Speyer et al., 1963; Khorana, 1968) confirmed the sensitivity and specificity of nucleotide base hybridization. The finding of the stop codons (UAA, UAG, UGA) (Watson et al., 1987) showed once again that oligonucleotide hybridization could act both in a positive sense, by promoting translation of the genetic code, and also in a negative sense, by terminating translation. Streptomycin was found to induce frameshift mutations by intercalation, inducing a crimp in the linear readout sequence, so that a single nucleotide was missed, and different tRNA antisense trinucleotides were chosen thereafter to add to the growing chain. Thus, both negative and positive metabolic effects were early found as features of oligonucleotide hybridization in readout of the genetic code.

In retrospect, it is not surprising that for the past two decades among prokaryotes naturally occurring oligonucleotides were occasionally reported, which from genetic evidence influenced the growth of microorganisms (Mizuno et al., 1983). More lately in eukaryotes also, naturally occurring oligonucleotides of unknown function have been reported (Kowalski and Denhardt, 1978; Kolodny, 1986; Plesner et al., 1987; Benner, 1988). It was also found that negatively sensed oligonucleotides were able to inhibit protein synthesis in cell-free systems (Paterson et al., 1977; Hastie and Held, 1978).

The relationship of hybridization competition to chemotherapeutics, however, grew out of somewhat different roots. The realization dawned in the early 1970s that oncogenic viruses were either DNA viruses or RNA viruses, and that the avian oncogenic viruses were predominantly the latter. It therefore became important to try to sequence an appropriate avian oncogenic virus.

The first of these to become available in sufficient amounts for sequencing was the Rous sarcoma virus (RSV), made available to investigators through the offices of the ill-fated National Cancer Institute Virus Cancer Program of the early 1970s. In our labora-

Prospects for Antisense Nucleic Acid Therapy of Cancer and AIDS, pages 1–6
©1991 Wiley-Liss, Inc.

tories Drs. Dennis Schwartz and H. Lee Weith struggled for 3 years to sequence a region just inside the poly(A) segment we had earlier found at the 3′-end of RSV. They finally deduced the sequence of 21 nucleotides, at which time we heard that Dr. Walter Gilbert's laboratory, across the river in another part of Harvard, had developed a new and faster nucleotide sequencing technique they would be pleased to share with us. They had in fact begun to sequence the 5′-end of RSV, unbeknown to us, and had deduced 101 residues in about 6 weeks. On comparing the 5′- and 3′-sequences of RSV thus clarified, both groups were astonished to find that 21 nucleotides at the 3′-end of RSV, just inside the poly(A) tail, were identical and in the same polarity as 21 nucleotides at the 5′-end, just inside the 5′-cap (Haseltine et al., 1977; Schwartz et al., 1977). It was already known at that time that RNase H hydrolyzed the 5′-end of the viral RNA as far in the 3′ direction as the primer tRNA^{Lys}3, once the complementary DNA strand had been synthesized by reverse transcriptase. Thus a piece of nascent single-stranded DNA was available for hybridization with the reiterated complementary piece of viral RNA still present at the 3′-end of the RSV. There was also a second identical linear strand of RSV packaged in the same virion, with its 3′-end available for hybridization.

Dr. Schwartz elected to pursue the sequencing of the entire 5000-odd base linear structure of RSV in the Gilbert laboratory. He and his colleagues achieved this by extraordinary efforts in the next 3 years (Schwartz et al., 1983) in a piece of work that served as a model for subsequent retroviral sequencing, including that of HTLV-I and HIV.

The other feature of the reiterated terminal sequences that fascinated us was the possibility that viral DNA reverse transcription might continue from the 3′-end of either the same viral strand or its virally encapsulated accompanying mate. There was in fact evidence available from electron microscopy of a circularized DNA. The meaning of this circularized intermediate is still obscure, since linear viral DNA appears at present to be the immediate precursor of the recipient cell integrated viral genome.

At any rate, it appeared logical to wonder whether the purpose of the reiterated sequences at the 5′- and 3′-termini of the RSV linear genome strand might be to permit continuation of reverse transcription. It had seemed curious to the writer that reverse transcription should begin close to the 5′-end of the genome rather than at the 3′-end. It was, however, pointed out that a mechanism must exist for copying those nucleotides complementary to the 3′-end of tRNA^{Lys}3, and located underneath this tRNA, which served as the specific natural primer for the reverse transcriptase. The mechanism chosen by RSV achieved the copying of this piece of the RSV genome. What a complicated arrangement, and why tRNA^{Lys}3, the latter bearing two unusual minor bases (Raba et al., 1979), one located next to the anticodon region?

It was generally thought that oligonucleotides were unable to traverse the outer cell membrane and enter living cells. Some of this information had come from studies of 2′,5′-oligonucleotides involved in the interferon story. In addition, the highly negative charge of nucleoside mono-, di-, and triphosphates had been known for years to create a barrier to their cell entry.

We found, however, that when a 13-mer oligodeoxyribonucleotide, complementary to a segment of the 3′-terminal sequence of RSV, was synthesized by solution chemistry, purified, and incubated with chick embryo fibroblasts in the presence of RSV added at the same time, that replication of RSV was inhibited (Zamecnik and Stephenson, 1978) and translation halted (Stephenson and Zamecnik, 1978). It was also clear to us from the work of Wickstrom (1986) and from our own (Zamecnik et al., 1986), that deoxyribooligomer was being rapidly degraded by nucleases. We therefore blocked the 5′- and 3′-hydroxyl groups of the oligomer as the isourea derivatives. This terminally blocked oligomer was 5–10 times as effective as an inhibitor on a molar basis as the unblocked, unmodified oligomer.

There had also been a sizeable literature

running back to the early 1970s that homo-oligomeric ribooligomers were able to enter cells. As in our case, however, the nagging doubt persisted that the oligomers were really not freely circulating in the cytosol or nuclear domain of the cell, but were only trapped in deep membranous crypts. One piece of genetic evidence supported our hope that genuine cell entry of deoxyoligomers occurred. We found that although RSV infected fresh chick embryo explants were regularly transformed into RSV transformants when planted in loose agar, that treatment with the 3′-complementary oligomer prevented transformation. This observation added support to the consideration that the complementary oligomer entered the chick fibroblast and prevented transformation by interfering with the viral replication process. We called an oligonucleotide capable of altering the genetics or metabolism of a living cell by means of complementary hybridization a "hybridon," a term that over the years has become politely buried.

There was little interest in this approach to therapeutics during the next half dozen years, and a stack of reprints of this work still gathers dust in our files. Around 1984 however an exciting development in the "antisense" field, as it came to be known by the molecular biologists, grew out of experiments of Izant and Weintraub (1984), Inouye (Green et al., 1986), and Melton (1985) in particular, who excised 1000-plus monomer length segments of double-stranded DNA from living organisms, religated them in reverse (or antisense) direction into plasmids, and by biological means achieved integration of these segments of DNA into the natural DNA of the accepting living organism. The antisense-oriented segments of DNA thus became part of the genetic equipment of the acceptor cell. With the correct promoter equipment in place, they could therefore be expressed along with other naturally occurring DNA. This antisense DNA therefore was endowed with inheritable properties, and could inhibit chosen sectors of the parent genome indefinitely. This feature of the antisense oligonucleotide inhibitory principle

was graphically displayed in changes of petunia color, due to inhibition of the flavanoid instruction gene of the petunia plant (van der Krol et al., 1988).

A confluence of two streams of genetic discoveries therefore occurred in the mid-1980s, one coming from advances in synthesis of oligonucleotides due to the new solid phase technique (Caruthers, 1989; Gait, 1984), involving strict nucleic acid synthetic chemistry, and the other as a continuation of the genetic revolution set in motion by Cohen et al. (1973) and Berg (1981) plus Nathans and Smith (1975).

Sharp distinctions exist between the potentialities of the two techniques. The synthetic small antisense oligomers are potential chemotherapeutic agents, with relatively short half lives, and with phenotypic effects. They can be supplied in $10^{-4}-10^{-9} M$ concentrations, which upon cell entry temporarily overwhelm by numbers the short segments of instructional or sense portions of the cell genetic apparatus with which they hybridize. The concentrations of these synthetic antisense oligomers may with some comfort be calculated to be two or more orders of magnitude greater than the usual intracellular viral or cellular genome targets. The assumption is made that virtually every cell of the target tissue is presented with this potential for inhibitory hybridization. The limitation is that due to degradation or excretion of oligomer over a period of time the potential for hybridization inhibition decreases. There is little or no published evidence of insertion into the cellular or viral genome of such oligomeric DNA fragments, although this possibility must not be overlooked.

On the other hand, a more lengthy piece of DNA, inserted in reverse orientation into a plasmid or retroviral vector, becomes built into the genome of a (generally) small proportion of the recipient cells, and remains a permanent part of the genetic equipment of the cell. It is thus, under favorable promoter circumstances, expressed indefinitely, and is part of the genotype of the cell. The synthetic oligomeric case, therefore, induces phenotypic changes, while the vector-induced genetic insertion results in genotypic changes. These two approaches may be expressed roughly as complementary potential forms of therapy of human disease: (1) oligonucleotide chemotherapy and (2) the potential cure of genetic disease by a genetic repair mechanism.

A host of potential targets for antisense therapy has become available during the past few years. As soon as a new piece of a genome of a living organism becomes known, a new potential target for antisense inhibition or modulation presents itself. These genetic targets may be divided into two general classes, as follows: (1) genetic instructions for replication or expression of the primary structure of microorganisms inimicable to human, animal, or plant development, and (2) replication instructions already present in the human, animal, or plant genome that may conceivably be modified by the antisense nucleic acid technique. In discussion of the first-mentioned class of target, some of the genetic instructions of viruses would be expected to be unique for them, or at least less likely to be present in the human organism. Thus, an antisense oligonucleotide, directed against a particular protein expressed by a virus, would less probably encounter an identical segment of nucleotides in the host genome. Thus, the viruses, protozoa, and other microorganisms pathogenic for humans have become desirable targets for chemotherapy.

As for the possibility of inhibiting human or animal genes that are overexpressed in certain diseases, the potential exists of "overinhibiting" a gene that has a useful function, to the detriment of the host organism. This approach nevertheless has attracted attention as a possible therapeutic aid against such diverse diseases as diabetes mellitus, Alzheimer's disease, rheumatoid arthritis, and transplantation immunity—to mention a few possibilities.

These are of course limitations to the antisense oligonucleotide approach: More experiments resulting in failures of antisense oligomers to inhibit a targeted segment of genetic expression have been performed than the successes reported in the literature. Three general reasons for these failures may be mentioned:

1. The oligonucleotide may not enter the cell. Our own experiments indicate that oligomers do traverse the living cell membrane, in eukaryotic cells, with the exception of adult human erythrocytes. There are few published data on permeability of bacterial cell walls, however; and the blood–brain barrier has been little explored. Considerable efforts have been expended on improving cell penetrability by means of liposomes or hydrophobic moieties attached to the oligomers (Letsinger et al., 1989; Boutorin et al., 1989).

2. The oligonucleotides may be promptly degraded either by nucleases outside the cell as in the case of tissue cultures with media fortified with fetal calf serum, or within the cell. Thus, it is generally agreed that nuclease resistant modified internucleoside phosphate linkages are advantageous (Matsukura et al., 1988; Agrawal et al., 1988, 1989), since they prolong the internal half-life of the oligomer and increase the likelihood of successful base hybridization. Against this advantage must be balanced the potentiality of enhanced toxicity.

3. The most troublesome imponderable in selection of an antisense synthetic oligomer is whether the region of primary structure selected is free of secondary and tertiary structure, or at least sufficiently free to make competitive hybridization possible. Computer realizations of probable secondary structure of single-stranded regions of DNA or RNA are useful guides to target selection (Wickstrom et al., 1988) and will be increasingly used in the future.

In general, the antisense oligonucleotides have an ID_{50} in tissue culture systems of 10^{-5}–$10^{-8} M$. Among nuclease resistant oligomers, the phosphorothioates have been most explored (Matsukura et al., 1988; Agrawal et al., 1988, 1989), and reach the lower therapeutic range. Cells already infected with a virus such as HIV are more resistant to tissue culture chemotherapy than freshly infected cells by a factor of around 5 (Agrawal et al., 1989). It would appear from a commercial viewpoint that enhancement of the therapeutic index by a factor of 10 would go far toward making the synthetic antisense oligomer approach economically feasible. Thus, much current effort is being devoted by a number of laboratories to efforts to attach to a matched 20–30-mer antisense oligomer a moiety that will withstand the rigors of deblocking, and serve to crosslink with the complementary DNA or RNA strand, cleave the complementary strand once hybridization has taken place, induce a stable intercalation, lie in the major or minor groove of the double helix and serve as a cork, preventing the passage down the groove of enzymes or protein factors, or, finally, my favorite, crosslink with the enzymes and proteins involved in splicing, chain initiation, and similar essential protein and nucleic acid interactions requisite for expression of a pathogenic viral or undesirable cellular protein. It is probable that the length of tethers and the use of model building will prove important in sharpening these future efforts. Details of the varied approaches to these "third generation" modifications of antisense oligomers are given in comprehensive reviews by Uhlmann and Peyman (1990), Goodchild (1990), and Cohen (1989).

ACKNOWLEDGMENTS

Supported by grants from the G. Harold and Leila Y. Mathers Foundation, NIH Grant AI-24846, and Cancer Center Grant P30-12708-17.

REFERENCES

Agrawal S, Goodchild J, Civeira MP, Thornton AH, Sarin PS, Zamecnik PC (1988): Oligodeoxynucleoside phosphoramidates and phosphorothioates as inhibitors of human immunodeficiency virus. Proc. Natl. Acad. Sci. U.S.A. 85:7079–7083.

Agrawal S, Ikeuchi T, Sun D, Konopka A, Maizel J, Zamecnik PC (1989): Inhibition of human immunodeficiency virus in early infected and chronically infected cells by antisense oligonucleotides and their phosphorothioate analogues. Proc. Natl. Acad. Sci. U.S.A. 86:7790–7794.

Allen DW, Zamecnik PC (1962): The effect of pur-

omycin on rabbit reticulocyte ribosomes. Biochim. Biophys. Acta 55:865–874.

Benner SA (1988): Hypothesis, extracellular 'communicator RNA'. FEBS 233:225–228.

Berg P (1981): Dissections and reconstructions of genes and chromosomes. Science 213:296–303.

Boutorin AS, Gus'kova LV, Ivanova EM, Kobetz ND, Zarytova VF, Ryte AS, Yurchenko LV, Vlassov VV (1989): Synthesis of alkylating oligonucleotide derivatives containing cholesterol or phenazinium residues at their 3'-terminus and their interaction with DNA within mammalian cells. FEBS Lett. 254:129–132.

Caruthers MH (1989): Synthesis of oligonucleotides and oligonucleotide analogues. In Cohen J (ed.): Oligodeoxynucleotides, Antisense Inhibitors of Gene Expression. Boca Raton, FL: CRC Press, pp 7–24.

Cohen JS (1989): Oligodeoxynucleotides. Antisense Inhibitors of Gene Expression. CRC Press, Inc., Boca Raton, FL.

Cohen SN, Chang ACY, Boyer HW, and Helling RB (1973): Construction of biologically functional bacterial plasmids *in vitro*. Proc. Natl. Acad. Sci. U.S.A. 70:3240–3244.

Gait MJ (ed.) (1984): Oligonucleotide Synthesis: A Practical Approach. Washington D.C.: IRL Press.

Green PJ, Pines O, Inouye M (1986): The role of antisense RNA in gene regulation. Annu. Rev. Biochem. 55:569–597.

Goodchild J (1990): Conjugates of oligonucleotides and modified oligonucleotides: A review of their synthesis and properties. Bioconjugate Chem. 1:165–187.

Haseltine WA, Maxam AM, Gilbert W (1977): Rous sarcoma virus is terminally redundant: The 5' sequence. Proc. Natl. Acad. Sci. U.S.A. 74:989–993.

Hastie ND, Held WA (1978): Analyses of mRNA populations by cDNA mRNA hybrid-mediated inhibition of cell-free protein synthesis. Proc. Natl. Acad. Sci. U.S.A. 75:1217–1221.

Hoagland MB, Stephenson ML, Zamecnik PC (1957): Intermediate reactions in protein synthesis. Biochim. Biophys. Acta 24:215–216.

Hoagland MB, Zamecnik PC, Stephenson ML (1959): A hypothesis concerning the roles of particulate and soluble ribonucleic acids in protein synthesis. In Zirkle RE (ed.): A Symposium on Molecular Biology. Chicago: University of Chicago Press, pp 105–114.

Izant JG, Weintraub H (1984): Inhibition of thymidine kinase gene expression by antisene RNA: A molecular approach to genetic analysis. Cell 36:1007–1115.

Khorana HG (1968): Nucleic acid synthesis in the study of the genetic code. In Nobel Lectures: Physiology or Medicine (1963–1970). Amsterdam: Elsevier.

Kolodny GM (1986): Very small molecular weight eukaryotic RNA. IRCS Med. Sci. 14:1123–1124.

Kowalski J, Denhardt D (1978): Most short DNA molecules isolated from 3T3 cells are not nascent. Nucl. Acids Res. 5:4355–4373.

Letsinger RL, Zhang G, Sun DK, Ikeuchi T, Sarin PS (1989): Cholesteryl-conjugated oligonucleotides: Synthesis, properties, and activity as inhibitors of replication of human immunodeficiency virus in tissue culture. Proc. Natl. Acad. Sci. U.S.A 86:6556.

Matsukura M, Zon G, Shinozuka K, Stein CA, Mitsuya H, Cohen JS, Broder S (1988): Synthesis of phosphorothioate analogues of oligodeoxyribonucleotides and their antiviral activity against human immunodeficiency virus. Gene 72:343–347.

Melton DA (1985): Injected anti-sense RNAs specifically block messenger RNA translation *in vivo*. Proc. Natl. Acad. Sci. U.S.A. 82:144–148.

Mizuno T, Chou M-Y, Inouye M (1983): A unique mechanism regulating gene expression: Translational inhibition by a complementary RNA transcript (micRNA). Proc. Natl. Acad. Sci. U.S.A. 81:1966–1970.

Nathans D, Smith HO (1975): Restriction endonucleases in the analysis and restructuring of DNA molecules. Annu. Rev. Biochem. 44:273–293.

Nirenberg MW, Matthaei JH (1961): The dependence of cell-free protein synthesis in E. coli upon naturally occurring or synthetic polyribonucleotides. Proc. Natl. Acad. Sci. U.S.A. 47:1588–1602.

Paterson BM, Roberts BE, Kuff EL (1977): Structural gene identification and mapping by DNA·mRNA hybrid-arrested cell-free translation. Proc. Natl. Acad. Sci. U.S.A. 74:4370–4374.

Plesner P, Goodchild J, Kalckar HM, Zamecnik PC (1987): Oligonucleotides with rapid turnover of the phosphate groups occur endogenously in eukaryotic cells. Proc. Natl. Acad. Sci. U.S.A. 84:1936–1939.

Raba M, Limburg K, Burghagen M, Katze JR, Simsek JRM, Heckman JE, Rajbhandary UL, Gross HJ (1979): Nucleotide sequence of three accepting lysine tRNAs from rabbit liver and SV40 transformed mouse fibroblasts. Eur. J. Biochem. 97:305–318.

Schwartz D, Zamecnik PC, Weith HL (1977): Rous sarcoma virus genome is terminally redundant: The 3′ sequence. Proc. Natl. Acad. Sci. U.S.A. 74:994–998.

Schwartz DE, Tizard R, Gilbert W (1983): Nucleotide sequence of Rous sarcoma virus. Cell 32:853–869.

Speyer JF, Lengyel P, Basilico C, Wahba AJ, Gardner RS, Ochoa S (1963): Synthetic polynucleotides and the amino acid code. Cold Spring Harbor Symp. Quant. Biol. 28:559–568.

Stephenson ML, Zamecnik PC (1978): Inhibition of Rous sarcoma viral RNA translation by a specific oligoribonucleotide. Proc. Natl. Acad. Sci. U.S.A. 75:285–288.

Uhlmann E, Peyman A (1990): Antisense oligonucleotides: A new therapeutic principle. Chem. Rev. 90:543–584.

van der Krol AR, Lenting PE, Veenstra J, Meer IM, van der Koes RE, Gerats AGM, Mol JNM, Stuitje AR (1988): Expression of an antisense CHS gene in transgenic plants inhibits flower pigmentation. Nature (London) 333:866–869.

Watson JD, Crick FHC (1953): Molecular structure of nucleic acids: A structure for desoxyribonucleic acids. Nature (London) 171:737–738.

Watson JD, et al. (1987): Molecular Biology of the Gene, Vol. 1, 4th ed. Reading, MA: Benjamin/Cummings, p 443.

Wickstrom E (1986): Oligonucleotide stability in subcellular extracts and culture media. J. Biochem. Biophys. Methods 13:97–102.

Wickstrom EL, Bacon TA, Gonzalez A, Freeman DL, Lyman GH, Wickstrom E (1988): Human promyelocytic leukemia HL-60 cell proliferation and c-*myc* protein expression are inhibited by an antisense pentadecadeoxynucleotide targeted against c-*myc* mRNA. Proc. Natl. Acad. Sci. U.S.A. 85:1028–1032.

Yarmolinsky MB, de la Haba GL (1959): Inhibition by puromycin of amino acid incorporation into protein. Proc. Natl. Acad. Sci. U.S.A. 45:1721–1729.

Zamecnik PC, Stephenson ML (1978): Inhibition of Rous sarcoma virus replication and cell transformation by a specific oligodeoxyribonucleotide. Proc. Natl. Acad. Sci. U.S.A. 75:280–284.

Zamecnik PC, Goodchild J, Taguchi Y, Sarin PS (1986): Inhibition of replication and expression of human T-cell lymphotropic virus type III in cultured cells by exogenous synthetic oligonucleotides complementary to viral RNA. Proc. Natl. Acad. Sci. U.S.A. 83:4143–4146.

Antisense DNA Treatment of HL-60 Promyelocytic Leukemia Cells: Terminal Differentiation and Dependence on Target Sequence

Eric Wickstrom

Departments of Chemistry, Biochemistry and Molecular Biology, and Surgery, University of South Florida, Tampa, Florida 33620

INTRODUCTION

Cancerous cells display overexpression or mutant expression of one or more of the genes normally used in cell proliferation. Such genes are called protooncogenes (Bishop, 1987), and are a logical target for antisense inhibition. The protooncogene c-*myc*, an evolutionarily conserved gene found in all vertebrates, has been found to be overexpressed in a wide variety of human leukemias and solid tumors (Klein and Klein, 1986), and is sufficient alone to transform cells to a neoplastic, tumorigenic state (Ramsay et al., 1990). The c-*myc* gene expresses a nuclear protein with an electrophoretic apparent molecular mass of 65 kDa (p65), plus a minor band 2–3 kDa larger (Hann and Eisenman, 1984; Persson et al., 1984; Spector et al., 1987). Overexpression of p65 promotes replication of SV40 DNA (Classon et al., 1987); hence, it appears likely that the c-*myc* gene product plays some direct or indirect role in replication. In agreement with this model, inhibition of c-*myc* p65 expression by an antisense oligodeoxynucleotide targeted against a predicted loop containing the primary ini-

tiation codon of the human c-*myc* mRNA was found to inhibit mitogen-stimulated human peripheral blood lymphocytes from entering S phase, in a sequence-specific, dose-dependent manner (Heikkila et al., 1987).

The HL-60 human promyelocytic leukemia cell line (Fig. 1) consists predominantly of rapidly dividing cells with promyelocytic characteristics (Collins et al., 1978). HL-60 cells are transformed by overexpression of c-*myc* protooncogene, mutational activation of N-*ras* protooncogene, and mutational inactivation of p53 suppressor gene (Collins, 1987). As with the peripheral lymphocytes, down-regulation of c-*myc* expression by treatment of HL-60 cells with the anti-c-*myc* initiation codon oligomer was observed to inhibit their proliferation, in a sequence-specific, dose-dependent manner (Wickstrom et al., 1988).

Untreated cells spontaneously differentiate (10–15%) into forms that exhibit characteristics of more mature granulocytic cells, while the promyelocytes continue to proliferate at a constant rate (Collins et al., 1977). Dimethyl sulfoxide (Me$_2$SO) inhibits the proliferation of this cell line, and induces the cells to terminally differentiate into slowly proliferating granulocytic cells that exhibit morphological and chemical properties similar to more mature myelocytes, metamyelocytes, and banded and segmented neutrophils; 1% Me$_2$SO (v/v) treatment for 5 days

Abbreviations: p65, the 65 kDa c-*myc* protein; FBS, fetal bovine serum; Me$_2$SO, dimethyl sulfoxide; NBT, nitroblue tetrazolium; nt, nucleotide(s); PMA, phorbol 12-myristate 13-acetate; VSV, vesicular stomatitis virus.

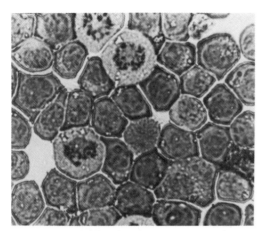

Fig. 1. Human promyelocytic leukemia HL-60 cells, which are transformed by overexpression of c-*myc* protooncogene, mutational activation of N-*ras* protooncogene, and mutational inactivation of p53 suppressor gene (Collins, 1987).

induces 65–78% differentiation (Collins et al., 1978). Phorbol 12-myristate 13-acetate (PMA) at 16 nM, on the other hand, induces HL-60 cells to terminally differentiate within 24 hr into nonproliferating macrophage/monocytic cells that exhibit morphological and biochemical properties similar to more mature promonocytes and monocytes (Rovera et al., 1979).

Overexpression of c-*myc* p65 usually correlates with the inability of the cells to differentiate (Coppola and Cole, 1986), while induction of HL-60 cell differentiation with Me$_2$SO coincides with a decline in c-*myc* mRNA (Westin et al., 1982) and the ability of the HL-60 cells to form colonies in semi-solid medium (Filmus and Buick, 1985). It has recently been shown that the domain of p65 necessary for transformation is also the domain necessary and sufficient for inhibition of differentiation (Freytag et al., 1990). A single dose of the anti-c-*myc* DNA that inhibited proliferation was also found to elicit a sequence-specific decrease in colony formation, and an increase in the number of HL-60 cells differentiating along the granulocytic line from the usual 10% to greater than 20% (Wickstrom et al., 1989). Expression of c-*myc* antisense RNA in HL-60 cells was re-

ported to induce differentiation along the monocytic line (Yokoyama and Imamoto, 1987), while expression of c-*myc* antisense RNA in F9 teratocarcinoma cells induced differentiation toward parietal endodermal cells (Griep and Westphal, 1988).

Hence, it seemed likely that reducing the level of c-*myc* mRNA translation further by daily addition of anti-c-*myc* oligomer might be sufficient to induce terminal differentiation of the entire population of *myc*-transformed cells. It was also observed that oligodeoxynucleotides are rapidly degraded, with a half-life of about 2 hr, in the culture medium used for growing HL-60 cells (Wickstrom et al., 1988), due to nucleases in fetal bovine serum (FBS) (Wickstrom, 1986). Therefore, in the studies reported here, a portion of the HL-60 cell line maintained in our laboratory was adapted to growth in a serum-free medium (Breitman et al., 1980) in order to minimize oligodeoxynucleotide degradation and maximize efficacy.

The serum-dependent and serum-free cultures were both treated with antisense oligodeoxynucleotides, Me$_2$SO, and PMA, and the effects of these treatments were compared. Daily addition of the anti-c-*myc* oligomer more effectively induced granulocytic differentiation, and inhibited proliferation and colony formation, in serum-free medium than in serum-containing medium. Daily addition was much more effective than a single addition, and, in the serum-free medium, was comparable in efficacy to 1% Me$_2$SO, yielding 80% or greater differentiation of the HL-60 population (Bacon and Wickstrom, 1991a). The results suggest that antisense DNA inhibition of c-*myc* p65 expression allows the induction of terminal granulocytic differentiation in HL-60 cells.

The original choice of the c-*myc* mRNA primary initiation codon region as an antisense target was based on calculation of a possible secondary structure (Fig. 2), in the absence of experimental evidence. Calculations of mRNA secondary structures often place the initiation codon in a single-stranded region following a highly basepaired 5'-untranslated leader (Konings et al., 1987).

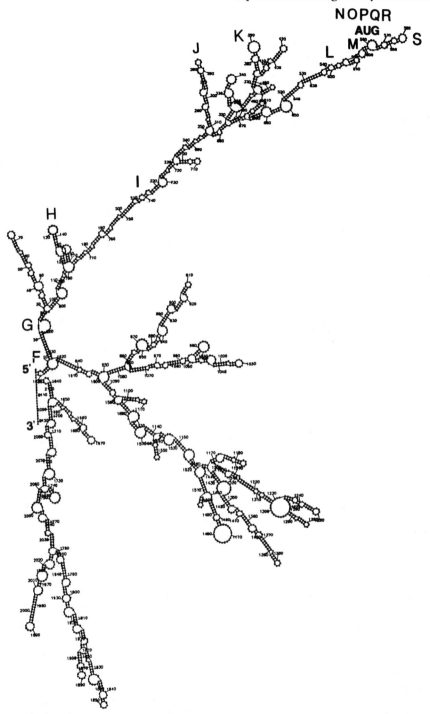

Fig. 2. Predicted secondary structure of the entire 2121-nt human c-*myc* mRNA from K562 cells (Watt et al., 1983) calculated with the RNAFLD program (Jacobson et al., 1984) using free energies of base pairing at 37°C (Freier et al., 1986). Target locations are indicated with capital letters. The principal initiation codon begins at nt 559, and the alternative CUG initiation codon for the larger minor product begins at nt 517 (Hann, et al., 1988).

This approach assumed that antisense DNA inhibition results from hybridization arrest of mRNA translation by ribosomes (Zamecnik and Stephenson, 1978), but such inhibition also clearly depends on RNase H attack on the RNA/DNA hybrid formed between the mRNA and the antisense oligodeoxynucleotide (Walder and Walder, 1988). Hence, one may not unequivocally predict the site on a message most sensitive to antisense DNA inhibition.

In an effort to find the most efficacious target in the cap to initiation codon domain of human c-*myc* mRNA, a series of antisense pentadecamers complementary to nine sites between the cap and initiation codon regions of c-*myc* mRNA, and one in the coding region, were synthesized and assayed for efficacy as antisense inhibitors. It was observed that the 5'-end of the message, the cap sequence, was the most sensitive target, at least twice as efficacious as the original initiation codon target (Bacon and Wickstrom, 1991b). The dependence of antisense inhibition at the initiation codon target on oligomer length was also examined, revealing modest improvement in efficacy by adding three deoxynucleotides, and modest reduction in efficacy by removing three residues, relative to the original pentadecamer.

MATERIALS AND METHODS

Cell Culture

HL-60 cells were grown and maintained in logarithmic phase with greater than 90% viability in RPMI 1640 (Sigma) supplemented with 10% heat-inactivated fetal bovine serum (FBS) (Sigma) at 37° C in a 5% CO_2 atmosphere saturated with water (HL-60–FBS). A serum-free culture was established by transferring 1 ml of RPMI 1640 with 10% FBS, containing 10^6 cells, into 9 ml of RPMI 1640 supplemented with 0.4% bovine serum albumin (BSA), 1:1000 insulin–transferrin–sodium selenite medium supplement (ITS) (Sigma) and gradually decreasing percentage of FBS over a 3-week period until cells were able to maintain greater than 80% viability without FBS (HL-60–ITS).

For growth in semisolid media, the latter media also contained 2% methyl cellulose (Fluka, 4000 mP·s) (Graf et al., 1981). All tissue culture media contained 10^5 units of penicillin, 0.1 g streptomycin, and 0.5 g gentamicin per liter. Cell titers and viability were determined by trypan blue (Gibco) dye exclusion. Error bars in cell counts represent one root mean square standard deviation for multiple determination, or $\sqrt{n}$ for single counts of large numbers. Adherence of cells to tissue culture plates was analyzed by washing the empty plate with warm medium, decanting, adding 0.1% trypsin for 10 min, inactivating the trypsin with an equal volume of warm medium, and counting 20 µl aliquots.

Treatment of Cells With Antisense DNA

HL-60 cells were sedimented, decanted, and resuspended in fresh RPMI 1640 with 10% FBS or RPMI 1640 with ITS-BSA to a titer of 10^5 cells/ml for long-term proliferation and differentiation assays, or 2×10^6 cells/ml for short-term radioimmunoprecipitation assays, and then incubated for 24 hr in the wells of culture plates in order to reestablish logarithmic growth before addition of the modulators: 1% (v/v) Me_2SO (Sigma), 16 nM PMA (Sigma), or antisense DNA oligomers. Antisense oligodeoxynucleotides (Table 1) were synthesized by the β-cyanoethyl phosphoramidite method (Sinha et al., 1983) on a Biosearch 8750 DNA synthesizer, purified as before (Wickstrom et al., 1988) on an ISCO liquid chromatograph, and added directly to cell suspensions at concentrations detailed in the text.

For experiments in semisolid media, aliquots of HL-60 cells (<10 µl) containing 10^4 cells were diluted to 1 ml in semisolid RPMI 1640 with 10% FBS or semisolid RPMI 1640 with BSA–ITS. The semisolid media contained no addition, 1% Me_2SO (Sigma), 16 nM PMA (Sigma), anti-c-*myc* oligomer, or anti-VSV oligomer at concentrations detailed in the text. For subsequent daily additions, the suspensions were mixed gently after each oligomer addition to assure distribution of the DNA.

TABLE 1. Antisense Oligodeoxynucleotide Target Sequences[a]

Target label	Target location	Antisense sequence
D	VSV-M	5'-d(TTG GGA TAA CAC TTA)-3'
E	−32 to −18	5'-d(CTC GCA TTA TAA AGG)-3'
F	1 to 14	5'-d(GCA CAG CTC GGG GGT)-3'
G	9 to 23	5'-d(CCG GCG GTG GCG GCC)-3'
H	127 to 141	5'-d(GCC CCG AAA ACC GGC)-3'
I	207 to 221	5'-d(CGC CCG GCT CTT CCA)-3'
J	277 to 291	5'-d(TGG GCC AGA GGC GAA)-3'
K	382 to 396	5'-d(GTG TTG TAA GTT CCA)-3'
L	536 to 550	5'-d(GAG GCT GCT GGT TTT)-3'
M	551 to 565	5'-d(GGG GCA TCG TCG CGG)-3'
N	559 to 570	5'-d(GTT GAG GGG CAT)-3'
O	559 to 573	5'-d(AAC GTT GAG GGG CAT)-3'
P	559 to 573	5'-d(AAC GTT GtG GaG CAT)-3'
Q	559 to 573	5'-d(CTG AAG TGG CAT GAG)-3'
R	559 to 567	5'-d(GCT AAC GTT GAG GGG CAT)-3'
S	574 to 588	5'-d(CCT GTT GGT GAA GCT)-3'

[a]Antisense oligodeoxynucleotide sequences targeted against predicted stems, loops, and bulges in the calculated secondary structure of the human c-*myc* mRNA (Fig. 2). Labels correspond to targets in Figure 2 and lanes in Figures 8, 9, and 10. Sequence O is the original AUG initiation codon region used previously (Heikkila et al., 1987; Wickstrom et al., 1988, 1989), and sequence L includes the exon 1/exon 2 splice junction. Sequences are numbered according to the c-*myc* mRNA sequence of Watt et al. (1983). Negative control sequences included nt 17–31 of VSV matrix protein mRNA (Rose and Gallione, 1981) (D), the initiation codon region of a possible upstream open reading frame transcribable from the P0 transcription start site (Gazin et al., 1984) (E), and a scrambled version (Q) (Heikkila et al., 1987) of the AUG initiation codon sequence (O). Lower case letters represent mismatches between sequence P and the initiation codon region target, 559–573.

Colony Formation in Semisolid Media

Tissue culture cell wells (Corning) containing 10^4 cells in 1 ml of semisolid RPMI 1640 with 10% FBS or semisolid RPMI 1640 with BSA–ITS media were visualized microscopically and colonies of two or more cells were counted and compared with single cell counts. Percentage of cells forming colonies were determined daily by counting 200 colonies/cells in the same quadrant of the well.

Cellular Differentiation and Mitotic Index

Treated and control cell suspensions were thoroughly mixed, and aliquots of 200–600 μl were sedimented onto microscope slides using a Shandon Cytospin II (Surrey, England). Cell monolayers were Wright–Giemsa stained (ASP), and cell differentiation and mitotic index were determined simultaneously under light microscopy on 1000 cells. Cells scored as myeloblasts/promyelocytes and dividing cells were considered undifferentiated. Those scored as myelocytes/metamyelocytes, banded neutrophils, segmented neutrophils, and hypersegmented neutrophils were considered differentiated.

Cytochemical Assays for Differentiation

Cell monolayers on slides, prepared as above, were used in cytochemical assays for uptake of sudan black (Sigma) (Sheehan and Storey, 1947), reduction of nitroblue tetrazolium (NBT) (Sigma) (Collins et al., 1979), and activity of α-naphthyl acetate esterase, and naphthol AS-D chloroacetate esterase (Sigma) (Yam et al., 1971), according to the published methods. Three hundred cells were evaluated in each case.

Radioimmunoprecipitation of c-*myc* p65 Protein

Samples of 2×10^6 cells were sedimented, decanted, washed in PBS, and resuspended in 1 ml of cysteine/methionine-free RPMI 1640 (Gibco) supplemented with 10% fetal bovine serum, L-cysteine (1066 Ci/mmol; 1 Ci = 37 GBq; du Pont/New England Nuclear) at 300 μCi/ml and L-methionine (1198 Ci/mmol; du Pont/New England Nuclear) at 300 μCi/ml. Each sample was grown for an additional 1.5 hr, after which the cells were sedimented, washed, lysed, immunoprecipitated, electrophoresed, fluorographed, and quantitated by densitometry as described (Wickstrom et al., 1988).

RESULTS

Development and Characterization of a Serum-Free HL-60 Cell Strain

An aliquot of the HL-60 cells growing in RPMI 1640 with 10% heat-inactivated FBS, designated HL-60–FBS, was introduced into RPMI 1640 medium supplemented with bovine serum albumin (BSA) and a mixture of insulin, transferrin, and sodium selenite (ITS) (Breitman et al., 1980), and grown continuously for over 6 months so that a steady state response to antisense oligomers could be investigated. The serum-free culture was designated HL-60–ITS.

The HL-60–ITS cells exhibited a slower growth rate than that of the parent cells, with a doubling time of 48 hr, 18 hr longer than HL-60–FBS cells, a longer lag time before returning to logarithmic growth, 2 days rather than 1 day, and the inability to grow colonies in semisolid medium, as reported before (Breitman et al., 1980). Both HL-60 cultures grew logarithmically to a maximum of 2.5×10^6 cells/ml before leveling off. Additionally, the HL-60–ITS cells displayed a lower mitotic index (0.35%) (Tables 2 and 3), a 6% increase in differentiated cell forms of the same morphology as seen in the HL-60–FBS cells (Tables 2 and 3), and the same percentages of sudan black positive cells (40–50%), NBT reducing cells (30%), and naphthyl AS-D chloroacetate esterase positive cells (30%).

Both strains of HL-60 cells reacted to sedimentation and resuspension in fresh medium by going into lag phase for a day. Therefore, daily additions of oligodeoxynucleotide were

TABLE 2. Differential Count Percentages of HL-60–FBS Cells After Incubation With Inducing Compounds and Antisense Deoxyoligonucleotides

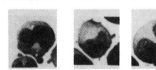

Inducer	Mp[a]	Mb/Pm	My/Me	Ba	Sn	HSn	Mo
None (0)[b]	2.0	88.8	4.6	0.5	3.7	0.4	0
None (5)[c]	3.4	79.9	6.1	2.4	6.1	2.1	0
PMA (5)[d]	0	0	0	0	0	0	100
Me₂SO (5)[c]	0.5	7.0	80.6	2.1	7.2	2.6	0
anti-VSV (5)[c]	3.6	79.4	7.1	2.7	5.3	1.9	0
anti-c-*myc* (5)[c]	1.8	34.3	48.2	6.6	6.8	2.3	0

[a]Mp, mitotic phase; Mb/Pm, myeloblasts/promyelocytes; My/Me, myelocytes/metamyelocytes; Ba, banded neutrophils; Sn, segmented neutrophils; HSn, hypersegmented neutrophils; Mo, monocytes.
[b]Values represent the analysis of 1000 cells each from duplicate samples of one experiment on day 0.
[c]Values represent the analysis of 1000 cells from a single experiment on day 5.
[d]Values represent the analysis of 200 cells. Fewer cells were recovered after treatment with PMA due to terminal differentiation of the cells within the first 24 hr of treatment.

TABLE 3. Differential Count Percentages of HL-60–ITS Cells After Incubation With Inducing Compounds and Antisense Deoxyoligonucleotides

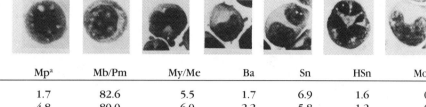

Inducer	Mp[a]	Mb/Pm	My/Me	Ba	Sn	HSn	Mo
None (0)[b]	1.7	82.6	5.5	1.7	6.9	1.6	0
None (5)[c]	4.8	80.0	6.0	2.2	5.8	1.2	0
PMA (5)[d]	0	0	0	0	0	0	100
Me$_2$SO (5)[c]	1.8	6.1	78.1	1.8	7.4	2.8	0
anti-VSV (5)[c]	4.9	79.7	5.9	1.8	6.2	1.5	0
anti-c-*myc* (5)[c]	2.2	8.3	79.8	3.7	6.2	1.8	0

[a]Mp, mitotic phase; Mb/Pm, myeloblasts/promyelocytes; My/Me, myelocytes/metamyelocytes; Ba, banded neutrophils; Sn, segmented neutrophils; HSn, hypersegmented neutrophils; Mo, monocytes.
[b]Values represent the analysis of 1000 cells each from duplicate samples of one experiment on day 0.
[c]Values represent the analysis of 1000 cells from a single experiment on day 5.
[d]Values represent the analysis of 200 cells. Fewer cells were recovered after treatment with PMA due to terminal differentiation of the cells within the first 24 hr of treatment.

done by adding new aliquots of oligomer to logarithmically growing cultures, without disturbing them further.

Antisense Inhibition of Cell Proliferation and Colony Formation

In agreement with our previous work (Wickstrom et al., 1988, 1989), sequence-specific dose-dependent inhibition of proliferation was observed after single additions of anti-c-*myc* oligomer to both the HL-60–FBS and HL-60–ITS cells, with greater inhibition noted in the HL-60–ITS cells (Fig. 3). The HL-60 cells used in the present work, the gift of Dr. Julie Djeu, did not respond as strongly to the antiproliferative effects of the anti-c-*myc* oligomer as did the culture used in the two previous studies (Wickstrom et al., 1988,1989). This characteristic probably reflects a lower copy number of c-*myc* mRNA, which varies widely among HL-60 cultures in different laboratories (J. Bresser, personal communication). No significant inhibition was noted in cells treated with anti-VSV oligomer, while cells treated with Me$_2$SO and PMA were inhibited in their proliferation by greater than 90% (Figs. 3 and 4). Adherence of these normally anchorage-independent cells was not

significant in any case examined, and never exceeded 3% of the cells in culture.

HL-60–FBS cells that were treated daily for 5 days with 10 nmol/ml of anti-c-*myc* oligomer, in order to reestablish a concentration of 10 μM in the culture medium every day, exhibited a titer at day 5 similar to HL-60–FBS cells treated with a single addition of 15 μM anti-c-*myc* oligomer (Figs. 3a and 4a) (Wickstrom et al., 1988, 1989). Similar results were seen with HL-60–ITS cells, with greater antiproliferative efficacy (Figs. 3b and 4b). No significant inhibition was noted in cells treated daily with anti-VSV oligomer, while the cells treated with a single addition of Me$_2$SO or PMA were inhibited by greater than 90% (Figs. 3ab and 4ab).

A second method for evaluating the growth potential of a cell line is to add the cells to a semisolid medium so that dividing cells will form colonies (Graf et al., 1981). HL-60–FBS cells grown in semisolid medium, and treated daily with anti-c-*myc* oligomer, showed a decrease in colony formation of greater than 50%, relative to untreated cells (Fig. 5a). However, no significant decrease in colony number was noted in cells treated with anti-VSV oligomer. By contrast, 12% or less of HL-

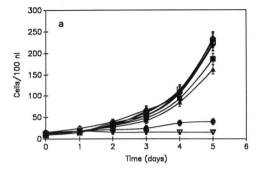

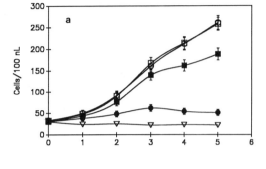

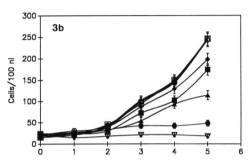

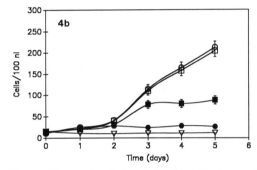

Fig. 3. Titers of cells treated with a single dose of antisense oligodeoxynucleotide Me$_2$SO or PMA in the presence or absence of serum. Cells were grown for 5 days in untreated medium (○), medium supplemented with 1% Me$_2$SO (●), 16 nM PMA (▽), 5 μM anti-VSV oligomer (◇), 10 μM anti-VSV oligomer (□), 15 μM anti-VSV oligomer (△), 5 μM anti-c-*myc* oligomer (◆), 10 μM anti-c-*myc* oligomer (■), or 15 μM antic-*myc* oligomer (▲). (**a**) With serum, HL-60−FBS cells; (**b**) without serum, HL-60−ITS cells.

Fig. 4. Titers of cells treated with daily doses of antisense oligodeoxynucleotides, or single doses of Me$_2$SO or PMA. Cells were grown for 5 days in untreated medium (○), medium supplemented with 1% Me$_2$SO (●), 16 nM PMA (▽), anti-VSV oligomer (10 nmol/ml/day)(□), or anti-c-*myc* oligomer (10 nmol/ml/day) (■). (**a**) With serum, HL-60−FBS cells; (**b**) without serum, HL-60−ITS cells.

60−FBS cells treated with Me$_2$SO and PMA formed colonies, none greater than 2 cells. Daily treatment with antisense oligomer treatment required that the semisolid medium be stirred sufficiently to achieve uniform distribution of the oligomer was assured. Despite our attempts to stir the medium gently, the existing colonies were measurably disturbed, such that the number of colonies, and colony sizes, were reduced by at least 70%, when compared with undisturbed colonies of untreated HL-60−FBS cells (Wickstrom et al., 1989). Similar experiments were attempted with HL-60−ITS cells, but they failed to grow in semisolid beyond

36 h, as observed previously (Breitman et al., 1980).

Upon counting colonies and cells in each cell well, it was found that 23 ± 5% of the untreated cells formed colonies (Fig. 5b). In contrast, only 12 ± 3% of cells treated with Me$_2$SO formed colonies, and only 5 ± 1% of those treated with PMA. However, colony formation by cells treated with anti-c-*myc* oligomer was reduced to 10 ± 3%, while 20 ± 5% of cells treated with the anti-VSV oligomer continued to form colonies.

Morphological Analysis of Cellular Differentiation

Analysis of Wright−Giemsa stained cells allows one to establish the overall degree of

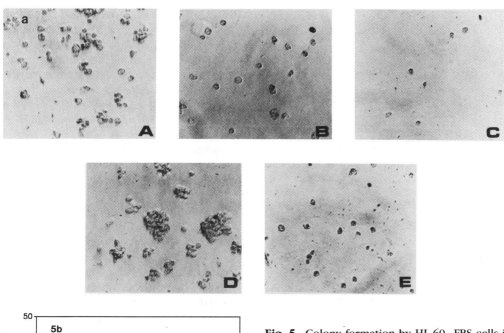

Fig. 5. Colony formation by HL-60−FBS cells in semisolid medium treated with daily doses of antisense oligodeoxynucleotides, or single doses of Me₂SO or PMA. Cells were grown for 5 days in untreated medium (**A**), medium supplemented with 1% Me₂SO (**B**), 16 nM PMA (**C**), anti-VSV oligomer (10 nmol/ml/day) (**D**), anti-c-*myc* oligomer (10 nmol/ml/day) (**E**). (**a**) Light micrographs of representative portions of cultures; (**b**) percentages of cells forming colonies.

differentiation in a heterogeneous cell population. Untreated HL-60−FBS cells, HL-60−FBS cells induced to differentiate into mature granulocytic forms with Me₂SO, and into mature monocytic forms with PMA, were compared with HL-60−FBS cells treated daily with anti-c-*myc* or anti-VSV oligomer over a 5-day period. Differential counts of untreated and treated cells are shown in Table 2. Light micrographs of cells representative of each category used in scoring cell types are shown at the head of the table.

Cells grown in culture at 10 μM anti-c-*myc* oligomer, replenished daily, exhibited a dramatic decrease in the number of promyelocytes and cells undergoing mitosis, a large increase in myelocytes/metamyelocytes and banded neutrophils, but little difference in

segmented neutrophils or hypersegmented neutrophils, compared with untreated cells. The overall extent of differentiation into the latter four cell types was $64 \pm 10\%$. No significant change in mature morphological forms was noted in cells treated with anti-VSV oligomer, while cells treated with Me₂SO exhibited even greater differentiation into granulocytic forms, to a total of $92 \pm 13\%$, and cells treated with PMA exhibited virtually complete differentiation into mature monocytic forms. The time course of differentiation over 5 days is shown in Figure 6a.

The same kind of analysis was carried out on HL-60−ITS cells, shown in Table 3 for the state of differentiation at 5 days, and the time course over 5 days appears in Figure 6b. The most significant difference seen in the ab-

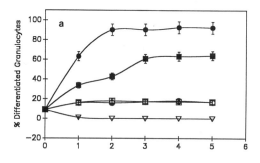

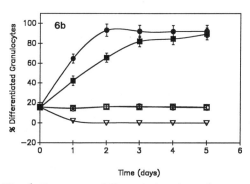

Fig. 6. Percentage differentiation along the granulocytic pathway of cells treated with daily doses of antisense oligodeoxynucleotides, or single doses of Me₂SO or PMA. Cells were grown for 5 days in untreated medium ($\bigcirc$), medium supplemented with 1% Me₂SO ($\bullet$), 16 nM PMA ($\triangledown$), anti-VSV oligomer (10 nmol/ml/day) ($\square$), or anti-c-*myc* oligomer (10 nmol/ml/day) ($\blacksquare$). (**a**) With serum, HL-60–FBS cells; (**b**) without serum, HL-60–ITS cells.

sence of serum is the greater efficacy of the anti-c-*myc* oligomer, nearly approximating that of 1% Me₂SO. Both in the presence and absence of serum, the predominant changes seen in Tables 2 and 3 occurred in the myelocyte/metamyelocyte category. That is, induction of differentiation by the anti-c-*myc* oligomer did not significantly drive the cells into the segmented or hypersegmented neutrophil categories.

Cytochemical Analysis of Cellular Differentiation

The positive results found by morphological analysis of cells treated with daily addi-

tions of antisense oligodeoxynucleotides required confirmation by cytochemical assays for granulocytic or monocytic differentiation. Sudan black stains lipophilic granules characteristic of granulocytic differentiation, while monocytic cells stain weakly (Sheehan and Storey, 1947). HL-60–ITS cells stayed in the range of 40–50% positive for Sudan black incorporation for 5 days, as did cells treated daily with anti-VSV oligomer (Fig. 7a). On the other hand, cells treated daily with anti-c-*myc* oligomer, or with Me₂SO, rose to over 80% positive by 5 days, while cells treated with PMA failed to take up Sudan black at all by 2 days. Similarly striking results were seen with HL-60–FBS cells (not shown).

NBT is reduced intracellularly to insoluble formazan by superoxide, which is generated by PMA induction of phagocytosis-associated oxidative metabolism in normal granulocytes (DeChatelet et al., 1976) and differentiated HL-60 cells (Collins et al., 1979). Hence, a high level of NBT reduction correlates with terminally differentiated mature granulocytic cells, a low level of NBT reduction correlates with immature or nondifferentiated cells, while the complete lack of NBT reduction is associated with monocytic or lymphocytic differentiation (Collins et al., 1979). Accordingly, untreated HL-60–ITS cells stayed in the neighborhood of 30% positive for NBT reduction over 5 days, as did cells treated daily with anti-VSV oligomer, while cells treated daily with anti-c-*myc* oligomer rose to about 60% positive by 5 days, and cells treated with Me₂SO rose to about 80% (Fig. 7b). In contrast, cells treated with PMA decreased by the first day to less than 5% positive. Comparable clear-cut results were seen with HL-60–FBS cells (not shown).

High levels of naphthol AS-D chloroacetate esterase activity have been observed in myeloblasts, promyelocytes, and granulocytes while little or no activity has been noted in monocytes. Conversely, high levels of α-naphthyl acetate esterase activity have been found in monocytes but not in myeloblasts, promyelocytes, or granulocytes (Yam et al., 1971). Hence, high levels of naphthol AS-D chloroacetate esterase correlate with imma-

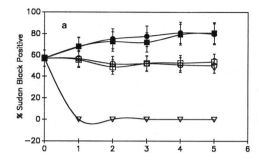

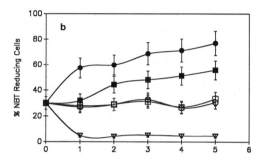

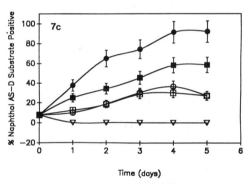

Time (days)

Fig. 7. Cytochemical assays of HL-60–ITS cells treated with daily doses of antisense oligodeoxynucleotides, or single doses of Me₂SO or PMA. Cells were grown for 5 days in untreated medium (○), medium supplemented with 1% Me₂SO (●), 16 n*M* PMA (▽), anti-VSV oligomer (10 nmol/ml/day) (□), or anti-c-*myc* oligomer (10 nmol/ml/day) (■). Three hundred cells were evaluated in each case. (**a**) Sudan black incorporation; (**b**) NBT reduction; (**c**) naphthol AS-D chloroacetate esterase activity.

ture cells and maturing cells differentiating along the granulocytic pathway while high levels of α-naphthyl acetate esterase activity correlate with cells maturing along the mon-

ocytic pathway. Untreated HL-60–ITS cells and those treated daily with anti-VSV oligomer leveled out by 3 days at about 30% positive for naphthol AS-D chloroacetate esterase while those treated with anti-c-*myc* oligomer rose to about 50% positive. In contrast, cells treated with Me₂SO were over 90% positive for chloroacetate esterase by 4 days, while those treated with PMA displayed no detectable activity after only 1 day of treatment (Fig. 7c). Untreated, Me₂SO, anti-VSV, and anti-c-*myc*-treated cells were found to be less than 2% positive for α-naphthyl acetate esterase while PMA-treated cells were greater than 90% positive (not shown). Virtually the same pattern was observed for HL-60–FBS cells (not shown).

Selection of Antisense Targets From a Calculated Secondary Structure

Calculation of one possible secondary structure for the entire 2121-nt human c-*myc* mRNA from K562 human erythroleukemia cells (Wickstrom et al., 1988) (Fig. 2) placed the main AUG initiation codon in a large bulge loop 559 nt downstream from the cap, and the minor CUG initiation codon (Hann, et al., 1988) in an inaccessible stem 517 nt 3′ of the cap. The predicted AUG-containing loop was chosen as the initial target for antisense inhibition, on the assumption that those nucleotides might be readily available for hybridization arrest. The efficacy of c-*myc* inhibition that we found by utilizing this target (Heikkila et al., 1987; Wickstrom et al., 1988, 1989) was consistent with its predicted accessibility, but is hardly a test of the secondary structure model.

While one might hypothesize that the antisense effectiveness of a particular oligomer might correlate with the extent of secondary structure at each corresponding mRNA target, and the number of G-C base pairs in the DNA–RNA hybrid, the levels of antisense inhibition may also depend on varying efficiencies of cellular uptake, transport, or degradation for oligomers of differing sequences. Hence, a series of antisense oligodeoxynucleotides was prepared against a series of predicted stems, loops, and bulges from the cap to the initiation codon region, including a predicted hairpin just downstream of the AUG (Table 1).

It is also expected that the effectiveness of antisense inhibition at each target in c-*myc* mRNA should increase with the length of the antisense oligomer, in proportion to the increased free energy of binding. If one literally assumed that an antisense 12-mer, 15-mer, and 18-mer directed against the predicted initiation codon loop could bind freely with all complementary residues in the mRNA target, then the association constant of a 15-mer would be orders of magnitude greater than that of a 12-mer, and the binding of an 18-mer would be orders of magnitude greater than that of the 15-mer. In order to test this simple model, antisense oligomers were synthesized against the initiation codon and the next three, four, and five codons (Table 1). The effects of mismatches were examined in one pentadecamer analog of the initiation codon antisense sequence that contained two nonadjacent mismatches.

Efficacy of Antisense Inhibition at Different Sites in c-*myc* mRNA

The efficacy of each oligomer was measured by radioimmunoprecipitation of c-*myc* p65 antigen from untreated HL-60 cells and cells treated with 10 μ*M* of each oligomer for 24 hr (Fig. 8). The minor p67 antigen was not apparent on these gels or in our earlier study (Wickstrom et al., 1988). Quantitation of p65 bands appears in Figure 9. The degree of variation in this method may be seen by comparing untreated control lanes B and U, as well as the VSV sequence control lane D.

Sequence E, the initiation codon region of the 20 kDa upstream open reading frame which might be transcribed from P0 (Gazin et al., 1984) was also expected to serve as a control; the apparent up-regulation of p65 was unexpected, and must be explored further. The 5' cap sequence (F), which appeared in the predicted structure as a weak stem and bulge region, was more than twice as sensitive a target as the original AUG ini-

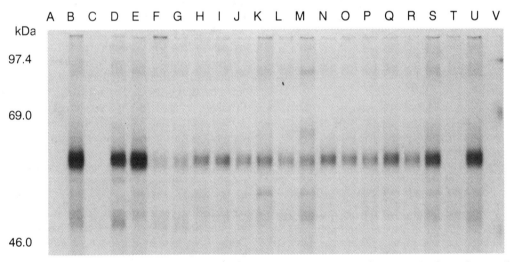

Fig. 8. Expression of c-*myc* p65 protein in treated and untreated HL-60 cells, measured by radioimmunoprecipitation and fluorography. Antisense oligodeoxynucleotides were added directly to the suspensions at a final concentration of 10 μ*M* for 12 hr. Lane A, molecular weight markers; lane B, no oligomer, anti-p65 antibody; lane C, no oligomer, rabbit IgG; lanes D–S, oligomers from Table 1, anti-p65 antibody; lane T, no oligomer, rabbit IgG; lane U, no oligomer, anti-p65 antibody; lane V, molecular weight markers.

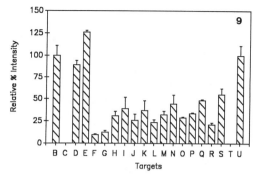

Fig. 9. Relative intensities of c-*myc* p65 protein expression from Figure 8. Each band was scanned at three different points with an LKB 2220 laser densitometer and the intensities were averaged. The entire scanning routine was repeated, and error bars represent the standard deviation of the six scans. B and U, no oligomer, anti-p65 antibody; C and T, no oligomer, rabbit IgG; D–S refer to oligomers in Table 1, with anti-p65 antibody.

tiation codon sequence (O), which was predicted to occur in an even weaker bulge and stem area. Even the oligomer directed against nt 9–23 (G), slightly downstream from the cap, was more effective than the initiation codon oligomer, despite the prediction that sequence G was included in a strong stem.

Within experimental error, sequence H, directed against a predicted hairpin loop and bulge beginning 127 nt downstream from the cap, was just as effective as our standard sequence O, as were targets I, supposedly a weak helical region, J, calculated to be a tight hairpin, K, a putative hairpin loop, and M, which began in a helical region upstream of the initiation AUG, and overlapped sequence O by 7 nt. The efficacy of sequence L, which overlaps the splice junction of exons 1 and 2, and was predicted to occur in a weak helix, was marginally better than the initiation codon sequence O. Sequence S, just downstream of the initiation codon sequence, showed the least inhibition of any sequence tested, except the VSV control, D.

Length Dependence of Antisense DNA Inhibition

Upon varying the length of the AUG initiation codon probes, it was found that se-

quence N, which was only 12 nt, was half as effective as the 15-mer O, while the 18-mer R was about twice as effective. The analog of sequence O containing two separated mismatches, P, was less effective than O, but more so than the 12-mer N or the scrambled 15-mer Q.

Concentration Dependence of Antisense DNA Inhibition

The strong efficacy at the 5'-end and the length dependence of inhibition at the initiation codon target were studied as a function of oligomer concentration, in order to test the validity of the observations at 10 µM (Fig. 10). Despite significant variability, it was apparent that antisense inhibition was dose dependent, and that the relative efficacies at 10 µM (Fig. 9) were consistent with those seen in the concentration ramps.

DISCUSSION

The observation that HL-60 cells grew in a serum-free medium implies that growth factors in serum are not absolutely essential for their proliferation, with the possible excep-

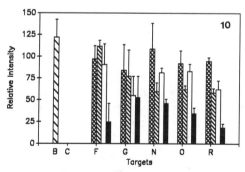

Fig. 10. Relative intensities of c-*myc* p65 protein expression in untreated cells (wide right crosshatch) or HL60 cells treated with 1 µM (narrow right and left crosshatch), 2.5 µM (narrow right crosshatch), 5 µM (blank box) or 10 µM (solid box) of each antisense oligodeoxynucleotide. Antigens were radioimmunoprecipitated, fluorographed, and scanned as in Figures 8 and 9. B, no oligomer, anti-p65 antibody; C, no oligomer, rabbit IgG; F, G, N, O, and R refer to oligomers in Table 1 and lanes in Figure 8, with anti-p65 antibody.

tions of insulin, transferrin, and sodium selenite. It is significant, however, that the doubling time increased in the absence of serum. Hence, one must assume that at least some of the missing serum factors play a stimulatory role in the HL-60 cell cycle. On the other hand, the extent of differentiation was the same under both culture conditions.

The greater antiproliferative efficacy of a single addition of anti-c-*myc* oligomer in the absence of serum is consistent with the hypothesis that serum nucleases limit the amount of oligomer that may be taken up by cells over the 5 days of the experiment. In both cases, inhibition of proliferation was dose dependent. Daily addition of anti-c-*myc* oligomer inhibited proliferation more than a single addition, consistent with the observation that oligodeoxynucleotides taken up by cells turn over within 1–2 days (Wickstrom et al., 1988). The greater efficacy of daily addition was observed both in the presence and absence of serum, with an even stronger effect in the absence of serum. Comparing Figure 3a with 4b, one sees that daily addition of 10 μM anti-c-*myc* oligomer in the absence of serum was roughly three times as effective as an antiproliferative as a single addition in the presence of serum. In no case did the control anti-VSV oligomer display any effect on proliferation, while the cells were similarly inhibited by Me$_2$SO and PMA in the presence and in the absence of serum.

Examination of colony formation in semisolid medium provided an answer to the question of whether the HL-60 cell culture contained a subpopulation resistant to antisense inhibition of c-*myc* expression. As the concentration of daily replenished anti-c-*myc* oligomer was increased, the number of cells forming colonies decreased, and in addition the size of colonies themselves decreased. This result reaffirms the observation that c-*myc* expression was inhibited significantly in all cells following single addition of antisense oligomer (Wickstrom et al., 1989).

Antisense inhibition of c-*myc* gene expression in HL-60 cells grown in the presence of serum, treated with a single addition of anti-c-*myc* oligomer, has been observed to enhance differentiation along the granulocytic pathway (Wickstrom et al., 1989). In the results presented above, the differential counts of cells treated daily with antisense oligomers provided a detailed profile of the extent of differentiation of each population. Daily addition of 10 μM anti-c-*myc* oligomer to HL-60 cells grown without serum was found to induce terminal differentiation almost as effectively as 1% Me$_2$SO.

Cytochemical assays of HL-60 cell differentiation confirmed the morphological analyses. Sudan black staining gave the strongest response, while NBT reduction and naphthol AS-D chloroacetate activity showed a greater response to Me$_2$SO than to anti-c-*myc* oligomer. In the latter two assays, differentiation induced by anti-c-*myc* oligomer was similar in both the presence and absence of serum. Perhaps uptake of Sudan black reaches a maximum as soon as promyelocytes enter the myelocyte/metamyelocyte stage, while superoxide and esterase levels may not maximize until the final granulocytic stages.

The pattern of antisense inhibition as a function of target sequence described above for human c-*myc* mRNA does not correlate with the predicted secondary structures of the message (Fig. 2). The most sensitive site in c-*myc* mRNA was the 5′ cap, followed by the first splice junction, and then the AUG initiation codon region and other targets in the 5′ untranslated leader. Similar results have been found for antisense oligodeoxynucleotide inhibition of rabbit globin mRNA translation in cell-free extracts (Blake et al., 1985; Goodchild et al., 1988a), and for c-Ha-*ras* oncogene (Daaka and Wickstrom, 1990). On the other hand, antisense inhibition of human immunodeficiency virus mRNA translation in infected cells revealed that the most efficacious targets in the viral system were the 3′ polyadenylation signal, 5′ leader sequences at nt 54–73 and 162–181, and the splice acceptor site at nt 5349–5368, rather than the cap or initiation codon, which were a little less sensitive (Goodchild et al., 1988b).

To the extent that antisense inhibition depends on genuine hybrid arrest of ribosomal translation of an mRNA, it is logical that those sites in a message that are recognized by initiation factors would be the most sensitive. To the extent that antisense inhibition depends on RNase H hydrolysis of antisense oligodeoxynucleotide/mRNA hybrids, the sensitivity of each site in a message should correlate with its accessibility, as determined by its secondary and tertiary structure. To the extent that antisense inhibition depends on interruption of RNA processing, splice sites should be particularly sensitive, if oligodeoxynucleotides enter the nucleus. From the observations presented above for c-*myc* mRNA in whole HL-60 cells, it seems likely that all of these modes of action may occur in this system. On the other hand, it is also possible that the most sensitive sites are those that are most exposed in the tertiary structure of c-*myc* mRNA, which would allow preferential binding to antisense oligomers, and resulting hydrolysis by RNase H.

The modest dependence of antisense effectiveness on oligomer length at the initiation codon target, and the residual efficacy of the pentadecamer with two mismatches, imply that none of the oligomers employed against the AUG initiation codon region hybridizes completely to the message. This is consistent with the prediction that the AUG is exposed in a bulge (Fig. 2), rather than a large loop, as predicted for a 400 nt portion of c-*myc* mRNA centered on the initiation codon (Wickstrom et al., 1988).

FUTURE DIRECTIONS

These observations suggest that it might be possible to keep c-*myc*-transformed leukemic cells under control by daily administration of an antisense oligomer concentration, 10 μM, which was not found to inhibit c-*myc* p65 expression in normal peripheral blood lymphocytes stimulated by phytohemagglutinin (Heikkila et al., 1987). On the other hand, the differential counts and cytochemical assays revealed that HL-60 cells treated daily with anti-c-*myc* oligomer did not differentiate into fully mature granulocytic forms, but were found primarily in intermediate forms along the granulocytic pathway. It is

worth asking whether continued antisense oligomer treatment beyond 5 days would differentiate the cells completely, and whether discontinuation of treatment would lead the cells to revert to promyelocytic forms.

While it is now clear that down-regulation of c-*myc* inhibits replication, it is not at all clear how a decrease in p65 stimulates differentiation along the granulocytic pathway, as opposed to the monocytic pathway. Elucidation of this question requires examination of the impact of anti-c-*myc* oligomer treatment on both transcription and translation of a panel of developmental and cell cycle genes, in a variety of normal and transformed cell lines.

The dependence of antisense inhibition on mRNA secondary and tertiary structure, and the relevance of cap and initiation codon sensitivity to the mechanism of translational initiation, should be tested in more detail. Elucidation of the actual secondary structure of c-*myc* mRNA in vivo is necessary. Additional targets should be probed, such as the upstream initiator CUG (Hann et al., 1988) between targets K and L, the other splice junctions in the original transcript, more sites in the coding region, and in the 3′ tail. At the most sensitive sites, sequence specificity and concentration dependence should be studied for each oligomer size over a broader range, in order to determine the optimum sequence and length for maximum efficacy and specificity.

The independence of the cap and AUG sites should be examined by assessing the effectiveness of combinations of their respective antisense oligodeoxynucleotides. Finally, the separation of the cap and initiation codon within the mRNA tertiary structure should be tested by preparing oligomers containing both antisense sequences in a single molecule, separated by variable length deoxynucleotide molecular rulers, and measuring their efficacy.

If antisense inhibition is usually most effective at cap and initiation codon sequences, and a more modest phenomenon in coding regions, then even in a human system a 12-mer may display sufficient statistical uniqueness, among the available targets, for gene-specific inhibition.

SUMMARY

Expression of the human protooncogene c-*myc* is necessary for replication, and may be inhibited in a sequence-specific, dose-dependent manner by an antisense oligodeoxynucleotide specific for the first five codons of c-*myc* mRNA. Antisense inhibition of c-*myc* inhibits the proliferation and enhances the differentiation of the HL-60 human promyelocytic leukemia cell line. In order to raise the efficacy of antisense oligomers, HL-60 cells were grown in a serum-free medium so as to minimize nuclease activity in the culture medium. Daily addition of anti-c-*myc* oligomer was then found to induce terminal granulocytic differentiation of 80% or more of HL-60 cells, and inhibit colony formation by greater than 50%, comparable to 1% Me_2SO. Since the dependence of antisense DNA inhibition on the target sequence within the c-*myc* mRNA had not yet been studied, a series of oligomers complementary to sequences between the cap and AUG initiation codon regions of c-*myc* mRNA were synthesized. HL-60 cells were treated with each oligomer, and c-*myc* p65 expression was measured by radioimmunoprecipitation. The sensitivity of the cap region sequences to antisense DNA was two to three times that of the original initiation codon antisense sequence. The other target sequences downstream of the cap, and up to the AUG initiation codon, were comparable to the initiation codon sequence, except that the first splice junction was slightly more sensitive. At the primary initiation codon target, a dodecamer was about half as effective as the original pentadecamer, while an octadecamer was about twice as effective. The observation of variation in antisense efficacy as a function of target location in c-*myc* mRNA may represent a combination of the effects of hybrid arrest, RNase H attack, and interference in RNA processing. Alternatively, the most sensitive targets might be those that are the most exposed in the secondary and tertiary structure of c-*myc* mRNA.

ACKNOWLEDGMENTS

Most importantly, I thank my co-worker on these studies, Thomas Bacon, for his dedicated efforts. I also thank Audrey Gonzalez, Kara Sessums, and Karen Mahovich for excellent technical assistance, Drs. Rosemary Watt, Grace Ju, and Robert Eisenman for samples of polyclonal antisera against c-*myc* p65 protein, Dr. Julie Djeu for a sample of HL-60 cells, Dr. Thomas Graf for discussions of cell growth in semisolid media, Drs. Danielle Konings, Dixie Goss, and Gary Grotendorst for valuable discussions and critical readings of the manuscript, and the NCI Advanced Scientific Computing Laboratory for computer time. This work was supported by grants from the National Cancer Institute, CA42960, the Leukemia Society of America, the Florida High Technology and Industry Council, and Milligen/Biosearch, Inc.

Figs. 3–7 are reproduced from Bacon and Wickstrom (1991a), and Figs. 2 and 8–10 from Bacon and Wickstrom (1991b), with permission of Harwood Academic Publishers GmbH.

REFERENCES

Bacon TA, Wickstrom E (1991a): Daily addition of an anti-c-*myc* DNA oligomer induces granulocytic differentiation of human promyelocytic leukemia HL-60 cells in both serum-containing and serum-free media. Oncogene Res. 6:21–32.

Bacon TA, Wickstrom E (1991b): Walking along human c-*myc* mRNA with antisense oligodeoxynucleotides: Maximum efficacy at the 5' cap region. Oncogene Res. 6:12–19.

Bishop JM (1987): The molecular genetics of cancer. Science 235:305–311.

Blake KR, Murakami A, Miller PS (1985): Inhibition of rabbit globin mRNA translation by sequence-specific oligodeoxynucleotides. Biochemistry 24:6132–6138.

Breitman TR, Collins SJ, Keene BR (1980): Replacement of serum by insulin and transferrin supports growth and differentiation of the human promyelocytic cell line, HL-60. Exp. Cell Res. 126:494–498.

Classon M, Henriksson M, Sümegi J, Klein G, Hammarskjöld ML (1987): Elevated c-*myc* expression facilitates the replication of SV40 DNA in human lymphoma cells. Nature (London) 330:272–274.

Collins SJ (1987): The HL-60 promyelocytic leukemia cell line: Proliferation, differentiation, and cellular oncogene expression. Blood 70:1233–144.

Collins SJ, Gallo RC, Gallagher RE (1977): Continuous growth and differentiation of human myeloid leukaemic cells in suspension culture. Nature (London) 270:347–349.

Collins SJ, Ruscetti FW, Gallagher RE, Gallo RC (1978): Terminal differentiation of human promyelocytic leukemia cells induced by dimethyl sulfoxide and other polar compounds. Proc. Natl. Acad. Sci. U.S.A. 75:2458–2462.

Collins, SJ, Ruscetti FW, Gallagher RE, Gallo RC (1979): Normal functional characteristics of cultured human promyelocytic leukemia cells (HL-60) after induction differentiation by dimethylsulfoxide. J. Exp. Med. 149:969–974.

Coppola JA, Cole MD (1986): Constitutive c-*myc* oncogene expression blocks mouse erythroleukemia cell differentiation but not commitment. Nature (London) 320:760–763.

Daaka Y, Wickstrom E (1990): Target dependence of antisense oligodeoxynucleotide inhibition of c-Ha-*ras* p21 expression and focus formation in T24-transformed NIH3T3 cells. Oncogene Res. 5:267–275.

DeChatelet LR, Shirley PS, Johnston RB (1976): Effect of phorbol myristate acetate on the oxidative metabolism of human polymorphonuclear leukocytes. Blood 47:545–554.

Filmus J, Buick RN (1985): Relationship of c-*myc* expression to differentiation and proliferation of HL-60 cells. Cancer Res. 45:822–825.

Freier SM, Kierzek R, Jaeger JA, Sugimoto N, Caruthers MH, Neilson T, Turner DH (1986): Improved free-energy parameters for prediction of RNA duplex stability. Proc. Natl. Acad. Sci. U.S.A. 83:9373–9377.

Freytag SO, Dang CV, Lee WMF (1990): Definition of the activities and properties of c-*myc* required to inhibit cell differentiation. Cell Growth Diff. 1:339–343.

Gazin C, de Dinechin D, Hampe A, Masson J-M, Martin P, Stehelin D, Galibert F (1984): Nucleotide sequence of the human c-*myc* locus: Provocation open reading frame within the first exon. EMBO J. 3:383–387.

Goodchild J, Carroll III E, Greenberg JR (1988a): Inhibition of rabbit β-globin synthesis by complementary oligonucleotides: Identification of mRNA sites sensitive to inhibition. Arch. Biochem. Biophys. 263:401–409.

Goodchild J, Agrawal S, Civeira MP, Sarin PS, Sun D, Zamecnik PC (1988b): Inhibition of human im-

munodeficiency virus replication by antisense oligodeoxynucleotides. Proc. Natl. Acad. Sci. U.S.A. 85:5507–5511.

Graf T, von Kirchbach A, Beug H (1981): Characterization of the hematopoietic target cells of AEV, MC29 and AMV avian leukemia viruses. Exp. Cell. Res. 131:331–343.

Griep AE, Westphal H (1988): Antisense *myc* sequences induce differentiation of F9 cells. Proc. Natl. Acad. Sci. U.S.A. 85:6806–6810.

Hann SR, Eisenman RN (1984): Proteins encoded by the human c-*myc* oncogene: Differential expression in neoplastic cells. Mol. Cell. Biol. 4:2486–2497.

Hann SR, King MW, Bently DL, Anderson CW, Eisenman RN (1988): A non-AUG translational initiation in c-*myc* exon 1 generates an N-terminally distinct protein whose synthesis is disrupted in Burkitt's lymphomas. Cell 52:185–195.

Heikkila R, Schwab G, Wickstrom E, Loke SL, Pluznik DH, Watt R, Neckers LM (1987): A c-*myc* antisense oligodeoxynucleotide inhibits entry into S phase but not progress from G_0 to G_1. Nature (London) 328:445–449.

Jacobson AB, Good L, Simonetti J, Zuker M (1984): Some simple computational methods to improve the folding of large RNAs. Nucleic Acids Res. 12:45–52.

Klein G, Klein E (1986): Conditioned tumorigenicity of activated oncogenes. Cancer Res. 46:3211–3224.

Konings DAM, van Duijn LP, Voorma HO, Hogeweg P (1987): Minimal energy foldings of eukaryotic mRNAs form a separate leader domain. J. Theor. Biol. 127:63–78.

Persson H, Hennighausen L, Taub R, DeGrado W, Leder P (1984): Antibodies to human c-*myc* oncogene product: Evidence of an evolutionarily conserved protein induced cell proliferation. Science 225:687–693.

Ramsay GM, Moscovici G, Moscovici C, Bishop JM (1990): Neoplastic transformation and tumorigenesis by the human protooncogene *MYC*. Proc. Natl. Acad. Sci. U.S.A. 87:2102–2106.

Rose JK, Gallione CJ (1981): Nucleotide sequences of the mRNAs encoding the vesicular stomatitis G and M proteins determined from cDNA clones containing the complete coding regions. J. Virol. 39:519–528.

Rovera G, Santoli D, Damsky C (1979): Human promyelocytic leukemia cells in culture differentiate into macrophage-like cells when treated with a phorbol diester. Proc. Natl. Acad. Sci. U.S.A. 76:2779–2783.

Sheehan HL, Storey GW (1947): An improved method of staining leucocyte granules with Sudan black B. J. Pathol. Bacteriol. 59:336–337.

Sinha ND, Biernat J, Köster H (1983): β-cyanoethyl N,N-dialkylamino/N-morpholinomonochloro phosphoamidites, new phosphitylating agents facilitating ease of deprotection and work-up of synthesized oligonucleotides. Tetrahedron Lett. 24:5843–5846.

Spector DL, Watt RA, Sullivan NF (1987): The v- and c-*myc* oncogene proteins colocalize in situ with small nuclear ribonucleoprotein particles. Oncogene 1:5–12.

Walder RY, Walder JA (1988): Role of RNase H in hybrid-arrested translation by antisense oligonucleotides. Proc. Natl. Acad. Sci. U.S.A. 85:5011–5015.

Watt R, Stanton LW, Marcu KB, Gallo RC, Croce CM, Rovera G (1983): Nucleotide sequence of cloned cDNA of human c-*myc* oncogene. Nature (London) 303:725–728.

Westin EH, Wong-Staal F, Gelmann EP, Dalla-Favera R, Papas TS, Lautenberger JA, Eva A, Reddy EP, Tronick SR, Aaronson SA, Gallo RC (1982): Expression of cellular homologues of retroviral *onc* genes in human hematopoietic cells. Proc. Natl. Acad. Sci. U.S.A. 79:2490–2494.

Wickstrom E (1986): Oligodeoxynucleotide stability in subcellular extracts and culture media. J. Biochem. Biophys. Methods 13:97–102.

Wickstrom EL, Bacon TA, Gonzalez A, Freeman DL, Lyman GH, Wickstrom E (1988): Human promyelocytic leukemia HL-60 cell proliferation and c-*myc* protein expression are inhibited by an antisense pentadecadeoxynucleotide targeted against c-*myc* mRNA. Proc. Natl. Acad. Sci. U.S.A. 85:1028–1032.

Wickstrom EL, Bacon TA, Gonzalez A, Lyman GH, Wickstrom E (1989): Anti-c-*myc* DNA oligomers increase differentiation and decrease colony formation by HL-60 cells. In Vitro Cell Dev. Biol. 25:297–302.

Yam LT, Li CY, Crosby WH (1971): Cytochemical identification of monocytes and granulocytes. Am. J. Clin. Pathol. 55:283–290.

Yokoyama K, Imamoto F (1987): Transcriptional control of the endogenous *MYC* protooncogene by antisense RNA. Proc. Natl. Acad. Sci. U.S.A. 84:7363–7367.

Targeting and Intracellular Delivery of Antisense Oligonucleotides Interfering With Oncogene Expression

Lee Leserman, Geneviève Degols, Patrick Machy, Jean-Paul Leonetti, Nadir Mechti, and Bernard Lebleu

UA CNRS 1191, Université Montpellier II, Sciences et Techniques du Languedoc, 34095 Montpellier Cedex 5 (G. D., J.-P. L., N. M., B. L.) and Centre d'Immunologie INSERM-CNRS de Marseille-Luminy, 13288 Marseille Cedex 9 (L. L., P. M.), France

INTRODUCTION

The antisense oligonucleotide (oligomers) field has undergone rapid development over the last few years primarily because of progress in nucleic acid chemistry. To the natural oligodeoxy- and oligoribonucleotides have been added an increasing number of analogues modified at their phosphodiester linkage, at their sugar residues, or conjugated to various reactive or photoactivatable groups (see Goodchild, 1990, for a recent review). In principle, such "second generation" antisense oligomers might overcome several of the drawbacks generally encountered in the approach, namely low metabolic stability and reversibility of the interaction between oligomers and their nucleic acid (or protein) targets. However, their behavior in intact cells has not been investigated in detail. Similarly, little is known about the potential toxicity of these synthetic molecules, their conjugated reactive groups, or their degradation products. Moreover, few of these approaches have dealt with the problem of cell penetration and targeting, which will obviously become of major concern when moving to in vivo experimentation.

A potential receptor-mediated endocytic pathway for oligomers has been recently described (Loke et al., 1989; Yakubov et al., 1989); however, its role in the uptake of antisense oligomers remains to be formally demonstrated. We have engaged in studies aimed at targeting oligonucleotides to cell surface determinants, at increasing the efficiency of oligonucleotides internalization in cells, and, more recently, at understanding their intracellular distribution (Leonetti et al., unpublished results). These studies were performed with unmodified or 3'-modified oligodeoxyribonucleotides. This renders them resistant to phosphodiesterases, which have the most active degradative activity encountered in serum.

In principle, however, our approaches could be extended without major alteration to the oligomer derivatives described above.

Initial studies have used virus [vesicular stomatitis virus (Lemaître et al., 1987) and more recently human immunodeficiency virus] infected cells in vitro (Stevenson and Iversen, 1989).

We are now attempting to adapt the same tools to control the expression of an oncogene as c-*myc* in search of a means to artificially control cell growth and differentiation. Controlling oncogene expression is a more difficult challenge than the antiviral approach that we and others have used previously. Oncogenes are altered or aberrantly expressed (deregulated level or inappropriate timing of expression) genes whose normal counterparts often play pivotal roles in cell growth and differentiation. Selective suppression of an abnormal allele, down-regulation of an abnormally expressed normal allele, or selective action in cancer cells has to be achieved to use the full potential of the antisense approach. Despite these conceptual limitations, c-myc protein is translated from short-lived mRNA and is a short-lived nucleoprotein, which increases its attractiveness as a target molecule.

Initial work by several groups has demonstrated the possibility of promoting in various cell lines an antiproliferative response and/or of initiating steps in terminal differentiation, either through short antisense oligodeoxynucleotides (generally targeted to the translation initiation site of c-myc mRNA), or through recombinant plasmids encoding antisense RNAs (Heikkila et al., 1987; Holt et al., 1988; Wickstrom et al., 1988). The exciting prospect of interfering with c-myc transcription through triple helix formation has been raised in experiments that have been confined to cell-free systems at the present time (Cooney et al., 1988). Finally, McManaway et al. (1990) demonstrated the possibility of a selective shut-off of the translocated c-myc allele transcript in lymphoma cells with an antisense oligomer inactive against its normal counterpart in lymphoid cells. A major limitation resides in the high concentrations (up to 100 μM depending upon cell culture conditions) of oligomers required to achieve meaningful phenotypic responses. Intriguingly, phosphorothioate analogues have not been successful in such systems, which is at variance with the case of virus-infected cells (Loke et al., 1988 and our own preliminary results); α-anomeric phosphodiester or α-anomeric phosphorothioate analogues also did not appear active (our unpublished data), although this comparative study is still at an early stage.

This rapid and nonexhaustive overview of the literature should also mention the demonstration of endogenous stable antisense N-myc transcripts in small-cell lung cancer H249 cells (Kristal et al., 1990) although the physiological meaning of this observation has not yet been evaluated.

CONTROL OF C-MYC EXPRESSION BY POLY(L-LYSINE) CONJUGATES OF ANTISENSE OLIGONUCLEOTIDES

As briefly outlined in the introduction, relatively large amounts of synthetic oligomers (5 to 100 μM in most experiments described so far) have to be used to achieve significant down-regulation of c-myc expression in cultured cell lines.

Previous work in our group has taken advantage of poly(L-lysine) (PLL) conjugation to circumvent one of the major problems encountered in the antisense oligomer approach, e.g., inefficient cell uptake (Lemaître et al., 1987). The chemical linkage of oligomers 3'-end to amino groups residues in PLL results in an N-morpholine terminal ring, which protects the oligomers from phosphodiesterase degradation. Steric hindrance by the PLL moiety (16,000 mean MW) might protect the oligomers from attack by other nucleases as well. Poly(L-lysine) conjugates are internalized by endocytosis, as initially described (Shen and Ryser, 1978), and confirmed by us for antisense–PLL (Leonetti et al., 1990a). The mechanism through which conjugated drugs are released from the endocytic pathway and separate from the polypeptide carrier to reach their intracellular target is still far from being understood, however.

As expected, antisense–PLL conjugates have been shown to be appreciably more active than unconjugated oligomers in inhibiting, in a sequence-specific way, the multiplication of vesicular stomatitis virus in various cell lines (Lemaître et al., 1987) and the development of cytopathic effects in HIV-1-infected cells (Stevenson and Iversen, 1989).

We have attempted to adapt similar tools in order to increase the antiproliferative activity of short oligomers (15-mer) complementary to the translation initiation site of c-myc mRNA (see Fig. 1 for sequences). Since poly(L-lysine) is not devoid of cytotoxicity above certain doses in various cell lines we have made use of ternary complexes in which the oligomer–PLL conjugates were mixed with polyanions as heparin; this reduces drasti-

```
oligomer complementary to the c-myc mRNA

c-myc            ACGTTGAGGGGCATCrA


control oligomer with the same base contents as c-myc

c-myc mismatch AGCTTAGGGGCGATCrA


control oligomer complementary to the VSV N protein mRNA

VSV              CATTTTGATTACTGTrA
```

Fig. 1. Oligonucleotides sequences used. Oligomers were synthesized on a riboadenosine derivatized support using a Biosearch cyclone automatic DNA synthesizer and were linked to poly(L-lysine) (PLL).

cally PLL cytotoxicity even at high concentration, as initially described by Morgan et al. (1988) and confirmed in our own work (Degols et al., unpublished results).

As summarized in Figure 2, ternary complexes comprising antisense oligomers of the appropriate sequence (Fig. 2B) specifically inhibit the multiplication of exponentially growing L929 cells when added to their culture medium. The antiproliferative activity is sequence specific and dose dependent, with a treshold level below 1 μM; heparin (or the other polyanions used throughout) or PLL–heparin complexes alone do not affect cell growth in these experimental conditions.

This is clearly different from the experiments described in Figure 2A where unconjugated oligomers have been added to the culture medium; within our experimental condition, e.g., with cells growing continuously in a culture medium supplemented with 10% (w/v) fetal calf serum, no significant antiproliferative activity can be attained at oligomers doses as high as 40 μM. Nonconjugated unmodified antisense oligomers do inhibit growth, however, when cells are transferred to serum-free conditions (for instance, in opti-MEM completely synthetic medium). Even in these conditions, however, unconjugated oligomers remain appreciably less active than PLL–conjugated ones (compare Fig. 2A and B). Poly(L-lysine) conjugation thus appears as a tool to increase the

antiproliferative activity of antisense oligomers in line with our initial data in an antiviral model.

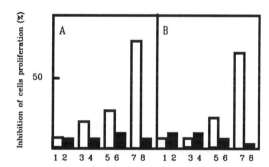

Fig. 2. (**A**) Antiproliferative effects of unconjugated antisense oligomers directed to c-*myc*. L929 cells were incubated for 4 hr in optiMEM or MEM in the presence or absence of 10% (V/V) fetal calf serum (this cell line is negligibly affected by a transient serum depletion). Unconjugated oligomers specific of c-*myc* (1,3,5,7) or control VSV oligomers (2,4,6,8) were then added in optiMEM (1,2,3,4) or MEM and sera (5,6,7,8) at 10 (1,2,5,6) or 40 (3,4,7,8) μM concentration. Cells were trypsinized 24 hr later and counted (vide supra). (**B**) Antiproliferative effects of antisense oligomers in ternary complexes with PLL and heparin. L929 cells (in 24-wells tissue culture plates) were incubated for 24 hr with oligomer–PLL conjugates specific of c-*myc* (1,3,5,7) or VSV (2,4,6,8) at 0.2 (1,2,5,6) or 2 (3,4,7,8) μM concentrations with (5,6,7,8) or without (1,2,3,4) heparin (100 μg/ml). The cells were then trypsinized and counted with a Coulter counter.

The mechanisms through which antisense oligomers act have not been fully investigated in most cases. Ternary complexes involving antisense oligomers with the relevant sequence reduce the amount of steady-state c-*myc* mRNA (Fig. 3) and c-*myc* protein. This might result from (1) nonspecific degrada-

tion of c-*myc* mRNA after translation arrest, (2) RNase H-mediated degradation of the RNA moiety in the hybrid formed between the antisense oligomer and its RNA target, or even (3) a blockade by the antisense oligomer of c-*myc* mRNA synthesis, pre-mRNA processing or nucleocytoplasmic mRNA transport (see Hélène and Toulmé, 1990, for a recent review). Further experiments will be required to distinguish these possibilities.

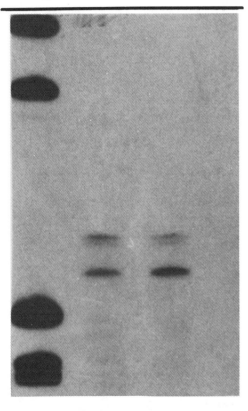

Fig. 3. Effects of antisense oligomers in ternary complexes with PLL and polyanions on c-*myc* RNA level. L929 cells growing in 24-well tissue culture plates [MEM 10% (v/v) fetal calf serum] were left untreated (track 2) or incubated for 8 hr with 2 μM c-*myc* (track 4) or c-*myc* mismatch (track 3) oligomer–PLL conjugates with 100 μg/ml heparin and the RNA samples were then isolated. The *Sst*I *Eco*RV (211 bp) restriction fragment from mouse c-*myc* gene cloned in pSP64 was used to generate an RNA probe synthesized using Sp6 RNA polymerase and [α-^{32}P]UTP (100 Ci/mmol). Hybridization of 2×10^5 cpm of probe with the RNA was in 40 μl of 80% (v/v) formamide, 40 mM PIPES,

CELL TARGETING OF ANTISENSE OLIGONUCLEOTIDES THROUGH ANTIBODY-BEARING LIPOSOMES

In spite of increased action of the oligodeoxyribonucleotides–poly(L-lysine) complex, the approach lacks specificity for particular cells, and the coupled oligodeoxyribonucleotide sequences may remain sensitive to degradation by nucleases. We are examining an alternative strategy for augmenting the cell association and enhancing the resistance to degradation of oligonucleotide sequences, namely the encapsulation of oligonucleotides in liposomes. Liposomes are synthetic vesicles that may contain a number of different kinds of molecules encapsulated in their aqueous spaces, including drugs, proteins, and polynucleotides. By virtue of their limited permeability, liposome membranes restrict access by the milieu, and protect their contents against enzymatic degradation. Liposomes may, in addition, be coupled covalently by well-established techniques to various ligands, including monoclonal antibodies or protein A, which permit their tar-

pH 6.8, 400 mM NaCl, 1 mM EDTA at 45°C for 12 hr. RNase A (100 μg/ml in 300 μl of 300 mM NaCl, 10 mM Tris-HCl, pH 7.4, 5 mM EDTA) was then added for 1 hr at 30°C. After treatment by proteinase K, solvent extraction and ethanol precipitation in the presence of 30 μg/ml yeast tRNA as carrier, probes were fractionated in a 5% (w/v) acrylamide–7 M urea sequencing gel. Their distribution in the gels was revealed by autoradiography on Kodak Xomat films. The size of the protected fragments was evaluated using φX174 RF digested with *Hae*III and labeled with [α-^{32}P] ddATP using terminal transferase (track 1) (fragments generated 873, 604, 311, 281–271 nt).

geting to specific cell populations. Depending on the target molecule and cell, these liposomes may be taken up and release the encapsulated product intracellularly (shown schematically in Fig. 4). We have previously shown intracellular delivery from liposomes of low-molecular-weight drugs (Leserman et al., 1989) and of polynucleotides such as $(I)_n \cdot (C)_n$ (Milhaud et al., 1989), and also the delivery of oligodeoxyribonucleotide sequences complementary to the 5'-end region of the N-protein of VSV RNA from antibody-targeted liposomes (Leonetti et al., 1990b).

In the present study we encapsulated in liposomes the same 15-mer anti-c-*myc* oligodeoxyribonucleotide sequence used in the poly(L-lysine) study discussed above. These liposomes were coupled covalently to protein A. Separation of nonencapsulated and encapsulated oligonucleotides was accomplished by gel-exclusion chromatography. When incubated (at nanomolar concentrations) with mouse L929 cells, together with antibodies directed at the expressed major histocompatibility complex-encoded class I molecule H-2K, inhibition of the proliferation of the L929 cells was suppressed (Fig. 5). Liposomes of the same composition but containing only buffer, or containing an antisense oligonucleotide specific for N-protein of VSV, incubated with the same antibody had no effect on the proliferation. As we have reported for liposomes transporting antiviral oligodeoxynucleotides (Leonetti et al., 1990b), the cell-binding antibody was necessary for the effects seen: when incubation occurred in the presence of antibodies of the same class that did not fix to the cells, no inhibition was observed. Incubation of the liposome preparation in DNase 1 at a concentration of

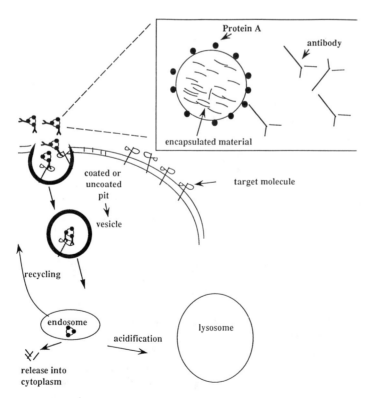

Fig. 4. Schematic presentation of the internalization of antibody-targeted liposomes.

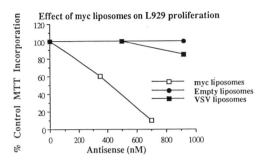

Fig. 5. Antiproliferative activity of liposome-encapsulated antisense oligomers directed to c-*myc*. Cells were incubated with H-2K-specific antibodies and various concentrations of *myc* or control liposomes, as indicated. Cell proliferation was monitored by an MTT assay.

50 μg/ml did not inhibit the potential of the liposomes to inhibit cell proliferation (data not shown).

Encapsulation in antibody-bearing liposomes thus permits protection of liposome contents from degradation and a certain degree of cell-specific targeting, and also permits their delivery into cells in an active form.

INTRACELLULAR DISTRIBUTION OF ANTISENSE OLIGONUCLEOTIDES

The experiments described so far in this and other papers have generally made use of oligomers targeted to c-*myc* mRNA translation initiation site. Their actual site of action within intact cells has not been firmly established since the same sequences are also present in the primary transcript and gene complementary DNA strand. Interference of such antisense oligomers with transcriptional events, pre-mRNA maturation steps, or nucleocytoplasmic transport could thus also be envisaged, as already emphasized for other genes. Oligonucleotides specifically designed to interfere at nuclear steps of c-*myc* gene expression would be worth testing for their relative efficiency as compared to the "antimessenger" ones.

As a preliminary step toward this goal, the intracellular fate of antisense oligomers was investigated. The problem of their internali-

zation pathway and escape from the endocytic compartment was obviated by microinjection with micropipets into the cytoplasm of somatic cells. Their conjugation to various probes allowed an assessment of their intracellular distribution by fluorescence microscopy of fixed or of living cells. As shown in Figure 6A for fixed cells, a 15-mer oligomer conjugated at its 5'-end to tetramethylrhodamine isothiocyanate (TRITC) has accumulated within cell nuclei as early as 1 min after its microinjection within the cytoplasm of REF-52 rat embryo fibroblasts. In contrast, coinjected FITC-labeled dextran has not diffused out of the cytoplasm in line with its high molecular weight. Oligomers of various sizes and lengths, including double-stranded 50-mer oligodeoxyribonucleotides, behaved similarly whether their fate was monitored by fluorescence microscopy of their associated fluorochromes (TRITC, FITC, AMCA), or indirect immunofluorescence with antibodies specific for bromodeoxyuridine-modified nucleotides. The direct observation of intracellular fluorescence distribution in nonfixed microinjected cells has demonstrated that oligomer nuclear translocation takes place within 1 min. Diffusion through the nuclear pores is probably involved, as opposed to active transport, since the process remains unaffected in its rate and magnitude upon ATP depletion, low temperature incubation, or with oligomer concentration.

First steps in the identification of oligomer nuclear binding sites have been made. Our ongoing experiments show that (1) the nuclear retention of the fluorescent signal is directly proportional to the metabolic stability of the conjugated oligomers, α-anomeric or phosphorothioates oligomers remaining in the nucleus for much longer periods of times than natural oligomers, and (2) photoactivatable oligomers bind to a restricted set of nonhistone, but as yet unidentified proteins, in intact nuclei.

Whatever the biological meaning and the nature of the nuclear structures concentrating oligomers, these observations might be worth being considered in target choice and in the design of the most appropriate oligomers or oligomer delivery system.

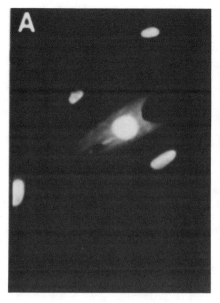 

Fig. 6. Nuclear accumulation of rhodamine-labeled oligomers. TRITC conjugated 15-mer oligomer (100 µg/ml) was coinjected with 2 mg/ml FITC-labeled dextran (MW 70,000) in the cytoplasm of rat embryo fibroblasts (REF52) cells. After incubation (30 min at 37°C) the cells were fixed with ethanol–acetone (70/30, v/v). Fluorescence photomicrographs have been taken at a 40 times enlargement and display the same field; (**A**) TRITC-labeled material and (**B**) FITC-labeled one.

PROBLEMS AND PERSPECTIVES

The published literature and experiments in progress in our group as well as in an increasing number of academic institutions and corporations have demonstrated the feasibility and the specificity of the antisense approach in in vitro experiments. It is our feeling however that we are still far from having explored all the potential of the antisense approach, or even optimized the existing procedures. Our knowledge of the basic mechanisms involved in internalization of oligomers by cells, intracellular traffic, interaction with their RNA, DNA, or protein targets, and mode of action is still rather fragmentary and requires additional study.

As a first example, these and previous studies have established the possibility of internalizing free PLL-conjugated, or liposome-encapsulated oligomers. Whether fluid-phase or receptor-mediated endocytosis is involved, oligomers (and their carriers) will transit through the endocytic compartment and ultimately will have to be released in a nondegraded form at their cytoplasmic or nuclear site of action. How and at which stage oligomers dissociate from their carrier and cross the bilayer membrane of the endocytic compartments remain unknown. Experiments aiming at associating oligo- and polynucleotides with lipid vesicles engineered to fuse with the plasma or endosome membrane are currently in progress in our group.

The rapid translocation of microinjected antisense oligomers to the cell nucleus and its accumulation therein offers new insights to target choice, as briefly discussed above, but raises potential problems as well. The control of nuclear events in c-*myc* expression through specific interference at transcription initiation, transcription elongation, splicing, or through triple helix formation could indeed be envisaged without involving the conjugation of synthetic oligomers to karyophilic peptides, which appears as a favorable situation. However, the nuclear binding sites described here might prevent the oligomers from interacting with their RNA or DNA targets, and therefore be detrimental to specific interference.

This represents but two examples of the many unknowns we face in this still new, but already promising field. Such problems obviously complicate the prospect of the utilization of antisense oligomers as pharmacological agents. This should, however, not be considered as a source of pessimism, since increasing complexity stimulates the research that often offers unexpected new avenues.

SUMMARY

Antisense oligonucleotides offer great promise for the control of viral genes and oncogenes. Studies using vesicular stomatitis virus genes as model target genes have been engaged in our group. They aimed at targeting antiviral oligonucleotides to cell surface determinants and at increasing the efficiency of oligonucleotides internalization in cells (Lemaître et al., 1987; Leonetti et al., 1990b).

These tools, oligonucleotide encapsidation into antibody-coated liposomes or chemical conjugation of oligonucleotides to poly(L-lysine), have now been adapted to c-*myc* oncogene. Both direct a sequence-specific down-regulation of this oncogene leading to an antiproliferative activity in various cell lines at micromolar or lower doses. These represent significant achievements toward the goal of controlling efficiently oncogene expression in specific cell types.

The direct microinjection in cell cytoplasm of antisense oligonucleotides tagged with fluorochromes has revealed their fast translocation and accumulation in nuclei; implications of this observation for target choice

and oligonucleotide design have yet to be evaluated.

REFERENCES

Cooney M, Czernuszewicz G, Postel EH, Flint SJ, Hogan ME (1988): Site-specific oligonucleotide binding represses transcription of the human c-myc gene in vitro. Science 241:456–459.

Goodchild J (1990): Conjugates of oligonucleotides and modified oligonucleotides: A review of their synthesis and properties. Bioconjugate Chem. 1:165–186.

Heikkila R, Schwab G, Wickstrom E, Loke SL, Pluznik DH, Watt R, Neckers LM (1987): A c-myc antisense oligodeoxynucleotide inhibits entry into S phase but not progress from G_0 to G_1. Nature (London) 328:445–449.

Hélène C, Toulmé JJ (1990): Specific regulation of gene expression by antisense and anti-gene nucleic acids. Biochem. Biophys. Acta 1049:99–125.

Holt JT, Redner RL, Nienhuis AW (1988): An oligomer complementary to c-myc mRNA inhibits proliferation of HL-60 promyelocytic cells and induces differentiation. Mol. Cell. Biol. 8:963–973.

Kristal GW, Armstrong BC, Battey JF (1990): N-myc mRNA forms an RNA-DNA duplex with endogenous antisense transcripts. Mol. Cell. Biol. 8:4180–4191.

Lemaître M, Bayard B, Lebleu B (1987): Specific antiviral activity of a poly(L-lysine)-conjugated oligodeoxyribonucleotide sequence complementary to vesicular stomatitis virus N protein mRNA initiation site. Proc. Natl. Acad. Sci. U.S.A. 84:648–652.

Leonetti JP, Degols G, Lebleu B (1990a): Biological activity of oligonucleotide-Poly(L-lysine) conjugates: Mechanism of cell uptake. Bioconjugate Chem. 1:149–153.

Leonetti JP, Machy P, Degols G, Lebleu B, Leserman L (1990b): Antibody-targeted liposomes containing oligodeoxyribonucleotides complementary to viral RNA selectivity inhibit viral replication. Proc. Natl. Acad. Sci. U.S.A. 87:2448–2451.

Leserman L, Langlet C, Schmitt-Verhulst AM, Machy P (1989): Positive and negative liposome-based immunoselection techniques. In Tartakoff A (ed): Methods in Cell Biology, Vol. 32, Vesicular Transport, Part B. New York: Academic Press, pp 447–471.

Loke SL, Stein C, Zhang X, Avignan M, Cohen J, Neckers IM (1988): Delivery of c-myc antisense phosphorothioate oligodeoxynucleotides to hematopoietic cells in culture by liposome fusion: Specific reduction in c-myc protein expression correlates with inhibition of cell growth and DNA synthesis Curr. Topics Microbiol. Immunol. 141:282–289.

Loke SL, Stein CA, Zhang XH, Mori K, Nakanishi M, Subasinghe C, Cohen JS, Neckers M (1989): Characterization of oligonucleotide transport into living cells. Proc. Natl. Acad. Sci. U.S.A. 86:3474–3478.

McManaway ME, Neckers LM, Loke SJ, Al-Nasser AA, Redner RL, Shiramizu BT, Goldschmidts, Huber BE, Batia K, Magkath IT (1990): Tumour-specific inhibition of lymphoma growth by an antisense oligodexynucleotide. Lancet 335:808–811.

Milhaud PG, Machy P, Lebleu B, and Leserman L (1989): Antibody-targeted liposomes containing poly (rI)·poly (rC) exert a specific antiviral and toxic effect on cells primed with interferons α/β or γ. Biochim. Biophys. Acta 987:15–20.

Morgan DML, Clover J, Pearson JD (1988): Effects of synthetic polycations on leucine incorporation, lactate dehydrogenase release and morphology of human umbilical vein endothelial cells. J. Cell. Sci. 91:231–238.

Shen WC, Ryser HJP (1978): Conjugation of poly-L-lysine to albumin and horseradish peroxidase: A novel method of enhancing the cellular uptake of proteins. Proc. Natl. Acad. Sci. U.S.A. 75:1872–1876.

Stevenson M, Iversen PL (1989): Inhibition of human immunodeficiency virus type 1-mediated cytopathic effects by poly(L-lysine)-conjugated synthetic antisense oligodeoxyribonucleotides. J. Gen. Virol. 70:2673–2682.

Wickstrom EL, Bacon TA, Gonzalez A, Freeman DL, Lyman GH, Wickstrom E (1988): Human promyelocytic leukemia HL-60 cell proliferation and c-myc protein expression are inhibited by an antisense pentadecadeoxynucleotide targeted against c-myc mRNA. Proc. Natl. Acad. Sci. U.S.A. 85:1028–1032.

Yakubov LA, Deeva EA, Zarytova VF, Ivanova EM, Ryte AS, Yurchenko LV, Vlassov VV (1989): Mechanism of oligonucleotide uptake by cells: Involvement of specific receptors? Proc. Natl. Acad. Sci. U.S.A. 86:6454–6458.

Transcriptional Regulation of c-*myc* Protooncogene by Antisense RNA

Kazushige Yokoyama

Gene Bank and Frontier Research Program, Tsukuba Life Science Center, RIKEN (The Institute of Physical and Chemical Research), Tsukuba, Ibaraki 305, Japan

INTRODUCTION

The promyelocytic leukemia cells line HL-60 can be differentiated terminally *in vitro.* Treatments with agents such as dimethyl sulfoxide (DMSO) (Collins et al., 1978; Collins and Groudine, 1982), N^6,O^2-dibutyryl adenosine $3',5'$-cyclic monophosphate (Chaplinski and Niedel, 1982), and retinoic acid (Breitman et al., 1980) result in the expression of granulocytic markers in HL-60 cells, while exposure to vitamin D (Bar-Shavit et al., 1983; McCarthy et al., 1983) or phorbol esters (Rovera et al., 1979) leads to the expression of macrophage-associated markers. HL-60 cells exposed to these inducers undergo growth arrest and ultimately reach a terminally differentiated state. The c-*myc* steady-state mRNA decreases to very low levels within a few hours after the addition of differentiating agents (Bentley and Groudine, 1986; Bishop, 1987; Dalla-Favera et al., 1982; Dyson et al., 1985; Eick and Bornkamm, 1986; Filmus and Buick, 1985; Grosso and Pitot, 1985; McCachren et al., 1986; Reitsma et al., 1983; Simpson et al., 1987; Trepel et al., 1987; Watanabe et al., 1985; Westin et al., 1982).

It has been suggested that this decrease may be a necessary step in cellular differentiation and growth arrest of terminally differentiating cells. It is unclear, however, whether decreased c-*myc* expression is a cause, or a closely linked consequence, of HL-60 differentiation.

The introduction of an antisense RNA complementary to a specific mRNA into cells can effectively create or mimic the null-mutant phenotype, and provide a cell line useful for investigating gene products of unknown physiological function (Green et al., 1986; Weintraub et al., 1985). We have recently applied the use of antisense RNA approach to inhibit c-*myc* expression to study its role in HL-60 cell differentiation (Yokoyama and Imamoto, 1987).

In this review we will examine the effects of antisense c-*myc* transcripts on the constitutive expression of the c-*myc* gene and show that antisense c-*myc* gene introduced into the human promyelocytic leukemia cell line HL-60 can inhibit not only c-*myc* protein synthesis but also transcription of the endogenous c-*myc* gene. A 74-kDa nuclear transcriptional regulatory protein is identified in the antisense c-*myc* transformant. This protein can decrease transcription of c-*myc* and commit HL-60 cells to monocytic differentiation without the help of a differentiation inducer. These results suggest that a 74-kDa nuclear protein is one of several negative repressors of c-*myc* gene transcription, and one of the factors directed to cell differentiation.

SELECTION OF THE ANTISENSE C-*MYC* TRANSFORMANTS

Antisense and sense c-*myc* plasmids pSVgpt C5−8 and pMMTVgpt C5−8 contain the c-*myc* coding sequences cloned in the antisense and sense orientation relative to a simian virus 40 promoter or mammary tumor virus promoter. These plasmids with a selective marker gene *Ecogpt* (*Escherichia coli*

Prospects for Antisense Nucleic Acid Therapy of Cancer and AIDS, pages 35–51

xanthine/guanine phosphoribosyltransfer-ase) are shown in Figure 1. Plasmid pC5–8 (S. Tonegawa, unpublished result), which contains a functional c-*myc* cDNA (2.4 kbp *Bam*HI fragment without exon 1), was the source of the c-*myc* sequence. The strategy of all constructions was to clone *Ecogpt* into pSVMdhfr (Lee et al., 1981) and to replace the dihydrofolate reductase gene with c-*myc* or other coding sequences (Nishioka and Leder, 1979; Weiss et al., 1983; Yokoyama and Imamoto, 1987).

Although the structural significance of the noncoding exon 1 of c-*myc* gene is not clear, possible roles have been described (Cole, 1986): (1) the pausing effect of transcription might occur in the end of exon 1, (2) the possible stem–loop structure between exon 1 and exon 2 might suppress the translation frequency of the c-*myc* protein, and (3) the rate-limiting region of mRNA stability of c-*myc* is located in exon 1. Given these possibilities, we did not use the sequence of exon 1 for the antisense plasmid construction. Initially we constructed an antisense c-*myc* plas-

mid using the cloned genomic DNA. However, the suppression efficiency of endogenous c-*myc* expression was quite low, possibly because of the complicated structure formed by the complementary sequence between exon 1 and intron 1 (Cole, 1986) and therefore we decided to use the cDNA sequence for plasmid construction.

These antisense and sense plasmid DNAs were introduced into the promyelocytic leukemia cell line HL-60 by the protoplast fusion method (Sandri-Goldin et al., 1981).

Short-Term Transfection Assay

We used two different assay systems to establish the antisense c-*myc* HL-60 transformants. One is the inducible expression system. We constructed antisense c-*myc* plasmids that were driven by mammary tumor virus (MMTV) LTR promoter or by mouse metallothionein promoter (in MTII$_a$) and transfected into HL-60 cells. The second method is the constitutive expression system. The antisense c-*myc* expression vectors were constructed with the promoter of simian virus

Inducible Expression

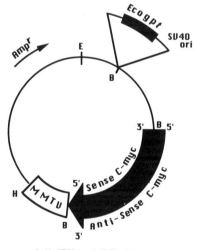

pMMTVgptC5-8

Constitutive Expression

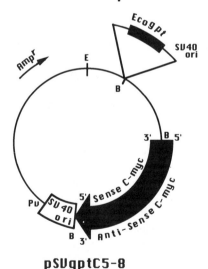

pSVgptC5-8

Fig. 1. Plasmid construction. Antisense and sense c-*myc* plasmids pMMTVgptC5–8, or pSVgptC5–8 contain the c-*myc* coding sequences cloned in the antisense and sense orientation relative to a

simian virus 40 promoter, or a mammary tumor virus promoter, plus the selective-marker gene *Ecogpt.*

40 (SV40) early region and of Rous sarcoma virus (RSV) LTR. However, none of the antisense c-*myc* clones established by the short-term transfection assay demonstrated the reduction of c-*myc* protein production.

Long-Term Transfection Assay

We have used two strategies to try to increase the level of antisense c-*myc* transcripts in the transformants (Yokoyama and Imamoto, 1987). First, the resistance to mycophenolic acid of a primary transformant (AM 93) was gradually increased by successive rounds of selection with increasing drug concentration over a 6-month period. Kim and Wold (1985) reported using the same strategy to increase the antisense transcript to a level sufficient to inhibit *tk* gene expression. In the second method we tried, secondary (AM 93-4) and tertiary (AM 93-4-12) transformants were established by successively retransfecting AM 93 and AM 93-4 with antisense c-*myc* DNA. These procedures resulted in elevated levels of antisense transcripts (Yokoyama and Imamoto, 1987).

A comparison of the amounts of antisense and sense c-*myc* RNA in various clones is given in Table 1. These data clearly show that selection for resistance to high concentrations of mycophenolic acid and repeated transfection with antisense c-*myc* plasmid DNA resulted in increased levels of antisense c-*myc* RNA. Examination of genomic DNA isolated from the clones demonstrated that amplification of the antisense c-*myc* gene had occurred, resulting in 100–250 copies in the tertiary transformants (Table 1). An RNase protection study confirmed the presence of antisense RNA-sense RNA-duplex in the nuclei (Yokoyama and Imamoto, 1987). The gpt control clone contained no detectable double-stranded c-*myc* RNA, whereas a 2.2-kb band was observed in the antisense clones. This protected band was found in the nuclear rather than in the cytoplasmic RNA fraction of the cells (Yokoyama and Imamoto, 1987).

The half lives of antisense c-*myc* transcript and the duplex RNAs between antisense and sense endogenous c-*myc* were quite long, about 3 to 4 hr, in contrast to the short half

life of sense c-*myc* transcript. The analysis of duplex formation between exogenous antisense and endogenous sense c-*myc* RNAs by a polyacrylamide gel showed 10-fold excess of antisense RNAs resulting in forming RNA duplex of all sense RNA.

C-*MYC* PROTEIN LEVELS IN ANTISENSE C-*MYC* TRANSFORMANTS

Antisense and sense c-*myc* transformants were incubated with [^{35}S]methionine for 8 hr to analyze c-*myc* protein synthesis under steady-state conditions. Six representative antisense transformants showed a significant reduction in the amount of c-*myc* gene product compared to control clones (Table 1). This reduction in the steady-state level of p 64 (c-*myc* protein) is probably not due to increased turnover of the protein in the antisense transformants because the pulse-chased experiment of c-*myc* protein demonstrates the half-life of p64 is about the same in the antisense and control clones. The relative amount of p64 in antisense and control clones was not significantly changed by correcting for differences in the half-life of p64 and in the pool sizes of [^{35}S]methionine (data not shown). Control experiments showed that the reduction in the amount of c-*myc* protein in the antisense transformants apparently depends specifically on the presence of antisense c-*myc* sequences. Control antisense clones were established by transfecting HL-60 cells with plasmids identical to pSVgpt C5–8 except that they contained antisense H-2K^b, antisense α-globin 68-mers, or antisense *Ecogpt*$^+$ in place of antisense c-*myc*. They were screened by the identical protocol as the antisense c-*myc* transformants. These clones exhibited no detectable reduction in p64 synthesis and showed no differentiation phenotypes, indicating that decreased c-*myc* expression requires antisense c-*myc* sequences and is not merely an artifact of mycophenolic acid selection or of culture conditions. A second control experiment showed that the level of actin protein synthesis is essentially unchanged in all transformants (Table 1).

TABLE 1. Summary of Sense and Antisense Transformants[a]

Cells	mRNA, cpm × 10⁻³			DNA, copy number		p64 protein, cpm × 10⁻³		Growth rate, hr per cell cycle
	S MYC	Actin	AS MYC	S MYC	AS MYC	MYC	Actin	
HL-60	20.5	18.1	0.1	25	—	2.7	4.2	33
Control pSV2gpt⁺	17.5	17.2	0.1	22	—	2.5	4.4	36
Sense transformant								
pSVgptC5-8	49.2	16.9	0.4	165	—	3.8	5.3	38
pSVH-2K[b]	16.9	16.3	—	25	—	2.5	4.8	36
pSVα-globin 68	18.5	17.0	—	20	—	2.6	4.4	34
Antisense transformant								
AM93 1 month	17.5	17.2	3.1	20	20	2.2	3.9	—
2 months	14.4	—	9.3	—	25	2.0	—	—
3 months	11.3	—	19.4	—	40	1.8	—	—
6 months	6.2	15.4	31.1	20	55	1.6	4.1	—
AM93-4	5.1	16.0	50.6	22	100	1.0	4.0	—
AM93-4-12	3.3	14.9	77.8	20	120	0.6	4.0	135
AM2-3-91	10.3	17.0	27.2	20	170	1.7	3.9	95
AM46-3-2	4.2	15.2	66.9	20	160	0.8	4.0	120
AM451-6-30	4.5	15.6	62.2	22	250	0.8	4.0	125
AM763-11-4	5.1	16.3	46.7	25	200	1.0	4.0	110
AM966-10-64	6.8	15.8	42.8	20	200	1.4	4.1	105
pSVH-2K[b]	17.5	17.1	—	20	—	2.6	4.1	36
pSVα-globin 68	18.1	16.0	—	25	—	2.6	4.5	38
pSV2gpt⁺	18.5	17.2	—	20	—	2.6	4.1	34

[a]Values are means of four experiments. Protein levels and mRNAs of antisense (AS) and sense (S) MYC and actin were measured by counting radioactivity in bands cut out of SDS₄ gels (protein) or out of RNA gel blot filters (mRNA). Copy number of the integrated DNA was determined by comparison with known amounts of MYC plasmid DNA (equivalent to 10⁻⁶ to 10⁻⁹ molecules).

According to the results of cell growth measurement, the inhibition of synthesis of the c-*myc* gene product might cause a substantial alteration in the growth machinery of the cell (Table 1).

C-*MYC* GENE EXPRESSION AND CELL DIFFERENTIATION

The HL-60 cell line, which consists predominantly of rapidly dividing cells with promyelocytic characteristics (Collins et al., 1978), contains multiple copies of the c-*myc* gene (Collins and Groudine, 1982; Dalla-Favera et al., 1982), and expresses 20- to 30-fold excesses of c-*myc* RNA per cell when compared with that of normal cells. Chemical reagents such as TPA and DMSO inhibited HL-60 cell proliferation and committed cell differentiation into two different pathways. Induction of HL-60 cells to differentiate with reagents leads to a decline in c-*myc* mRNA (Bentley and Groudine, 1986; Dyson et al., 1985; Eick and Bornkamm, 1986; Grosso and

Pitot, 1985; McCachren et al., 1986; Reitsma et al., 1983; Simpson et al., 1987; Trepel et al., 1987; Watanabe et al., 1985; Westin et al., 1982). In order to see whether decreased c-*myc* expression is a cause or a closely linked consequence of HL-60 cell differentiation, we analyzed the phenotype of antisense c-*myc* transformants.

Immunofluorescence staining and cytochemical studies demonstrated a close correlation between a decrease in the amount of c-*myc* protein in cells and an increase in the number of cells with a monocytic phenotype (Table 2). This correlation suggests that a reduction in c-*myc* expression may be the initial event in the differentiation of HL-60 cells to commit to monocytes. Although reduced c-*myc* expression may not be sufficient to commit HL-60 cells to monocytic differentiation, the presence of antisense c-*myc* appears to change the developmental potential of HL-60 cells biasing the cells toward the monocytic pathway in preference to granulocytic development.

TABLE 2. Comparative Human MYC Protein Level and Relative Cell Number of the Differentiated Phenotypes[a]

Cells	S *MYC* expression,[b] ratio × 10^{-3}	% Reactive cells			
		M[c]	G[d]	OKM-1	OKM-5
HL-60	1.11	0	0	0	0
HL-60 + PMA	—	99	0	95	99
HL-60 + Me₂SO	—	0	98	3	4
Control pSV2gpt⁺	1.18	12	15	5	6
Transformants					
Sense pSVgptC5–8	1.45	15	18	8	10
Antisense					
AM2-3-91	0.22	27	10	32	29
AM46-3-2	0.10	56	13	61	63
AM93-4-12	0.08	72	12	80	88
AM451-6-30	0.08	58	14	67	74
AM763-11-4	0.22	34	15	49	53
AM966-10-64	0.10	41	17	54	69

[a]Values are means of four experiments. Phorbol 12-myristate 13-acetate (PMA) and dimethyl sulfoxide (Me₂SO) were used as specific inducers of monocyte and granulocyte, respectively. S, sense.
[b]The ratio of the levels of MYC protein and mRNA per cell were calculated from data such as in Table 1, correcting for differences in cell number and in incorporation rate of methionine.
[c]Percent of positive cells by staining of α-naphthyl acetate esterase.
[d]Percent of positive cells by staining of naphthol AS-D chloroacetate esterase.

TRANSCRIPTIONAL CONTROL OF THE C-*MYC* GENE

During the isolation of antisense c-*myc* transformants, we found that most of the antisense c-*myc* transformants showed decreased levels of endogenous c-*myc* mRNA (approximately 1/3- to 1/10-fold). This finding prompted us to analyze the effect of antisense c-*myc* on the promoter activity of the endogenous c-*myc* gene.

Transcriptional Expression of Endogenous c-*myc* Gene by Antisense RNA

The effect of antisense c-*myc* RNA on constitutive transcription of c-*myc* gene was analyzed by measuring promoter activity. Two fragments of the c-*myc* promoter, the *Hin*dIII–*Pvu*II fragment (3 kbp) and the *Pst*I–*Pvu*II fragment (920 bp), were tested in a promoter CAT vector. In parental HL-60 cells and in the pSV2gpt$^+$ clone, weak CAT activity was observed with the larger fragment of the c-*myc* promoter (*Hin*dIII–*Pvu*II), while the smaller fragment (*Pst*I–*Pvu*II) produced strong CAT activity.

Surprisingly, in antisense c-*myc* transformants, the enhancer carrying promoter fragment (*Pst*I–*Pvu*II) gave less than 10% of CAT activity when compared to that of the *Hin*dIII–*Pvu*II fragment (Yokoyama and Imamoto, 1987).

A study of nuclear run-on assay also confirmed that this regulation of promoter activity was at the transcriptional level (Yokoyama and Imamoto, 1987). In order to determine the precise nucleotide sequences required for this inhibitory activity we constructed various deletion mutants with *Exo*III. These constructs were transfected into both HL-60 cells and antisense c-*myc* transformants (AM93-4-12) and their relative CAT activities were measured.

Results shown in Figure 2 indicate that the target sequence of antisense c-*myc* induced suppression is localized within a 56-bp fragment of the c-*myc* gene promoter region (lanes 1 and 2). An internal deletion between − 158 and − 102 displayed no effect on negative regulation (lane 7). These data suggest the presence of a target sequence for antisense c-*myc* induced gene suppression located between − 158 and − 102.

Specific Factors in the Nuclear Extracts From Antisense c-*myc* Transformants

To investigate whether nuclear factors in extracts derived from AM93-4-12 were capable of binding to sequences in the − 158 to − 102 region, DNase I footprinting was carried out. Double-stranded 56-bp oligodeoxynucleotides corresponding to this sequence were prepared and used as a probe for DNase I footprinting experiment. The protected region, using a nuclear extract from antisense c-*myc* transformant AM93-4-12, was identified at position − 151 to − 117 in the promoter region (Fig. 3A). However, this region was not protected with extracts from HL-60 cells that were not transfected.

We next synthesized 45-bp oligodeoxynucleotides duplex, TCTCCTCCCCACCTTCCCCACCCTCCCCACCCTCCCCATAAGCGC to be used as a probe in a gel retardation assay (Fig. 3B). The shifted band was detected in the extract from AM93-4-12 (lane 5) and the binding was completely inhibited with the three tandem repeated oligodeoxynucleotides containing CACC-TCC sequence, (CACCCTCCC)$_3$ (lane 6). The nuclear extracts from HL-60 contained much lower amounts of the binding proteins which bind to this sequence (lanes 2 and 4).

Thus, in the nuclear extracts from AM93-4-12, specific DNA binding proteins were induced and found to bind to the nucleotide sequence CACC-TCC repeat in the promoter region of c-*myc* gene. Thus the evidence suggests that this binding protein might be a negative regulator of c-*myc* gene transcription.

To characterize the nuclear proteins that interact with the CACC-TCC elements, southwestern blotting (Miskimins et al., 1985; Fainsod et al., 1986) was carried out. The nuclear extracts from HL-60 and AM93-4-12 cells were fractionated on a heparin agarose column. The proteins eluted from this column by 0.4 *M* KCl were applied to a DEAE-

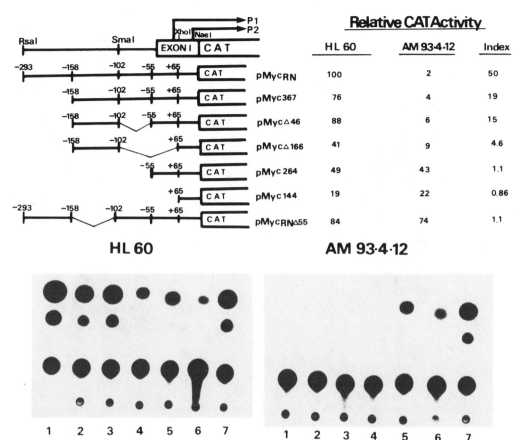

Fig. 2. Expression of the pC-myc-CAT deletion constructs in HL-60 and antisense c-*myc* clone (AM93-4-12). CAT activity was measured by thinlayer chromatography as described (Yokoyama and Imamoto, 1987). P1 and P2, transcriptional start sites of c-*myc* gene. Left panel shows CAT activities in HL-60 and right panel shows CAT activities in AM93-4-12. Lane 1, pMyc RN; lane 2, pMyc 367; lane 3, pMyc Δ46; lane 4, pMyc Δ166; lane 5, pMyc 264; lane 6, pMyc 144; lane 7, pMyc RNΔ55.

Sepharose column and eluted with a stepwise gradient of KCl. These fractionated proteins were then electrophoresed on an SDS polyacrylamide gel. The nylon filters, after transfer of proteins by western blotting, were incubated with ^{32}P-radiolabeled probe containing the CACC-TCC element in order to detect specific binding proteins. As shown in Figure 4A, 74-kDa proteins were seen in nuclear fractions from both HL60 and AM93-4-12 cells. However, the DNA binding activity of 74-kDa protein in AM93-4-12 cells was much higher than in HL60 cells. The probes not containing the CACC-TCC sequence did

not show specific bands (Fig. 4B, C). The proteins eluted from DEAE Sepharose were directly applied to an affinity column containing CACC-TCC repeated sequences. The bound fraction was recycled twice. The eluted fractions were analyzed by two-dimensional SDS polyacrylamide gel. As shown in Figure 4D distinct spots of 74-kDa protein were observed in the eluted fraction of AM93-4-12 nuclear extracts in addition to the bands of 110-kDa protein. In the fraction of HL60 nuclear extracts, a faint band of 74-kDa protein was observed (Fig. 4E). These results indicate that the synthesis of both 74-kDa and

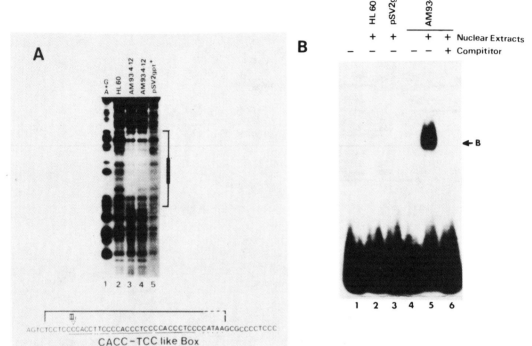

Fig. 3. DNase I footprint and gel retardation assays of CACC-TCC binding protein. (**A**) A radioactively labeled promoter fragment (-158 to -102) was incubated with nuclear extracts prepared from HL-60, antisense c-*myc* transformant, and pSV2gpt$^+$ transformant and analyzed by the DNase I footprint (Katoh et al., 1990). Lane 1, G + A; lane 2, 65 μg of nuclear extract from HL-60; lane 3, 35 μg of nuclear extract from AM93-4-12; lane 4, 50 μg of nuclear extract from AM93-4-12; lane 5, 65 μg of nuclear extract from pSV2gpt$^+$ clone. The location of the protected region is in-

dicated to the right and the protected nuclear sequence is indicated below. (**B**) T4 kinase-treated oligodeoxynucleotides duplex (45-mers) containing the binding site of CACC-TCC protein (lane 2) was incubated with the nuclear extracts (5 μg) from HL-60, pSV2gpt$^+$ clone (lane 3) and AM93-4-12 (lanes 5, 6) by the method described elsewhere (Katoh et al., 1990). The competitor oligodeoxynucleotides of three tandem repeat of CACCCTCCC motif were added in the incubation mixture (lane 6). The shifted band was indicated as B.

110-kDa proteins is specifically induced in the antisense c-*myc* transformant, and these proteins might be candidates for suppressors of the endogenous c-*myc* gene expression. Given the results of southwestern blot, it may be unlikely that the 110-kDa molecule is a DNA-binding protein. It is unknown how this protein might couple with 74-kDa protein and how it might modulate the transcriptional activity of the endogenous c-*myc* gene.

We next attempted to purify this 74-kDa negative regulator from AM93-4-12 cells using a combination of column chromatogra-

phy and DNA-affinity chromatography. The nuclear extracts from cultured AM93-4-12 cells (200 liters) were applied to a heparin-agarose column. The fraction eluted by 0.4 *M* KCl was then applied to a DEAE-Sepharose column and S-300 column chromatography. Each fraction eluted from the column was assayed by gel retardation using CACC-TCC DNA fragment as a probe. The final product (1.5 μg) was analyzed for homogeneity by SDS gel electrophoresis as shown in Figure 5.

We also found a 38-kDa protein on a gel under reducing conditions. The gel bands

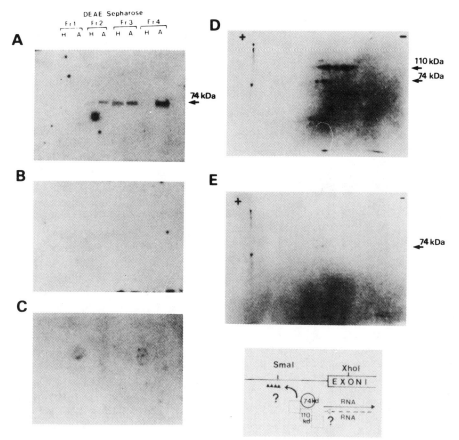

Fig. 4. Binding of various ^{32}P-labeled DNAs to nuclear proteins from HL-60 and AM93-4-12. A crude nuclear extract from cells of HL-60 (indicated as H) and AM93-4-12 (indicated as A) was fractionated on a heparin-agarose column and both 0.4 *M* KCl-eluted fractions were applied to DEAE-Sepharose column. Both fractions were eluted by stepwise elution of NaCl and applied to a SDS-PAGE. Fraction 1, 0.1 *M* NaCl; fraction 2, 0.25 *M* NaCl; fraction 3, 0.3 *M* NaCl; fraction 4, 0.4 *M* NaCl. Binding of [32]p-labeled DNA fragments (− 158 to − 102, **A**), (*Pst*I–*Pvu*II fragment, **B**) and (pUC 19 *Bam*H1 fragment, **C**) were carried out as described (Fainsod et al., 1986; Miskimins et al., 1985). HL-60 and AM93-4-12 cells were radiolabeled by [35S]methionine and fractionated as described above. The proteins of fraction 4 on DEAE-Sepharose column were then applied to DNA-affinity column (CACC-TCC oligonucleotides-agarose) and recycled twice (Kadonaga and Tjian, 1986). The bound proteins were analyzed by 2D-gel electrophoresis. (**D**) Bound fraction of extracts from AM93-4-12; (**E**) bound fraction of extracts from HL-60.

representing the 38-kDa protein were cut out, eluted and renatured by successive treatment with 6 *M* guanidine-HCl and dialysis against 10 m*M* Tris-buffered saline in the presence of protease inhibitors. The binding activity to the CACC-TCC DNA fragment was found only in the 38-kDa protein fraction. In the absence of a reducing reagent, the major band observed was the 74-kDa protein (data not shown). Thus it is highly possible that the 74-kDa protein is composed of a dimer of the 38-kDa subunit. Southwestern blotting indicated that the DNA-binding activity was observed only in the fraction of 74-kDa protein. Other fractions did not bind to DNA with a CACC-TCC sequence. Thus the spe-

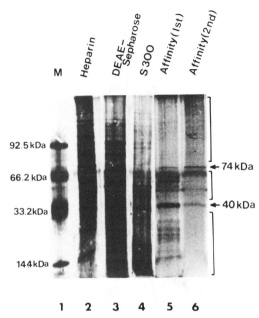

92.5 kDa

66.2 kDa   ← 74 kDa

33.2 kDa   ← 40 kDa

14.4 kDa

1   2   3   4   5   6

Fig. 5. Sodium dodecyl sulfate-polyacrylamide gel electrophoresis (SDS-PAGE) analysis of a 74-kDa protein. Lane 1, molecular weight marker; phosphorylase B (92.5 kDa), bovine serum albumin (66.2 kDa), carbonic anhydrase (33.2 kDa), and lysozyme (14.4 kDa); lane 2, 0.4 M KCl fraction; lane 3, DEAE-Sepharose fraction; lane 4, S-300 fraction; lane 5, DNA-affinity column (CACC-TCC fraction) (first); Lane 6, DNA-affinity column (CACC-TCC fraction) (second).

cific negative regulator may be a 74-kDa protein that is probably a dimer of a 38-kDa protein subunit.

Function of c-*myc* Negative Repressor

In an attempt to see whether the 74-kDa protein complex affected the promoter activity of the c-*myc* gene, we performed a CAT assay of mutant constructs using nuclear extracts in the presence or absence of the affinity purified 74-kDa protein.

As shown in Figure 6, the CACC-TCC box is revealed to be a negative element of c-*myc* promoter (lane 3). Without addition of CACC-TCC binding protein (74-kDa protein), we did not detect any decreased activity of c-*myc* gene promoter (lane 2). By contrast, the nuclear extract from AM93-4-12 containing 74-kDa protein caused a significant reduction

of c-*myc* promoter activity. This negative activity is evident even when the enhancer—promoter (GC-box) binding factors of c-*myc* gene like E2F or Sp1 (Nishikura, 1986; Thalmeier et al., 1989) were added to the nuclear extracts as shown in Figure 6 (*in vitro* promoter assay). The presence of 74-kDa protein in nuclei was shown in other antisense c-*myc* transformants besides AM93-4-12 (unpublished result). These results indicate that negative regulation of the c-*myc* gene is mediated by the induced 74-kDa nuclear protein factor in the antisense c-*myc* transformant.

In order to identify the biological activity of 74-kDa protein liposome-mediated protein transfer was used to introduce 74-kDa protein into HL-60 cells. This was done to determine whether transcription of endogenous c-*myc* gene could be decreased by this protein and result in the triggering of cell differentiation. As shown in Table 3, the 74-kDa protein can commit HL-60 cells to differentiate into macrophages as detected by monoclonal antibody or esterase staining. Both levels of endogenous c-*myc* protein and RNA in these cells were also significantly reduced. Although the concentration of 74-kDa protein in cells was estimated to be less than 10^{-6} M, it is clear that the overintroduction of this 74-kDa protein results in the decreased expression of c-*myc* gene and the commitment of cells to differentiate into either monocytes or macrophages.

DISCUSSION

We have shown here that high levels of cellular antisense c-*myc* RNA can reduce the accumulation of sense c-*myc* RNA and the synthesis of p64 c-*myc* protein. Plasmids capable of producing *Ecogpt* and antisense c-*myc* RNA were introduced into HL-60 cells by DNA transfer. Resistance to high concentration of mycophenolic acid usually involves overexpression of *Ecogpt*, and, in the clones isolated in this study, it also resulted in an increase in the amount of antisense c-*myc* RNA. Selection for resistance to increasing concentrations of mycophenolic acid followed by repeated transfections with plas-

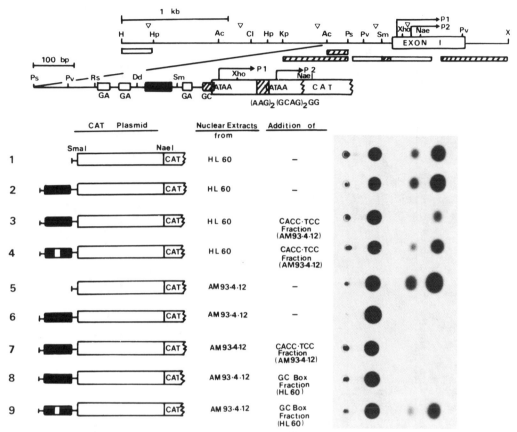

Fig. 6. Effect of 74-kDa protein on the level of CAT expression directed by pC-myc-CAT$_{SN}$ (CACC-TCC)$_3$. A CAT plasmid, pC-*myc*-CAT$_{SN}$ (CACC-TCC)$_3$ was constructed by adding three tandem repeats of 45-mer including CACC·TCC sequence (■) or mutated sequence (■□■) to pC-*myc*-CAT$_{SN}$ vector and transfected into either HL-60 or AM93-4-12. Nuclear extract (50 µg) was mixed with the partially purified 74-kDa protein (1.2 µg) or GC box binding proteins (2.4 µg). A CAT assay followed as previously described (Yokoyama and Imamoto, 1987).

mid DNA caused an increase in antisense RNA resulting from gene amplification (Table 1).

Gene amplification of *dhfr* sequence was first observed in *CHOdhfr$^-$* cells (Huberman and Riggs, 1968). Our study was the first to show gene amplification using HL-60 cells. The exact mechanism of this gene amplification of *Ecogpt* gene is not known. One possible explanation is the c-*myc* gene, which was already amplified to 20 to 30 copies, is localized to replicating submicroscopic circular DNA molecules in HL-60 cells (Von Hoff et al., 1988). The plasmid driven by SV40 promoter can be active enough to increase

the copy number of the integrated plasmid DNA (unpublished result). A previous study by Ariga et al. (1987a) showed that the c-*myc* protein can substitute for the function of large T antigen in cells to directly promote DNA replication. Thus it might be that c-*myc* protein can bind to the large T binding site of the SV40 promoter region and substitute for the function of large T antigen. Given the above findings, the SV40 promoter may replicate autonomously in cells to high copy numbers. The double selection, by increasing concentrations of mycophenolic acid, and by the positive regulation of c-*myc* protein itself

TABLE 3. Comparative Studies of the Human c-*myc* Protein Production and the Differentiated Phenotype of HL-60 Transformants[a]

Cells		MYC expression p64/m RNA/cell (× 10³)	DNA synthesis in vitro (%)	% Reactive cells				
				α-Naphthyl acetate esterase	Monoclonal antibody			Naphthol AS-D chloroacetate esterase
					OKM-1	OKM-5	Anti-neutrophil	
HL-60		1.11	100	0	0	0	0	0
(induced by)								
TPA		—	—	99	95	99	8	0
DMSO		—	—	0	3	4	98	98
Control pSV2gpt⁺		1.18	85	12	5	6	3	15
Sense MYC (pSV2gptC5–8)		1.45	187	15	8	10	6	18
Antisense MYC								
AM93-4-12		0.08	11	72	80	88	9	12
AM451-6-30		0.08	10	58	67	74	8	14
HL-60 treated with µg/ml								
liposomes containing	(0.3)	0.61	60	19	21	27	2	10
partially purified 74-	(2.6)	0.57	44	22	26	21	10	9
kDa protein complex	(8.9)	0.40	35	34	37	49	9	13
	(28.0)	0.16	18	68	73	82	7	12
HL-60 cultured with the cultured medium from AM93-4-12		0.98	95	17	10	14	13	14

[a]Values are means of three experiments. Protein and mRNA levels were measured as described in Table 1. Percent of positive cells by staining of esterase was shown as described in Table 2. In vitro DNA synthesis was measured as reported elsewhere (Studzinski et al., 1986).

to replicate the SV40 promoter, might increase the copy numbers of antisense c-*myc* DNA and result in the higher expression of antisense c-*myc* RNA.

Other promoters besides SV40 early promoter (metallothionein promoter, MMTV promoter) do not function in HL-60 cells. The selection by G418 to yield neomycin-resistant clones can cause reduced survival of cells. Selection by G418 may be toxic for HL-60 cells and also may commit cell differentiation by the selecting drug itself (unpublished data).

Growth inhibition was observed in cells that contained 100–250 copies of the antisense c-*myc* DNA sequence. The amount of sense c-*myc* DNA in the antisense transformants is 20–25 copies per cell. We found that a minimum ratio of 10:1 antisense to sense DNAs is required for inhibition of c-*myc* expression in transfection studies; at this DNA ratio, there is 15- to 25-fold more antisense c-*myc* RNA than sense c-*myc* RNA. The steady-state level of c-*myc* protein in the antisense c-*myc* clones is reduced by >90% per cell compared to that in HL-60 cells. This suggests that antisense RNA may exert an inhibitory effect on p64 protein synthesis by the formation of a stable antisense-sense RNA hybrid *in vivo* (Yokoyama and Imamoto, 1987).

Reduced expression of c-*myc* transcripts correlates with the triggering of HL-60 cell differentiation. c-*myc* mRNA is no longer present in HL-60 cells rendered granulocytic or monocytic differentiation by exposure to chemical inducers (Westin et al., 1982). Still unknown, however, is whether modulation in oncogene expression is required for cell-cycle progression or terminal differentiation of hematopoietic cell types (Heikkila et al., 1987).

The results in Table 2 show a strong correlation between lowered c-*myc* transcriptions and monocytic phenotype, suggesting that antisense c-*myc* RNA directs HL-60 cells into the monocytic pathway in preference to the granulocytic pathway.

However, a recent study by Wickstrom et al. (1989) demonstrated reduction of c-*myc* expression by antisense c-*myc* DNA oligo-nucleotides that might be sufficient to allow terminal granulocytic differentiation. This discrepancy may be due to our use of constitutive inhibition of c-*myc* gene expression while Wickstrom et al. (1989) used a transient inhibition of c-*myc* gene expression. Alternatively, their use of short DNA oligomers against the translational initiation domain may not have induced the same proteins as were induced by sense–antisense RNA duplexes. Anyhow, the down-regulation of c-*myc* expression is a critical determinant of differentiation in HL-60 cell (Holt et al., 1988).

In vitro run-on assays and the promoter-specific CAT assays demonstrated that the antisense transformants were clearly defective in promoter activity of the endogenous c-*myc* gene and in the elongation of c-*myc* RNA in isolated nuclei. These results suggest that c-*myc* RNA transcription in the antisense clones is regulated by the formation of an RNA–RNA duplex in the nucleus. The regulatory sequence that seems to be recognized by antisense RNA was localized to the CACC-TCC repeated sequence of the c-*myc* gene promoter, which was confirmed by DNase I footprinting assays and gel retardation assays (Fig. 3A and B).

Although the molecular mechanism of the repression of the c-*myc* gene is not known, the synthesis of *trans*-acting negative regulatory elements that bind to the CACC-TCC repeats might be induced by RNA–RNA base paring, or by additional regulatory factors induced by the antisense RNA (Fig. 7).

Hogan and his colleagues have demonstrated that a 27-base-long complementary DNA oligonucleotide designed to bind to duplex DNA at −115 base pairs upstream from the transcription origin P1 (−115 to −142) of c-*myc* gene forms a colinear triplex, and repressed the transcription of c-*myc* gene in an *in vitro* assay system (Boles and Hogan, 1987; Cooney et al., 1988).

In contrast, Kinniburgh and his colleagues showed that the *cis*-acting DNA sequence element (two ACCCT sequence motif; −117 to −141) of c-*myc* gene promoter was assumed to form the H-DNA conformation and acted as a positive transcriptional regulator.

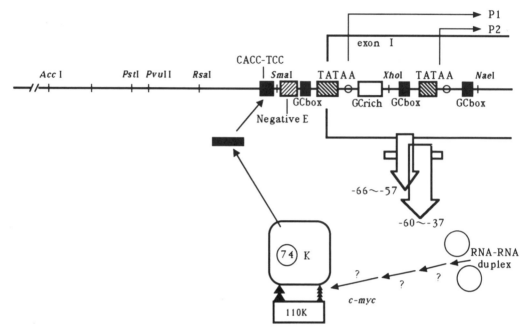

Fig. 7. Schematic model of transcriptional repression of c-*myc* protooncogene by antisense RNA. The duplex formation by antisense RNA and endogenous sense RNA commits the increased levels of 74-kDa protein and 110-kDa protein. The 74-kDa protein can bind the sequence of CACC-TCC in the c-*myc* gene promoter region and shut off the transcription of endogenous c-*myc* gene.

Both ribonucleoprotein and protein factors bound to this ACCCT repeated sequence (Davis et al., 1989). We do not know why the same *cis*-element of this repeated sequence is apparently involved in both positive and negative regulation of c-*myc* gene expression.

We have identified 74-kDa and 110-kDa proteins in the nuclei of antisense transformants that could be candidates for negative regulators of c-*myc* gene expression in our system. While the 74-kDa protein's action as a DNA binding protein has been characterized, the biological function of the 110-kDa protein is not yet known.

The results of CAT assays of c-*myc*-CAT constructs, by addition of 74-kDa protein purified by affinity chromatography, show that this 74-kDa protein causes inhibition of c-*myc* promoter activity. This suppressive activity is evident even when the endogenous E2F, or Sp1-like proteins as the enhancer—promoter binding factors of c-*myc* gene are present (Nishikura, 1986; Thalmeier et al.,

1989). The introduction of 74-kDa protein into HL-60 parental cells using liposome-mediated protein transfer resulted in decreased activity of c-*myc* and the commitment of cells into the macrophage pathway. This protein seems essential for regulation of the c-*myc* promoter and the monocytic differentiation pathway of HL-60 cells. We do not know how the RNA duplex induced this 74-kDa protein. One possible explanation is that nuclear events induced by interferon might be involved in c-*myc* gene regulation via RNA–RNA duplex formation (Stewart, 1979). However, the increased level of the 74-kDa protein was also observed in HL-60 cells treated with TPA (unpublished result). Thus we can speculate that the 74-kDa protein might be related to the phenomenon of cell differentiation, not directly correlated with antisense RNA introduction.

Ariga et al. (1987b) reported that c-*myc* protein is involved in DNA synthesis. Therefore, by interfering with c-*myc* gene expression antisense c-*myc* transformants might inhibit DNA synthesis by autoregulation of c-*myc* protein. The 74-kDa protein (and/or 110-kDa protein) might be closely linked to this autoregulatory mechanism of c-*myc* protein on c-*myc* gene transcription. The production of 74-kDa protein was paralleled by a decreased level of c-*myc* protein (unpublished result). Further study is required to determine which functions, the inhibition of DNA synthesis or the commitment to differentiation, is the direct target of 74-kDa protein complex induced gene regulation. The biological function of 74-kDa protein should be characterized to understand the negative regulatory mechanism of c-*myc* gene expression in the differentiation pathway of HL-60 cells.

SUMMARY

A plasmid carrying antisense human c-*myc* DNA and the gene encoding *Escherichia coli* xanthine-guanine phosphoribosyltransferase (*Ecogpt*) was introduced into human promyelocytic leukemia cell line HL-60 by protoplast fusion. High-level expression of antisense c-*myc* RNA was obtained by selecting cells resistant to progressively higher levels of mycophenolic acid over a period of >6 months. The constitutive production of c-*myc* protein in clones producing high levels of antisense c-*myc* RNA was reduced by 90% compared to parental HL-60 cells. Inhibition of c-*myc* expression was not only at the translation but also at the transcriptional level, implying that antisense RNA can regulate transcription of the c-*myc* gene. The CACC-TCC repeats in the c-*myc* leader sequence are the primary transcriptional target of the antisense RNA. A 74-kDa nuclear protein was able to bind to this sequence in the antisense c-*myc* transformant. The suppression of endogenous c-*myc* gene expression by either antisense RNA or this 74-kDa protein decreases cell proliferation and triggers monocytic differentiation.

ACKNOWLEDGMENTS

We thank Drs. R. DiNicolantonio and G. Gachelin for critical reading of the manuscript and Ms. T. Yamauchi for secretarial work.

This work was supported by grants from the Life Science Research Project, and from the Frontier Research Program of RIKEN and Agency of Science Technology in Japan.

REFERENCES

Ariga SMM, Itani T, Yamaguchi M, Ariga H (1987a): C-myc protein can be substituted for SV40 T antigen in SV40 DNA replication. Nucl. Acids Res. 15:4889–4899.

Ariga SMM, Itani T, Kiji Y, Ariga H (1987b): Possible function of the c-myc product: Promotion of cellular DNA replication. EMBO J. 6:2365–2371.

Bar-Shavit Z, Teitelbaum SL, Reitsma, Hall A, Pegg LE, Trial J, Kahn AJ (1983): Induction of monocytic differentiation and bone resorption by 1,25-dihydroxyvitamin D$_3$. Proc. Natl. Acad. Sci. U.S.A. 80:5907–5911.

Bentley DL, Groudine M (1986): A block to elongation is largely responsible for decreased transcription of c-myc in differentiated HL-60 cells. Nature (London) 321:702–706.

Bishop JM (1987): The molecular genetics of cancer. Science 235:305–311.

Boles TC, Hogan ME (1987): DNA structure equilibria in the human c-myc gene. Biochemistry 26:367–376.

Breitman TR, Selonick SE, Collins SJ (1980): Induction of differentiation of the human promyelocytic leukemia cell line (HL-60) by retinoic acid. Proc. Natl. Acad. Sci. U.S.A. 77:2936–2940.

Chaplinski TJ, Niedel JE (1982): Cyclic nucleotide induced maturation of human promyelocytic leukemia cells. J. Clin. Invest. 70:953–964.

Cole MD (1986): The c-myc oncogene: Its role in transformation and differentiation. Annu. Rev. Genet. 20:361–384.

Collins SJ, Groudine M (1982): Amplification of endogenous c-myc-related DNA sequences in a human myeloid leukemia cell line. Nature (London) 298:679–681.

Collins SJ, Ruscetti FW, Gallagher RE, Gallo RC (1978): Terminal differentiation of human promyelocytic leukemia cells induced by dimethylsulfoxide and other polar compounds. Proc. Natl. Acad. Sci. U.S.A. 75:2548–2462.

Cooney M, Czernuszewicz G, Postel EH, Flint SJ, Hogan ME (1988): Site-specific oligonucleotide binding represses transcription of the human c-myc gene *in vitro*. Science 241:456–459.

Dalla-Favera R, Wong-Staal F, Gallo RC (1982): Oncogene amplification in promyelocytic leukemia cell line HL-60 and primary leukemic cells of the same patient. Nature (London) 299:61–63.

Davis TL, Firulli AB, Kinniburgh AJ (1989): Ribonucleoprotein and protein factors bind to an H-DNA-forming c-myc DNA element: Possible regulators of the c-myc gene. Proc. Natl. Acad. Sci. U.S.A. 86:9682–9686.

Dyson PJ, Littlewood TD, Forster A, Rabbits TH (1985): Chromatin structure and transcriptionally active and inactive human c-myc alleles. EMBO J. 4:2885–2891.

Eick D, Bornkamm GW (1986): Transcriptional arrest within the first exon is a fast control mechanism in c-myc gene expression. Nucl. Acids Res. 14:8331–8346.

Fainsod A, Bogorad LD, Ruusala T, Lubin M, Crothers DM, Ruddle FH (1986): The homeo domain of a murine protein binds 5' to its own homeo box. Proc. Natl. Acad. Sci. U.S.A. 83:9532–9536.

Filmus J, Buick RN (1985): Relationship of c-myc expression to differentiation and proliferation of HL-60 cells. Cancer Res. 45:822–825.

Green PJ, Pines O, Inouye M (1986): The role of antisense RNA in gene regulation. Annu. Rev. Biochem. 55:569–597.

Grosso LE, Pitot HC (1985): Transcriptional regulation of c-myc during chemically induced differentiation of HL-60 cultures. Cancer Res. 45:847–850.

Heikkila R, Schwab G, Wickstrom E, Loke SL, Pluznik DH, Watt R, Neckers LM (1987): A c-myc antisense oligodeoxynucleotide inhibits entry into S phase but not progress from G_0 to G_1. Nature (London) 528:445–449.

Holt JT, Redner RL, Nienhuis AW (1988): An oligomer complementary to c-myc mRNA inhibits proliferation of HL-60 promyelocytic cells and induces differentiation. Mol. Cell Biol. 8:963–973.

Huberman JA, Riggs AD (1968): On the mechanism of DNA replication in mammalian chromosomes. J. Mol. Biol. 32:327–337.

Kadonaga JT, Tjian R (1986): Affinity purification of sequence-specific DNA binding proteins. Proc. Natl. Acad. Sci. U.S.A. 83:5889–5893.

Katoh S, Ozawa K, Kondoh S, Soeda E, Israel A, Shiroki K, Fujinaga K, Itakura K, Gachelin G, Yokoyama K (1990): Identification of sequences responsible for positive and negative regulation by E1A in the promoter of H-2K^{bm1} class I MHC gene. EMBO J. 9:127–135.

Kim SK, Wold BJ (1985): Stable reduction of thymidine kinase activity in cells expressing high levels of anti-sense RNA. Cell 42:129–138.

Lee F, Mulligan R, Berg P, Ringold G (1981): Glucocorticoids regulate expression of dihydrofolate reductase cDNA in mouse mammary tumor virus chimeric plasmids. Nature (London) 294:228–232.

McCachren SS, Nicols J Jr, Kaufman RE, Niedel JE (1986): Dibutyryl cyclic adenosine monophosphate reduces expression c-myc during HL-60 differentiation. Blood 68:412–416.

McCarthy DM, San Miguel JF, Freake HC, Green PM, Zola H, Catovsky D, Goldmar JM (1983): 1,25-

Dihydroxyvitamin D_3 inhibits proliferation of human promyelocytic leukemia (HL60) cells and induces monocyte-macrophage differentiation in HL60 and normal human bone marrow cells. Leukemia Res. 7:51–55.

Miskimins WK, Roberts MP, McClelland A, Ruddle FH (1985): Use of a protein-blotting procedure and a specific DNA probe to identify nuclear proteins that recognize the promoter region of the transferrin receptor gene. Proc. Natl. Acad. Sci. U.S.A. 82:6741–6744.

Nishikura K (1986): Sequences involved in accurate and efficient transcription of human c-myc gene microinjected into frog oocytes. Mol. Cell Biol. 6:4093–4098.

Nishioka Y, Leder P (1979): The complete sequence of a chromosomal mouse α-globin gene reveals elements conserved throughout vertebrate evolution. Cell 18:875–882.

Reitsma DH, Rothberg PG, Astrin SM, Trial J, Bar-Shavit Z, Hall A, Teitelbaum SL, Kahn AJ (1983): Regulation of c-myc gene expression in HL60 leukemia cells by a vitamin D metabolite. Nature (London) 306:492–495.

Rovera G, O'Brien TG, Diamond L (1979): Induction of differentiation in human promyelocytic leukemia cells by tumor promoters. Science 204:868–870.

Sandri-Goldin RM, Golin AL, Levine M, Glorioso JC (1981): High-frequency transfer of cloned herpes Simplex virus type 1 sequences to mammalian cells by protoplast fusion. Mol. Cell Biol. 1:743–752.

Simpson RV, Hsu T, Begley DA, Mitchell BS, Alizadeh BN (1987): Transcriptional regulation of the c-myc protooncogene by 1,25-dihydroxyvitamin D_3 in HL-60 promyelocytic leukemia cells. J. Biol. Chem. 262:4104–4108.

Stewart WE II (1979): The Interferon System. New York: Springer-Verlag.

Studzinski GP, Brelvi ZS, Feldman SC, Watt RA (1986): Participation of c-myc protein in DNA synthesis of human cells. Science 234:467–470.

Thalmeier K, Synovzik H, Mertz R, Winnacker E-L, Lipp M (1989): Nuclear factor E2F mediates basic transcription and trans-activation by E1a of the human c-myc promoter. Genes Dev. 3:527–536.

Trepel JB, Colamonici OR, Kelly K, Schwab G, Watt RA, Sausville EA, Jaffe ES, Neckers LM (1987): Transcriptional inactivation of c-myc and the transferrin receptor in dibutyryl cyclic AMP-treated HL-60 cells. Mol. Cell Biol. 7:2644–2648.

Von Hoff DD, Needham-VanDevanter DR, Yucel J, Windle BE, Wahl GM (1988): Amplified human c-myc oncogenes localized to replicating submicroscopic circular DNA molecules. Proc. Natl. Acad. Sci. U.S.A. 85:4804–4808.

Watanabe T, Sariban E, Mitchell T, Kufe D (1985):

Human c-myc and H-ras expression during induction of HL-60 cellular differentiation. Biochem. Biophys. Res. Commun. 126:999–1005.

Weintraub H, Izant JG, Harland RM (1985): Antisense RNA as a molecular tool for genetic analysis. Trends Genet. 1:22–25.

Weiss E, Golden L, Zakut R, Mellor A, Fahrner K, Kvist S, Flavell RA (1983): The DNA sequence of the H-2K^b gene: evidence for gene conversion as a mechanism for the generation of polymorphism in histocompatibility antigens. EMBO J. 2:453–462.

Westin EH, Wong-Staal F, Gelmann EP, Dalla-Favera R, Pappas TS, Lautenberger J, Eva A, Reddy EP, Tronick SR, Aaronson SA, Gallo RC (1982): Expression of cellular homologues of retroviral oncogenes in human hematopoietic cells. Proc. Natl. Acad. Sci. U.S.A. 79:2490–2494.

Wickstrom EL, Bacon TA, Gonzalez A, Lyman GH, Wickstrom E (1989): Anti c-myc DNA increases differentiation and decreases colony formation by HL-60 cells. *In Vitro* Cell. Dev. Biol. 24:297–362.

Yokoyama K, Imamoto F (1987): Transcriptional control of the endogenous MYC protooncogene by antisense RNA. Proc. Natl. Acad. Sci. U.S.A. 84:7363–7367.

The c-myb Protooncogene: Its Role in Human Hematopoiesis and the Therapeutic Implications of Perturbing Its Function

Alan M. Gewirtz and Bruno Calabretta

Hematology Sections, Departments of Pathology and Internal Medicine, University of Pennsylvania School of Medicine, Philadelphia, Pennsylvania 19104 (A.M.G.) and Department of Pathology and Fels Cancer Research Institute, Temple University School of Medicine, Philadelphia, Pennsylvania 19140 (B.C.)

INTRODUCTION

The c-*myb* protooncogene is the normal cellular homologue of the avian myeloblastosis virus (AMV) transforming gene, v-*myb*, which causes myeloblastic leukemia in chickens and transforms myelomonocytic hematopoietic cells in culture [1,2]. The avian leukemia virus E26 also contains the v-*myb* oncogene, in addition to a second oncogene, *ets*. E26, like AMV, can transform hematopoietic cells [3], and in association with the *ets* oncogene can induce erythroleukemia in chicken erythroid precursor cells [4–6]. MYB protein localizes to the cell nucleus [7,8] where it binds to DNA [9], specifically recognizing the nucleotide sequence pyAACG/TG [10]. The DNA-binding domain of the v-*myb* protein is composed of two imperfectly conserved 52 amino acid direct repeats located near the amino terminus [9], and corresponds to a truncated version of that found in chicken and mammalian c-*myb* proteins [11–15]. Concatemers of this consensus sequence is able to confer v-*myb*-dependent inducibility on otherwise unresponsive promoters suggesting that the v-*myb* protein acts as sequence-specific DNA binding transcription factor [16].

In the hematopoietic cell system, the c-*myb* protooncogene is thought to play a particularly important role in governing cell proliferation [17], and certain maturation functions as well [18]. For example, c-*myb* is preferentially expressed in primitive hematopoietic cell tissues and hematopoietic tumor cell lines of several species [19]. Further, when primitive hematopoietic cells are stimulated to undergo maturation, cellular differentiation is associated with a major decline in c-*myb* expression [17]. Similarly, constitutive expression of exogenously introduced c-*myb* inhibits the ability of a murine erythroleukemia cell line (MEL) to undergo erythroid differentiation in response to known inducing agents [20,21]. Finally, it has also recently been shown that v-*myb* directly regulates the expression of a cellular gene in chicken myeloblasts infected with a temperature sensitive AMV mutant [18]. This cellular gene, designated *mim*-1 [18], encodes a secretable protein contained in the granules of both normal and v-*myb*-transformed promyelocytes. The promoter of the *mim*-1 gene contains three closely spaced binding sites of v-*myb* protein and is strongly activated by v-*myb* in a cotransfection assay [18].

Those data summarized above clearly suggest that the c-*myb* gene product is an important regulator of hematopoietic cell development. Nevertheless, it is only recently that more direct experimentation has critically strengthened this conclusion. In our laboratories, for example, we have employed antisense oligodeoxynucleotides to specifically impair the function of this gene in order to directly examine its function in normal and malignant hematopoiesis. The results of these investigations, described below, suggest that inhibition of both normal and malignant hematopoiesis can be brought about by perturbing c-*myb* function, and this being so, may allow the antisense approach to be exploited for therapeutic purposes.

METHODS

Cells. Light density mononuclear cells (MNC) were obtained from consenting normal donors by Ficoll–Hypaque density gradient centrifugation. MNC were further enriched for hematopoietic progenitors by first removing adherent and phagocytic elements and then T lymphocytes as previously described [22]. MNC depleted of adherent/phagocytic and T cells (A⁻T⁻MNC) were plated in semisolid medium or further enriched for immature progenitors by immunorosetting with anti-HPCA-1 MAb (Becton Dickinson, Mountain View, CA), which recognizes the MY10 or CD34 antigen [22]. Briefly, marrow cells were incubated for 60 min at 4°C (6×10^6 cells/ml) in the presence of 1:50 anti-HPCA-1 MAb. After extensive washing, $CrCl_3$ antibody-treated SRBC (0.4 ml/10^6 cells) were added to cells, pelleted, and incubated for 1 hr at 4°C. Rosetted cells were then separated on a Ficoll gradient and SRBC were subsequently eliminated with a lysing reagent (Ortho Pharmaceutical, Raritan, NJ).

Human leukemia cell lines HL-60, ML-3, KG-1, and KG-1a were obtained from the American Type Culture Collection or as described [23].

Colony assays. Hematopoietic cell colonies were cloned in plasma clot culture [24] or in methylcellulose [22] as described. For myeloid colony stimulation, granulocyte/macrophage colony-stimulating factor (GM-CSF; 5 ng/ml), IL-3 (20 U/ml), or granulocyte colony-stimulating factor (G-CSF; 10% of

transfected Chinese hamster ovary cell supernatant) was used. CFU erythroid (CFU-E) colonies were obtained in the presence of recombinant human erythropoietin alone (3 U/ml; Amgen Biologicals, Thousand Oaks, CA). Burst-forming units-erythroid (BFU-E) colonies were obtained in the presence of combination of IL-3 (10 U/ml), GM-CSF (5 ng/ml), and erythropoietin (3 U/ml). Colonies were scored after 14–16 days of culture in a humidified 5% CO_2 incubator. CFU-E and G-CSF-stimulated granulocytic colonies were scored at 7–9 days of culture. To determine colony size, 30 consecutive individual colonies from each plate were plucked from methylcellulose, dispersed in tissue culture medium, and then counted in a hemocytometer. Cells in smaller colonies were counted directly in the plates.

Oligodeoxynucleotides. Unmodified, 18-base oligodeoxynucleotides were made on a DNA synthesizer (model 380B, Applied Biosystems Inc., Foster City, CA) by means of β-cyanoethylphosphoramidite chemistry. Oligodeoxynucleotides were purified by ethanol precipitation and multiple washes in 70% ethanol. They were subsequently lyophilized to dryness and redissolved in culture medium at a concentration of 1 mg/ml. The primary oligodeoxynucleotide c-*myb* sequences employed for our studies [22–24] are as follows: 5'-GCC CGA AGA CCC CGG CAC-3' (MYB S: codons 2–7; sense); 5'-GTG CCG GGG TCT TCG GGC-3' (MYB AS: codons 2–7; antisense).

Oligomer treatment of cells. Normal MNC or leukemia cells were incubated in tissue culture medium in polypropylene tubes (Falcon Plastics, Cockeysville, MD) in the presence of optimal concentrations of the relevant hematopoietic growth factor [22,24,25]. Some cultures were also supplemented with either sense or antisense oligomers for 18 hr at final concentrations up to ~ 17.5 μM. Cells placed into semisolid medium were not washed before plating. Control cultures were left untreated.

Detection of c-*myb* mRNA in bone marrow cells by reverse transcriptase–polymerase chain reaction (RT-PCR)

analysis. Expression of c-*myb* mRNA in oligomer-treated and control cells was performed as published [22,25]. In brief, cells were collected at the end of the required time of culture and total RNA was extracted in the presence of 20 mg of *Escherichia coli* ribosomal RNA. RNA was then reverse-transcribed and the resulting cDNA fragments were amplified with 7.5 U of *Thermus aquaticus* (*Taq*) polymerase in the presence of c-*myb* mRNA sequence-specific synthetic primers. Of the 50-ml polymerase chain reaction 10 ml was separated in a 4% Nusieve agarose gel and transferred to a nitrocellulose filter. The resulting blot was hybridized with a synthetic [γ-^{32}P]ATP end-labeled oligomer complementary to the amplified c-*myb* DNA.

Detection of c-*myb* encoded (MYB) protein. Expression of MYB protein was assessed by indirect immunofluorescence as described [25]. A sheep anti-human MYB serum (Cambridge Research Biochemicals, Valley Stream, NY) was utilized at a 1:40 dilution.

RESULTS

Role of c-*myb* Encoded Protein in Normal Human Hematopoiesis

Exposure of MNC to c-*myb* antisense oligomers not only resulted in the generation of small colonies but also in a statistically significant decrease in the total numbers of colonies formed. The decrease in granulocyte−macrophage colony forming units (CFU-GM) in four studies was 86% ($P = .018$), 70% ($P = .040$), 43% ($P = .001$), and 74% ($P = .001$), respectively. Similar data were generated when the effect of c-*myb* antisense oligomers on erythroid and megakaryocyte progenitor cell growth was examined. In contrast, the c-*myb* sense oligomer had no consistent effect on the formation of hematopoietic colonies of any lineage when compared to growth in control cultures ($P = .778, P = .796, P = .375$ for CFU-GM, CFU-E, and CFU-Meg, respectively).

We also determined if inhibition of colony formation correlated with oligomer concentration utilized. After progenitor cells were exposed to 0.3, 1.4, 7, and 14 μM of c-*myb*

sense and antisense oligodeoxynucleotides, erythroid and megakaryocytic colony formation was examined. As expected, the c-*myb* sense oligodeoxynucleotide had no significant effect on colony formation. In contrast, there was a direct association between increasing amounts of the c-*myb* antisense oligonucleotide and the degree of colony inhibition. Exposure to c-*myb* antisense oligomers at the above concentrations led respectively to the formation of 21 ± 3, 17 ± 1, 10 ± 2, and 4 ± 2 megakaryocyte colonies; exposure to 1.4, 7, and 14 μM concentrations of the antisense oligodeoxynucleotide led to the formation of 775 ± 127, 515 ± 9, and 244 ± 25 myeloid colonies.

To provide evidence that inhibition of hematopoietic colony formation was due to a specific effect of the c-*myb* oligomer, we determined whether it was taken up by target cells. Demonstrating direct uptake into hematopoietic progenitor cells was not feasible due to cell rarity. However, we reasoned that if cells took up a myeloperoxidase (MPO) antisense oligodeoxynucleotide, they should take up an identically sized (18-nt) c-*myb* oligomer. Accordingly, we examined the intracellular synthesis of MPO protein after prolonged exposure to the MPO antisense oligodeoxynucleotide. MPO synthesis by individual cells within colonies was monitored by indirect immunofluorescence with an MPO antibody [24]. The exposure of cells to either the c-*myb* sense or antisense oligodeoxynucleotide did not affect the synthesis of MPO expression [24]. Also, despite the fact that the antisense oligonucleotide inhibited cell proliferation, maturation, at least as demonstrated by an ability to synthesize MPO, was unaffected. In contrast, MPO anti-sense oligomer exposure resulted in a ~70% decrease ($P < .001$) in synthesis of this protein as determined by immunochemical means [24].

More recently, we determined if MYB protein is preferentially required during specific stages of normal human hematopoiesis we incubated normal marrow mononuclear cells (MNC) with c-*myb* antisense oligodeoxynucleotides [22]. Treated cells were cultured in semisolid medium under conditions designed to favor the growth of specific progenitor cell types. In comparison to untreated controls, CFU-GM-derived colonies decreased 77% when driven by recombinant human (rH) IL-3, and 85% when stimulated by rH GM-CSF; BFU-E and CFU-E derived colonies decreased 48 and 78%, respectively. In contrast, numbers of G-CSF-stimulated granulocyte colonies derived from antisense-treated MNC were unchanged from controls through the numbers of cells comprising these colonies were decreased ~90% (Fig. 1). Similar results were obtained when MY10$^+$ cells were exposed to c-*myb* antisense oligomers. When compared to untreated controls, numbers of CFU-GM- and BFU-E-derived colonies were unchanged, but the numbers of cells comprising these colonies was reduced ~75% and >90%, respectively, in comparison to controls. c-*myb* sense and c-*myc* antisense oligomers were without significant effect in these assays. These results in aggregate suggest that c-*myb* is required for proliferation of intermediate-late myeloid and erythroid progenitors, but is less important for lineage commitment and early progenitor cell amplification. They also suggest that c-*myb* may be of more importance than c-*myc* in regulating hematopoiesis, at least in this in vitro system.

Role of c-*myb* Encoded Protein in Leukemic Hematopoiesis

The effect of c-*myb* sense and c-*myb* antisense oligomers on the growth of four myeloid leukemia cell lines HL-60, K562, KG-1, and KG-1a was determined [23]. The results with HL-60, and KG-1 are shown in Figure 2. The antisense oligomer inhibited the proliferation of each leukemia cell line, although the effect was most pronounced on HL-60 cells. Specificity of this inhibition was demonstrated by the fact that the sense oligomer had no effect on cell proliferation, nor did "antisense" sequences with 2 or 4 nucleotide mismatches [23].

To determine whether the treatment with c-*myb* antisense modified cell cycle distribution of HL-60 cells, we measured the DNA content in exponentially growing cells exposed to either sense or antisense *myb* oligomers. Control cells and cells treated with c-*myb* sense had essentially equivalent DNA content values, which were approximately twice those observed in HL-60 cells treated with an-

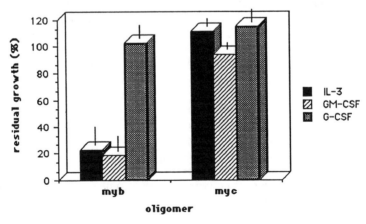

Fig. 1. Effect of antisense vs. sense and c-*myc* oligomers on in vitro colony formation of human myeloid progenitors. Bars express percent of colony number of antisense- vs. sense-treated A$^-$T$^-$MNC. Values are mean ± SD of four different experiments from duplicate plates. Colonies were scored after 14 days of culture (7–9 days for G-CSF-stimulated cultures). Solid bars, CFU-GM (IL-3); striped bars, CFU-GM (GM-CSF); stippled bars, CFU-G (G-CSF). (Adapted from [22], with permission of the publisher.)

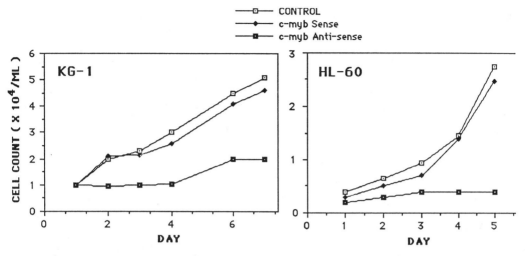

Fig. 2. Effect of sense and antisense oligomers on human myeloid leukemia cell line growth.

tisense oligomers. The majority of these cells appeared to reside either in the G_1 compartment, or were blocked at the G_1/S boundary.

It has been reported that an oligomer complementary to the first five codons of c-*myc* mRNA inhibits proliferation of HL-60 cells and induces differentiation [26–28]. We therefore compared the effect of two 18-nt oligomers complementary to the same region (codons 2–7) in HL-60, KG-1, and KG-1a cells. In each case, c-*myb* antisense inhibited cell proliferation to a much greater extent than the c-*myc* antisense oligomer. For example, untreated HL-60 cells had a cell doubling time of 20–21 hr, c-*myc* antisense-treated cells had a doubling time of 30–31 hr, while the c-*myb* antisense-treated cells had a doubling time of 50–52 hr. Of note, the c-*myc* antisense effect we noted is similar to that observed by other investigators. Finally, we determined if HL-60 cells treated with c-*myb* antisense oligomer would be induced to differentiate along the granulocytic pathway as has been reported to occur when these cells are exposed to a c-*myc* antisense oligomer. Incubation of HL-60 with a c-*myc* antisense oligomer for 5 days was followed by the appearance of 15–20% of the cells showing NBT reduction and morphological changes associated with a more differentiated phenotype than that of untreated cells.

On the contrary, less than 5% of HL-60 cells exposed to the c-*myb* antisense oligomer were NBT positive (Table 1).

These findings indicate that reduction of the c-*myb* protein level in a number of leukemic cell types is accompanied by a substantial inhibition in their growth rate. HL-'60 cells appeared to be most sensitive to the c-*myb* antisense effect, suggesting that the growth advantage conferred on these cells by c-*myc* amplification, and by activation of N-*ras*, is not enough to maintain increased cell proliferation in the absence of c-*myb* protein.

Evidence That Normal and Leukemic Progenitor Cells Rely Differentially on c-*myb* Function

If leukemic cells depended more on MYB protein than their normal counterparts inhibition of c-*myb* function might prove ther-

TABLE 1. Effect of *myc* and *myb* Antisense Oligodeoxynucleotides on HL-60 Cell Differentiation

Culture condition	% Positive cells
Control	<5
c-*myc* AS	~15–20
AS	<5

apeutically useful. To test this possibility, we incubated phagocyte and T cell-depleted normal human marrow mononuclear cells (MNC), human T lymphocyte leukemia cell line blasts (CCRF-CEM), blasts from acute myelogenous leukemia (AML) patients, or 1:1 mixtures of these cells with sense or antisense oligodeoxynucleotides to codons 2–7 of human c-*myb* mRNA. Oligomers were added to liquid suspension cultures at time 0 and at time +18 hr (25% 0 hr dose). Control cultures were untreated. In controls, or in high dose sense oligomers, CCRF-CEM proliferated rapidly, whereas MNC numbers and viability decreased <10%. In contrast, when CCRF-CEM were incubated for 4 days in c-*myb* antisense DNA, cultures contained $4.7 \pm 0.8 \times 10^4$ cells/ml (mean ± SD; $n = 4$) compared to $285 \pm 17 \times 10^4$/ml in controls. At the effective antisense dose, MNC were largely unaffected. After 4 days in culture, remaining cells were transferred to methylcellulose supplemented with recombinant hematopoietic growth factors. Myeloid colonies/clusters were enumerated at day 10 of culture inception. Depending on cell number plated, control MNC formed from 31 ± 4 to 274 ± 18 colonies. In dishes containing equivalent numbers of untreated or sense oligomers exposed CCRF-CEM, colonies were too numerous to count (TNTC). When MNC were mixed 1:1 with CCRF-CEM in antisense oligomer concentrations ≤ 5 μg/ml, only leukemic colonies could be identified by morphologic, histochemical, and immunochemical analysis. However, when antisense oligomer exposure was intensified, normal myeloid colonies could now be found in the culture while leukemic colonies could no longer be identified with certainty using the same analytical methods. At equivalent antisense DNA doses, AML blasts from 18 of 23 patients exhibited ~75% decrease in colony and cluster formation compared to untreated or sense oligomer treated controls. When 1:1 mixing experiments were carried out with primary AML blasts and normal MNC, we were again able to preferentially eliminate AML blast colony formation while normal myeloid colonies continued to form.

DISCUSSION

The results presented herein strongly suggest that the growth and survival of both lymphoid and myeloid leukemic blast cells are highly dependent on c-*myb* gene function. In this regard, we have shown that proliferation and cloning efficiency of cells from human myeloid and T cell leukemia cell lines, and primary blast cells from patients with acute myelogenous leukemia were significantly inhibited by exposure to antisense oligodeoxynucleotides to the c-*myb* protooncogene. In addition, we have also provided suggestive data that leukemic hematopoietic cells are much more dependent than their normal counterparts on c-*myb* gene function by demonstrating that c-*myb* antisense oligomers can preferentially inhibit the survival and cloning efficiency of leukemic cells at doses which permit the survival and continued clonal proliferation of normal hematopoietic progenitor cells.

The inhibitory mechanism(s) leading to preferential inhibition of leukemic cell growth remain undefined. Studies from our laboratories have demonstrated that MYB protein is required for G_1/S transition by normal PHA stimulated lymphocytes [25], and by normal hematopoietic progenitor cells that are actively progressing through the cell cycle [22]. Accordingly, c-*myb* function is certainly important in cells that are synthesizing DNA. Such a hypothesis might explain oligomer induced inhibition of the T leukemic cell line but not that of primary blast cells, many of which are not in active cycle [29]. In this regard, Kastin et al. [30] have recently demonstrated that MYB protein can be detected in cells that are not actively proliferating and that protein levels do not necessarily correlate with cell cycle activity [30]. Consistent with these data are the recent observations that a v-*myb*-encoded protein functions as a transcriptional activator for *mim*-1, a newly described gene that encodes a maturation-related myeloid granule protein [31], and that an alternatively spliced c-*myb* mRNA species can enhance hematopoietic cell differentiation [32]. Therefore, it appears likely that MYB protein exerts important cellular functions other than those related to cell cycle activity. The greater sensitivity of leukemia cells to MYB deprivation as compared to their normal counterparts may reflect a generally lower tolerance for pertubation of these as yet undefined functions in leukemic cells.

The ability of antisense oligodeoxynucleotides to block a specific gene's function is proving to be powerful research tool for investigating the biologic importance of the target gene's function [33,34]. That such synthetic nucleic acids might also have therapeutic utility was first suggested by Zamecnick and Stephenson [35], who demonstrated that a 13-mer oligodeoxynucleotide complementary to a reiterated Rous sarcoma virus sequence could inhibit viral replication in infected fibroblasts. Reports of inhibition of the replication of several viruses, including the human immunodeficiency virus, have followed [36,37]. Since overexpression of protooncogenes may be an important mechanism in the development and maintenance of solid tumors [38,39], as well as malignancies of the hematopoietic forming elements of the marrow [40], antisense oligomers could have significant therapeutic potential in the treatment of human neoplastic diseases if they prove to be practical agents for specifically blocking gene function in such tissues. In addition, because they can be specifically targeted, toxic effects should be limited to those tissues that are actively expressing the targeted gene. In this regard, a number of significant technical problems remain to be solved before the in vivo use of antisense oligomers can be contemplated. These include lack of effective in vivo delivery systems, oligomer digestion by serum or plasma nucleases, and efficiency of target cell uptake. Our results are intriguing because they suggest that antisense oligomers might find immediate utility as ex vivo bone marrow purging agents. Problems related to delivery and potential systemic toxicity are thereby avoided. Further, since the oligomers appear to be less toxic to normal hematopoietic progenitor cells at effective purging doses than standard chemotherapeutic purging agents improved engraftment rates of oligomer purged marrow would be expected to result [41,42]. Accordingly, antisense DNA may prove a useful addition to the treatment of human leukemias.

SUMMARY

The c-*myb* protooncogene is the normal cellular homologue of the avian myeloblastosis virus (AMV) transforming gene, v-*myb*. In the avian system, v-*myb* is thought to play an important causative role in the transformation of myelomonocytic hematopoietic cells in culture, and in the development of myeloblastic leukemia in vivo. In contrast, the significance of c-*myb* gene expression in the process of hematopoiesis has only recently been more clearly defined. One way in which this has been accomplished has been to employ antisense oligodeoxynucleotides to specifically impair the function of the c-*myb* gene in normal and in malignant hematopoietic cells. This approach has demonstrated that c-*myb* gene function is most important for cell proliferation. In addition, it has also been shown that perturbing c-*myb* function with antisense oligomers can inhibit both normal and malignant hematopoiesis. Interestingly, it appears that malignant cells are more sensitive to deprivation of MYB protein than are their normal counterparts. This being so, it is possible that antisense oligodeoxynucleotides may be exploited for therapeutic purposes in the future.

ACKNOWLEDGMENTS

This work was supported in part by National Institutes of Health (NIH) Grants CA-36896, CA-01324, and CA-54384 (A.M.G.), and CA-46782 (B.C.), Grant CH-455 from the American Cancer Society (B.C.), a Grant-in-Aid from the National Center of the American Heart Association (A.M.G.), and the WW Smith Charitable trust (A.M.G.). Dr. Calabretta is a Scholar of the Leukemia Society of America. Dr. Gewirtz is the recipient of a Research Career Development Award from the National Cancer Institute, NIH.

REFERENCES

1. Baluda MA, Goetz IE (1961): Morphological conversion of cell cultures by avian myeloblastosis virus. Virology 15:185–199.

2. Moscovici C (1975): Leukemia transformation with avian myeloblastosis virus: Present status. Curr. Top. Microbiol. Immunol. 71:79–96.

3. Moscovici C, Gazzolo L (1982): Transformation of hemopoietic cells and avian leukemia viruses. In Advances in Viral Oncology, Vol. 1, Oncogene Studies. New York: Raven Press, pp 83–106.

4. Radka K, Beug H, Kornfeld S, Graf T (1982): Transformation of both erythroid and myeloid cells by E26, an avian leukemia virus that contains the myb gene. Cell 31:643–653.

5. Golay J, Introna M, Graf T (1988): A single point mutation in the v-eta oncogens affects both erythroid and myelomonocytic cell differentiation. Cell 55:1147–1158.

6. Nunn MF, Hunter T (1989): The ets sequence is required for induction of erythroblastosis in chickens by avian retrovirus E26. J. Virol. 63:398–402.

7. Klempnauer K-H, Symonds G, Evan GI, Bishop JM (1984): Subcellular localization of proteins encoded by oncogenes of avian myeloblastosis virus and avian leukemia virus E26 and by the chicken c-myb gene. Cell 37:537–547.

8. Boyle WJ, Lampert MA, Lipsick JS, Baluda MA (1984): Avian myeloblastosis virus and E26 virus oncogene products are nuclear proteins. Proc. Natl. Acad. Sci. U.S.A. 81:4265–4269.

9. Klempnauer KH, Sippel AE (1987): The highly conserved amino-terminal region of the protein encoded by the v-myb oncogene functions as a DNA-binding domain. EMBO J. 6:2719–2725.

10. Biedenkapp H, Borgmeyer U, Sippel AE, Klempnauer KH (1988): Viral myb oncogene encodes a sequence-specific DNA-binding activity. Nature (London) 335:835–837.

11. Rosson D, Reddy P (1986): Nucleotide sequence of chicken a-myb complementary DNA and implications for myb oncogene activation. Nature (London) 319:604–606.

12. Gerondakis S, Bishop JM (1986): Structure of the protein encoded by the chicken proto-oncogene c-myb. Mol. Cell. Biol. 6:3677–3684.

13. Gonda TJ, Gouch NM, Dunn AR, De Blaguers J (1985): Nucleotide sequence of cDNA clones of the murine myb proto-oncogens. EMBO J. 4:2003–2008.

14. Bender TP, Kuehl WM (1986): Murine myb proto-oncogene mRNA: cDNA sequence and evidence for 5' heterogeneity. Proc. Natl. Acad. Sci. U.S.A. 83:3204–3208.

15. Majello B, Kenyon LC, Dalla-Favera R (1986): Human c-myb proto-oncogene: Nucleotide sequence of cDNA and organization of the genomic locus. Proc. Natl. Acad. Sci. U.S.A. 83:9616–9640.

16. Weston K, Bishop JM (1989): Transcriptional activation by the v-myb oncogene and its cellular progenitor c-myb. Cell 58:85–93.

17. Duprey SP, Boettiger D (1985): Developmental regulation of c-myb in normal myeloid leukemia cells. Proc. Natl. Acad. Sci. U.S.A. 82:6937–6941.

18. Ness SA, Marknell A, Graf T (1989): The v-myb oncogene product binds to and activates the promyelocyte-specific MIM-1 gene. Cell 59:1115–1125.

19. Westin EH, Gallo RC, Arya SE, Eva A, Souza LM, Baluda MA, Aaronson SA, Wong-Staal F (1982): Differential expression of the AMV gene in human hematopoietic cells. Proc. Natl. Acad. Sci. U.S.A. 79:21994–2198.

20. Clarke MF, Kukowska-Latallo JF, Westin E, Smith M, Procownick E (1988): Constitutive expression of a c-myb cDNA blocks Friend during erythroleukemia cell differentiation. Mol. Cell. Biol. 8:884–892.

21. Todokoro K, Watson RJ, Higo H, Amanuma H, Kuramuchi S, Yanagisawa H, Ikawa Y (1988): Down-regulation of c-myb gene expression is a requisite for erythropoietin-induced erythroid differentiation. Proc. Natl. Acad. Sci. U.S.A. 85:8900–8904.

22. Caracciolo D, Venturelli D, Valtieri M, Peachle C, Gewirtz AM, Calabretta B (1990): Stage-related proliferative activity determines c-myb functional requirements during normal human hematopoiesis. J. Clin. Invest. 85:55–61.

23. Anfossi G, Gewirtz AM, Calabretta B (1989): An oligomer complementary to c-myb encoded mRNA inhibits proliferation of human myeloid leukemia cell lines. Proc. Natl. Acad. Sci. U.S.A. 86:3379–3383.

24. Gewirtz AM, Calabretta B (1988): A c-myb antisense oligodeoxynucleotide inhibits normal human hematopoiesis in vitro. Science 242:1303–1306.

25. Gewirtz AM, Anfossi G, Venturelli D, Valpreda D, Sims R, Calabretta B (1989): G1/S transition in normal human T-lymphocytes requires the nuclear protein encoded by c-myb. Science 245:180–183.

26. Wickstrom EL, Bacon TA, Gonzalez A, Freeman DL, Lyman GH, Wickstrom E (1988): Human promyelocytic leukemia HL-60 cell proliferation and c-myc protein expression are inhibited by an antisense pentadecadeoxynucleotide targeted against c-myc mRNA. Proc. Natl. Acad. Sci. U.S.A. 85:1028–1032.

27. Wickstrom EL, Bacon TA, Gonzalez A, Lyman GH, Wickstrom E (1989): Anti-c-myc DNA increases differentiation and decreases colony formation

by HL-60 cells. *In vitro* Cell Dev. Biol. 25:297–302.

28. Holt JT, Redner RL, Nienhuis AW (1988): An oligomer complementary to c-myc mRNA inhibits proliferation of HL-60 cells and induces differentiation. Mol. Cell Biol. 8:963–973.

29. Minden M (1985): Stem cells in acute leukemia. In Golde DW, Takaku F (eds): Hematopoietic Stem Cells. New York and Basel: Marcel Dekker, pp 273–90.

30. Kastan MB, Stone KD, Civin CI (1989): Nuclear oncoprotein expression as a function of lineage, differentiation stage and proliferative status of normal human hematopoietic cells. Blood 74:1517–1524.

31. Ness SA, Marknell A, Graf T (1989): The v-myb oncogene product binds to and activates the promyelocyte-specific *mim*-1 gene. Cell 59:1115–1125.

32. Weber BL, Westin EH, Clarke MF (1990): Differentiation of mouse erythroleukemia cells enhanced by alternatively spliced mRNA. Science 249:1291–1293.

33. Weintraub H, Izant JG, Harland RM (1985): Antisense RNA as a molecular tool for genetic analysis. Trends Genet. 1:22–25.

34. van der Krol AR, Mol JNM, Stuitje AR (1988): Modulation of eukaryotic gene expression by complementary RNA or DNA sequences. BioTechniques 6:958–976.

35. Zamecnik PC, Stephenson ML (1978): Inhibition of Rous sarcoma virus replication and cell transformation by a specific oligodeoxynucleotide. Proc. Natl. Acad. Sci. U.S.A. 75:280–284.

36. Zerial A, Thuong NT, Helene C (1987): Selective inhibition of the cytopathic effect of type A influenza viruses by oligodeoxynucleotides covalently linked to an intercalating agents. Nucl. Acids Res. 15:9909–9919.

37. Matsukura M, Shinozuka K, Zon G, Mitsuya H, Reitz M, Cohen JS, Broder S (1987): Synthesis of phosphorothioate analogues of oligodeoxynucleotides: Inhibitors of replication and cytopathic effects of human immunodeficiency virus. Proc. Natl. Acad. Sci. U.S.A. 84:7706–7710.

38. Weinberg RA (1985): The action of oncogenes in the cytoplasm and nucleus. Science 230:770–776.

39. Slamon D (1987): Protooncogenes and human cancers. N. Engl. J. Med. 317:955–957.

40. Calabretta B, Venturelli D, Kaczmarek L, Narni F, Talpaz M, Anderson B, Beran M, Baserga R (1986): Altered expression of G_1-specific genes in human malignant myeloid cells. Proc. Natl. Acad. Sci. U.S.A. 83:1495–1498.

41. Auber ML, Horwitz LJ, Blaauw A, Khorana S, Tucker S, Wood T, Warmuth M, Dicke KA, McCredie KB, Spitzer G (1988): Evaluation of drugs for elimination of leukemic cells from the bone marrow of patients with acute leukemia. Blood 71:166–172.

42. Spitzer G, Verna DS, Fisher R, Zander A, Vellekoop L, Litan J, McCredie KB, Dicke KA (1980): The myeloid progenitor cell—its value in predicting hematopoietic recovery after autologous bone marrow transplantation. Blood 55:317–323.

Antisense *fos* Oligodeoxyribonucleotides Suppress the Generation of Chromosomal Aberrations

S. van den Berg, A. Schönthal, A. Felder, B. Kaina, P. Herrlich, and H. Ponta

Kernforschungszentrum Karlsruhe, Institut für Genetik und für Toxikologie, D-7500 Karlsruhe 1, Federal Republic of Germany (P.H., H.P.), Institute of Genetics, University of Karlsruhe, D-7500 Karlsruhe 1, Federal Republic of Germany (S.v.d.B., A.S., B.K., P.H.), and Ciba-Geigy AG, Division Pharma, CH-4002 Basel, Switzerland (A.F.)

THE NUCLEAR PROTEINS FOS AND JUN AS SUBUNITS OF A TRANSCRIPTION FACTOR

The discovery of ever more cellular proto-oncogenes, all of which can be altered to transform cells, has generated the idea that the encoded proteins are part of a cellular network of signal transduction involved in the regulation of proliferation and differentiation (reviewed in Heldin and Westermark, 1984; Hunter, 1985; Berridge, 1986, Rozengurt, 1986; Herrlich and Ponta, 1989). Nuclear oncogene products play crucial roles in the putative network in that they convert the incoming signals into changes of the genetic program. We have studied the role of the nuclear oncoproteins FOS and JUN in signal transduction pathways using collagenase as a target gene. The collagenase promoter is addressed by a variety of agents such as growth factors, tumor promoters, or DNA-damaging agents (Angel et al., 1987a; Herrlich et al., 1988). The promoter is also induced by overexpression of the cytoplasmic and membrane-bound oncoproteins RAS, MOS, and SRC and of the nuclear oncoproteins FOS and JUN (Schönthal et al., 1988a,b). The activation of the collagenase promoter depends largely on a single DNA element localized between positions -72 and -66, the TPA-responsive element (TRE; Angel et al., 1987b).

The TRE is necessary and sufficient for the response. It is recognized by the transcription factor AP-1, a heterodimer of FOS and JUN (reviewed in Curran and Franza, 1988). This suggests that serum, phorbol ester tumor promoters, DNA damage, and several oncoproteins feed into at least one common cellular pathway which converges onto the nuclear oncoproteins FOS and JUN (which in turn control the activity of the collagenase promoter).

DEPLETION OF CELLS FOR SHORT-LIVED PROTEINS BY ANTISENSE RNA

To prove the involvement of c-*fos* and c-*jun* in the activation of the collagenase promoter we eliminated their gene products using the antisense RNA technique. Antisense c-*fos* or c-*jun* RNA was transcribed from gene constructs comprising the ribosome-binding site and surrounding sequences (Schönthal et al., 1988b). The gene constructs were either driven by a constitutive (in transient transfection assays) or by an inducible promoter (in stable transfectants). The antisense RNA blocks FOS or JUN protein synthesis, respectively, and the cells lose presynthesized FOS or JUN at a $t_{1/2}$ of about $3-4$ hr. These experiments have demonstrated that both proteins, FOS and JUN, are absolutely necessary for the activation of the collagenase pro-

Prospects for Antisense Nucleic Acid Therapy of Cancer and AIDS, pages 63–70
©1991 Wiley-Liss, Inc.

moter by various agents: induction of colla-genase did not occur under conditions where synthesis of either FOS or JUN was blocked by antisense RNA expression. It has been shown using the same technique that the pro-moter of the stromelysin gene coding for an-other metalloprotease is also controlled by the FOS protein (Kerr et al., 1988).

The success of the antisense RNA tech-nique in elucidating the role of FOS and JUN in signal transduction pathways is largely due to the short lifetime of these proteins. Over the period of antisense action cells indeed lose substantial amounts of the two proteins. With long-lived proteins this is not to be ex-pected. The limitation is the maintenance of high levels of antisense RNA over longer pe-riods of time. In transient transfections, expression of antisense RNA is rapidly lost with the turnover of plasmid. In stable trans-fectants with inducible promoters available at present promoter activity is by and large transient and often rapidly down-modulated. In addition, the inducers of the conditional promoter may interfere with the process to be studied. A technique has been developed that overcomes at least some of these limi-tations and may be useful in combination with the above mentioned promoter-driven anti-sense techniques: treatment of cells with an-tisense oligodeoxyribonucleotides and oli-goribonucleotides (Weintraub et al., 1985; Heikkila et al., 1987; Holt et al., 1988; Frei et al., 1988; Wickstrom et al., 1988). The oli-gonucleotides are taken up rapidly by all cells (Heikkila et al., 1987; Holt et al., 1988; Frei et al., 1988; Wickstrom et al., 1988) and they can be added to cells at chosen points in time as well as repeatedly. Instead of depending on continuous promoter activity, oligonu-cleotide levels are determined by their up-take and stability. It has therefore been sug-gested to prolong their half-life by chemical modifications (Ott and Eckstein, 1987). We will show here that monothiophosphate bridges between deoxynucleotides increase the half-life by a factor of two as measured in a functional assay, that is the repression of serum-induced FOS protein synthesis. We then use these conditions to analyze a complex cellular phenotype, the generation of chro-

mosomal mutations by either UV irradiation or oncogene overexpression. Aberrations are introduced by a genetic program that de-pends on the presence of FOS protein.

PROLONGATION OF OLIGODEOXYRIBONUCLEOTIDE HALF-LIFE BY MONOTHIOPHOSPHATE–INTERNUCLEOTIDE BONDS

The effect of modification of the backbone of oligodeoxyribonucleotides was studied by comparing the influence of unmodified and modified 16-mers on serum-induced FOS protein synthesis in 3T3 cells. The 16-mers were complementary to the ribosome-bind-ing site of the murine c-fos RNA, extending 11 bases 5' and 5 bases 3' of the A in the ATG initiator codon (see legend to Fig. 1). FOS protein was determined by immunoflu-orescence. The cells were starved prior to serum induction. After 48 hr of serum dep-rivation preexisting FOS protein was de-pleted and no new synthesis was detectable by FOS-specific polyclonal antibodies fol-lowed by treatment with swine anti-rabbit antibodies conjugated to rhodamine. Serum treatment of cells led to a rapid induction of FOS protein synthesis: 1 hr after treatment the majority of nuclei (80–90%) showed flu-orescence labeling (Fig. 1A). This is consis-tent with previous observations (Verrier et al., 1986). Using preimmune serum as con-trol of specificity of the FOS polyclonal anti-bodies, no fluorescence staining upon serum treatment was observed (not shown). Addi-tion of antisense fos oligodeoxyribonucleo-tides (at 10 μM) together with serum treat-ment completely repressed FOS protein synthesis, independent of the use of unmo-dified or modified oligodeoxyribonucleo-tides (Fig. 1B). The specificity of repression is documented by the use of oligodeoxyri-bonucleotides without sequence comple-mentarity to fos mRNA and by the use of c-fos sense oligodeoxyribonucleotides (Fig. 1A). These control oligodeoxyribonucleotides did not influence the induction of FOS protein synthesis by serum. To determine the stabil-ity of the antisense c-fos oligodeoxyribonucleotides, cells were preincubated for various periods of time with the antisense oligodeoxyribonucleotide. Then, the degree of FOS synthesis in response to serum is a measure of the antisense oligodeoxynucleotide level at the time of serum addition. Antisense *fos* oligodeoxyribonucleotide added together with serum repressed completely, as already mentioned (Fig. 1B). Extension of the preincubation period up to 6 hr led to progressive release of repression of serum-induced FOS protein synthesis (Fig. 1B). Already after 4 hr, induction of FOS protein was indistinguishable from control cells (80% of the cells being FOS-positive). A similar set of experiments using backbone-modified oligodeoxyribonucleotides gave a qualitatively similar result. Quantitatively, however, there was a clear difference: With 4-hr preincubation less than 20% of the cells and with 6-hr preincubation about 80% of the cells were induced. These data can be plotted to yield a functional half-life. In summary our experiments demonstrate efficient repression of FOS protein synthesis with 16-mers of antisense *fos* oligodeoxyribonucleotides at an extracellular concentration of 10 μM. Unmodified oligodeoxyribonucleotides are inactivated with a functional half-life of about 2 hr; a modification of the backbone of the oligomer by phosphothioester bonds increases their functional half-life to more than 4 hr.

These data document the functional half-life of an oligonucleotide under conditions of low complementary mRNA concentration. It is possible that an oligonucleotide could be stabilized when hybridized to RNA. Stabilization could account for the longer physical half-lives that have been reported for MYC and RAS antisense oligonucleotides (Wickstrom et al., 1988; Daaka and Wickstrom, 1990). The functional lifetime has, however, not been determined.

UV RADIATION AND ONCOGENES INDUCE CHROMOSOMAL ABERRATIONS AND FOS PLAYS A DECISIVE ROLE IN THIS PROCESS

It is generally accepted that cancerous transformation occurs in several steps. On-

cogene expression certainly is one decisive step toward transformation, but it is obvious that genetic changes in addition to oncogene expression need to occur, as, e.g., suggested by the long delay in vivo between the onset of oncogene expression and tumor formation (reviewed by Hanahan, 1988; Balmain and Brown, 1988). In an attempt to elucidate the mechanisms leading to additional mutations, we tested the hypothesis whether oncogene expression itself increases the rate of mutations. The oncogenes *ras*, *mos*, and *fos* were expressed in NIH3T3 cells under control of the steroid hormone-inducible mouse mammary tumor virus promoter (Jaggi et al., 1986; Schönthal et al., 1988a). As an easily measurable parameter for mutations we followed chromosomal aberrations by cytohistochemical techniques. To elucidate the sensitivity of our test system (cell strain and detection method for aberrations) we also analyzed the induction of chromosomal aberrations by UV irradiation (UV).

UV increased the number of detectable aberrations in a dose-dependent manner (Table 1). Also induction of oncogene expression upon treatment of cells with the glucocorticoid dexamethasone enhanced the number of chromosomal aberrations (about 2-fold, Table 2). This enhancement is significant at the 5% confidence levels and is reproducible in all transfected clones tested. Dexamethasone treatment of mock-transfected NIH3T3 cells and of transfectants with a mutated c-*fos* gene had no influence on the number of chromosomal aberrations observed (Table 2).

As discussed above the activation of the collagenase gene by *ras* and *mos* overexpression depends on FOS protein (Schönthal et al., 1988a; Herrlich and Ponta, 1989). To investigate whether the same hierarchical order also holds for oncogene-dependent induction of chromosomal aberrations, we blocked, prior to and during the glucocorticoid-induced activation of oncoprotein synthesis, the synthesis of the FOS protein by use of antisense-thiooligodeoxyribonucleotide (Table 2). Possible unspecific effects were measured by using control oligonucleotides, e.g., an oligodeoxyribonucleotide of the same length but without any complementarity to

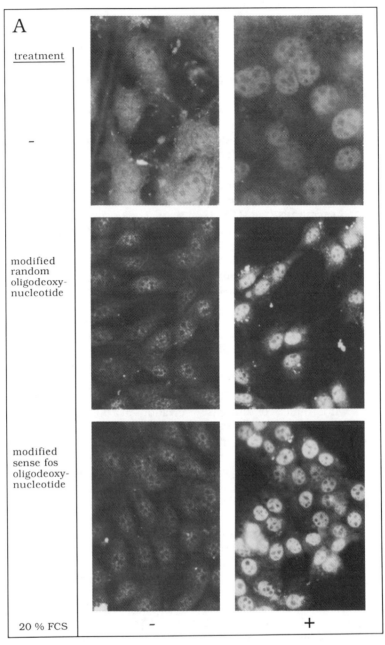

Fig. 1. A: Comparison of specific antisense oligonucleotide with the random oligonucleotide. **B:** Comparison between modified and unmodified antisense. Modified antisense oligodeoxyribonucleotides have a prolonged functional half-life. Cells (4×10^4) were plated on coverslips in microtiter dishes and, after 24 hr, starved in medium containing 0.5% serum. After 48 hr the cells were treated with medium containing 20% serum for 1 hr; 10 μM modified or unmodified antisense *fos* oligodeoxyribonucleotide was added at 6, 4, 2, and 1 hr before or together (0 hr) with serum treatment, as indicated. After serum treatment cells were fixed in 3% paraformaldehyde, permeabilized in 1% Triton X-100, and exposed to a polyclonal rabbit antiserum (1:30 dilution) directed against the FOS protein. Binding of these antibodies was visualized by rhodamine-conjugated swine anti-rabbit antibodies. The sequence of the oligodeoxyribonucleotides were sense, 5'-AAACTTCGACCATGAT-3'; random, 5'-TAGTCGTCGCAGCCT-3'; antisense, 5'-ATCATGGTCGAAGTTT-3'.

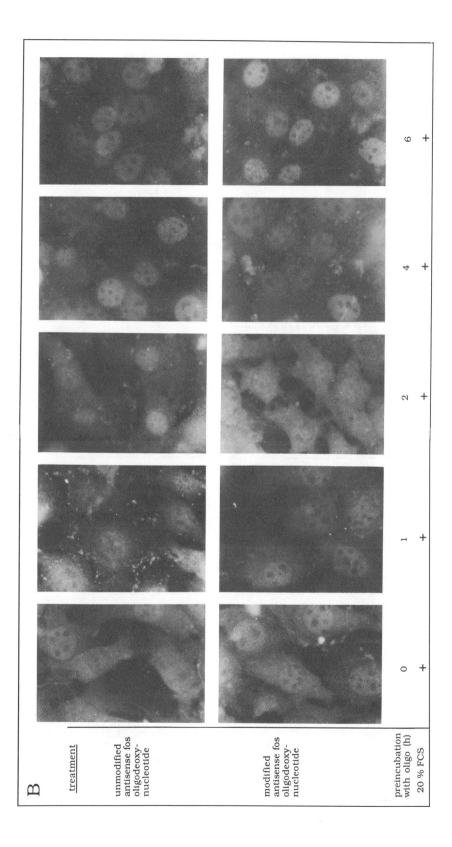

B

<u>treatment</u>

unmodified
antisense fos
oligodeoxy-
nucleotide

modified
antisense fos
oligodeoxy-
nucleotide

preincubation
with oligo (h)   0   1   2   4   6

20 % FCS   +   +   +   +   +

TABLE 1. UV Induction of Chromosomal Aberrations Requires FOS[a]

Oligo	UV (J/m^2)	Cells analyzed	Cells with aberrations (%)	Number of aberrations per 100 cells
—	—	300	11 ± 2	17
Random	—	300	11 ± 2	16
sof	—	300	14 ± 3	27
—	3	300	47 ± 4	154
Random	3	300	47 ± 7	187
sof	3	300	34 ± 1	92
—	5	300	72 ± 6	368
Random	5	300	72 ± 5	297
sof	5	300	50 ± 1	186

[a]Cells (4×10^4) were plated in 5-cm petri dishes and irradiated with UV (254 nm) after 48 hr; 10 μM modified antisense *fos* and random oligodeoxyribonucleotides were added 3 and 0.5 hr before and 3, 6, and 9 hr after UV irradiation. Sixteen hours after UV irradiation the cells were fixed and spreads of metaphase chromosomes were prepared. The number of aberrant metaphases and aberrations represents the average of three independent experiments. The reduction of UV-induced chromosomal aberrations by antisense *fos* oligodeoxyribonucleotides was statistically significant at the 5% level.

TABLE 2. A Key Role of FOS in the Induction of Chromosomal Aberrations by Extranuclear Oncoproteins[a]

Cells	Oligo (10 μM)	Dex ($2 \times 10^{-7} M$)	Cells analyzed	Cells with aberrations (%)	Number of aberrations per 100 cells
LTR-*fos*	—	−	400	14 ± 2	22
	—	+	400	26 ± 4	35
LTR-*ras*	—	−	300	7 ± 1	7
	—	+	250	12 ± 2	15
	Random	−	250	10 ± 0	14
	Random	+	200	18 ± 2	26
	sof	−	250	7 ± 0	9
	sof	+	250	8 ± 1	9
LTR-*mos*	—	−	400	8 ± 1	9
	—	+	400	13 ± 1	15
	Random	−	400	9 ± 2	10
	Random	+	400	14 ± 2	17
	sof	−	400	7 ± 1	8
	sof	+	400	10 ± 1	11

[a]Cells (4×10^4) per well were plated in microtiter dishes and treated after 48 hr with the modified antisense *fos* or the random oligodeoxyribonucleotide or not treated. Dexamethasone (Dex) was given immediately after addition of the oligodeoxyribonucleotide and, after another 2 hr, the oligodeoxyribonucleotide treatment was repeated. After 16 hr the cells were fixed and spreads of metaphase chromosomes were prepared. The number of aberrant cells and aberrations represents the average of three independent experiments. The reduction of oncogene-induced chromosomal aberrations by antisense *fos* oligodeoxyribonucleotide was statistically significant at the 5% level.

c-*fos* mRNA. This "random" oligonucleotide did not influence the induction of chromosomal aberrations, either in RAS or in MOS expressing cells (Table 1). In contrast, the *fos* antisense oligonucleotide reduced the induction of aberrations by RAS and MOS drastically (Table 2). This suggests that RAS and MOS increase the aberration frequency through FOS protein. Also UV-induced chromosomal aberrations were reduced in cells pretreated with *fos* antisense oligodeoxyribonucleotides, but not in control cells treated with the "random" oligonucleotide (Table 1). The repression was not as complete as with RAS or MOS expressing cells. This may suggest that a fraction of the UV-induced chromosomal aberrations is generated independently of FOS, or that UV-induced FOS synthesis is so fulminant that the level of antisense oligodeoxynucleotide does not suffice.

CONCLUSIONS

The analysis of control mechanisms regulating the proliferation and differentiation of mammalian cells rests on techniques that knock out specifically individual gene functions. The technique described offers this option. Antisense oligodeoxyribonucleotides block the function of specific genes. We have shown that the addition of small amounts of short antisense oligodeoxyribonucleotides extending over the start of translation for a specific protein knocks out completely the synthesis of this protein, in all cells of a culture. By chemically modifying the backbone of the oligodeoxyribonucleotides it is possible to prolong its functional integrity in this specific case by a factor of about 2. The technique is suitable to help in dissecting relatively complex cellular processes that require periods of time far exceeding the half-life of the oligonucleotide, provided a limiting function is addressed whose synthesis or activity is needed over the time of antisense blockage. In the example presented here, we have provided evidence that FOS protein plays a decisive role in oncogene and UV-driven mutagenesis. The experiments suggest that FOS acts only in an early time interval, which could

be explained if FOS acted as the subunit of a transcription factor or of a transcriptional repressor that turns on mutator genes (Walker, 1985) essential in the generation of chromosome aberrations. Future analyses will attempt to confirm this hypothesis, e.g., will investigate the other subunits FOS interacts with, and examine whether, indeed, the transactivating domain of FOS is required for induction of mutations.

SUMMARY

The fast induction of the FOS nuclear oncoprotein by serum treatment of starved cells was used to test the functional stability of antisense oligodeoxyribonucleotides. Unmodified oligodeoxyribonucleotides lost their blocking effect with a half-life of about 2 hr, modification of the backbone by thioesters extended the half-life to about 4 hr. The modified oligodeoxyribonucleotides were used to unravel a decisive role of FOS in a complex physiologic event: The induction of chromosomal aberrations upon overexpression of oncogenes like *ras* and *mos* and upon irradiation of fibroblasts with UV light.

REFERENCES

Angel P, Baumann I, Stein B, Delius H, Rahmsdorf HJ, Herrlich P (1987a): 12-O-Tetradecanoyl-phorbol-13-acetate induction of the human collagenase gene is mediated by an inducible enhancer element located in the 5'-flanking region. Mol. Cell. Biol. 7:2256–2266.

Angel P, Imagawa M, Chiu R, Stein B, Imbra RJ, Rahmsdorf HJ, Jonat C, Herrlich P, Karin M (1987b): Phorbol ester-inducible genes contain a common cis element recognized by a TPA-modulated transacting factor. Cell 49:729–739.

Balmain A, Brown K (1988): Oncogene activations in chemical carcinogenesis. Adv. Cancer Res. 51:147–182.

Berridge MJ (1986): Growth factors, oncogenes and inositol lipids. Cancer Surv. 5:413–430.

Curran T, Franza BR Jr (1988): Fos and Jun: The AP-1 connection. Cell 55:395–397.

Daaka Y, Wickstrom E (1990): Target dependence of antisense oligodeoxynucleotide inhibition of c-Ha-ras p21 expression and focus formation in T24-transformed NIH3T3 cells. Oncogene Res. 5:267–275.

Frei E, Schuh R, Baumgartner S, Burri M, Noll M, Jürgens G, Seifert E, Nauber U, Jäckle H (1988): Molecular characterization of spalt, a homeotic gene required for head and tail development in the Drosophila embryo. EMBO J. 7:197–204.

Hanahan D (1988): Dissecting multistep tumorigenesis in transgenic mice. Annu. Rev. Genet. 22:479–519.

Heikkila R, Schwab G, Wickstrom E, Loke SL, Pluznik DH, Watt R, Neckers LM (1987): A c-myc antisense oligodeoxynucleotide inhibits entry into S phase but not progress from G_0 to G_1. Nature (London) 328:445–449.

Heldin C-H, Westermark B (1984): Growth factors: Mechanism of action and relation to oncogenes. Cell 37:9–20.

Herrlich P, Ponta H (1989): "Nuclear" oncogenes convert extracellular stimuli into changes in the genetic program. Trends Genet. 5:112–116.

Herrlich P, Jonat C, Rahmsdorf HJ, Angel P, Haslinger A, Imagawa M, Karin M (1988): Signals and sequences involved in the ultraviolet- and 12-O-tetradecanoylphorbol-13-acetate (TPA)-dependent induction of genes. In Colburn NH, Moses HL, Stanbridge EJ (eds): Growth Factors, Tumor Promoters, and Cancer Genes. UCLA Symposia on Molecular and Cellular Biology, New Series, Vol. 58. New York: Alan R. Liss, pp 249–256.

Holt JT, Redner RL, Nienhuis AW (1988): An oligomer complementary to c-myc mRNA inhibits proliferation of HL-60 promyelocytic cells and induces differentiation. Mol. Cell. Biol. 8:963–973.

Hunter T (1985): Oncogenes and growth control. Trends Biol. Sci. 10:275–280.

Jaggi R, Salmons B, Muellener D, Groner B (1986): The v-mos and H-ras oncogene expression represses glucocorticoid hormone-dependent transcription from the mouse mammary tumor virus LTR. EMBO J. 5:2609–2616.

Kerr LD, Holt JT, Matrisian LM (1988): Growth factors regulate transin gene expression by c-fos-dependent and c-fos-independent pathways. Science 242:1424–1427.

Ott J, Eckstein F (1987): Protection of oligonucleotide primers against degradation by DNA polymerase I. Biochemistry 26:8237–8241.

Rozengurt E (1986): Early signals in the mitogenic response. Science 234:161–166.

Schönthal A, Herrlich P, Rahmsdorf HJ, Ponta H (1988a): Requirement for fos gene expression in the transcriptional activation of collagenase by other oncogenes and phorbol esters. Cell 54:325–334.

Schönthal A, Gebel S, Stein B, Ponta H, Rahmsdorf HJ, Herrlich P (1988b): Nuclear oncoproteins determine the genetic program in response to external stimuli. Cold Spring Harbor Symp. Quant. Biol. 53:779–787.

Verrier B, Müller D, Bravo R, Müller R (1986): Wounding a fibroblast monolayer results in the rapid induction of the c-fos proto-oncogene. EMBO J. 5:913–917.

Walker GC (1985): Inducible DNA repair systems. Annu. Rev. Biochem. 54:425–457.

Weintraub H, Izant JG, Harland RM (1985): Antisense RNA as a molecular tool for genetic analysis. Trends Genet. 1:22–25.

Wickstrom EL, Bacon, TA, Gonzalez A, Freeman DL, Lyman GH, Wickstrom E (1988): Human promyelocytic leukemia HL-60 cell proliferation and c-myc protein expression are inhibited by an antisense pentadecadeoxynucleotide targeted against c-myc mRNA. Proc. Natl. Acad. Sci. U.S.A. 85:1028–1032.

The Use of Antisense Oligonucleotides to Dissect the Role of c-*fos* Expression in HL-60 Cell Differentiation

Adrienne L. Block, John P. Doucet, Susan L. Morrow, and Stephen P. Squinto

Department of Biochemistry and Molecular Biology, Louisiana State University Medical Center, New Orleans, Louisiana 70119

INTRODUCTION

Understanding of the molecular events that guide the regulation of eukaryotic cellular proliferation and differentiation should ultimately lead to the rational design of specifically targeted drugs for the treatment of several diseases including cancer, immunodeficiency, and neurodegeneration. Recently, a powerful experimental tool has emerged that allows for a selective and efficient means for inhibiting the expression of key gene products known to be involved in the control of both eukaryotic proliferation and differentiation. The technique involves the use of antisense DNA or RNA molecules designed to provide translation arrest of these key cellular regulatory proteins. This new technology has allowed for the direct assessment of the specific function of a single gene product during important cellular transitional periods.

Currently there are two primary approaches to achieving antisense-directed translational arrest. The first method makes use of gene constructs designed to synthesize complementary antisense mRNA sequences resulting in hybridization arrest of protein translation. Possible mechanisms include impaired nuclear processing or the inability of the RNA duplex to be efficiently translated (van der Krol et al., 1988). An alternative approach to antisense translational arrest involves the synthesis of short 5' or 3' synthetic antisense oligodeoxyribonucleotides (antisense DNA). It has recently been suggested that RNase H may play a role in this mechanism by cleaving the RNA/DNA duplexes formed between mRNA and antisense DNA (Walder and Walder, 1988).

Several reports have recently appeared that demonstrate the ability of antisense DNA to arrest the translation of selected RNAs when added to reticulocyte lysate or wheat germ translation systems in vitro (Hastie and Held, 1978), when microinjected into *Xenopus* oocytes (Kawasaki, 1985) or when incubated directly with eukaryotic cells in vitro (Marcus-Sekura et al., 1987). We have chosen to use this latter approach to directly assess the role of the c-*fos* protooncogene in the differentiation of human HL-60 leukemia cells and to determine if c-*fos* expression is required for either monocytic or granulocytic differentiation.

HL-60 leukemia cells represent a useful model for examining the role of protooncogene expression during cellular differentiation. For example, the induction of both the monocytic and granulocytic HL-60 cell differentiation pathways is temporally correlated with a marked reduction of c-*myc* mRNA, which is apparently controlled at the level of transcription (Grosso and Pitot, 1985). Using c-*myc* antisense RNA constructs (Yokoyama and Imamoto, 1987) and c-*myc* an-

Prospects for Antisense Nucleic Acid Therapy of Cancer and AIDS, pages 71–82

tisense DNA (Wickstrom et al., 1988; Holt et al., 1988) it has been convincingly demonstrated that c-*myc* inhibition is necessary for the monocytic or granulocytic differentiation of HL-60 cells, respectively.

Chemical agents capable of inducing monocyte or macrophage differentiation [i.e., phorbol esters (PMA), interferon-γ (IFN-γ), and vitamin D_3] are also capable of rapidly and transiently inducing the levels of c-*fos* mRNA (Mitchel et al., 1984; Muller et al., 1985). Induction of granulocytic differentiation by some chemical agents [i.e., retinoic acid (RA) and dimethyl sulfoxide (DMSO)] is not temporally correlated with the rapid and transient increases of c-*fos* expression observed during monocytic differentiation. This has led to the hypothesis that c-*fos* expression is lineage-specific. Recently we demonstrated that the early changes in c-*fos* mRNA that occur during PMA- and TNF-α-induced HL-60 cell differentiation are temporally correlated with a nuclear accumulation of the c-*fos* protooncoprotein pp55^{fos} in discrete nuclear substructures, including semicondensed chromatin and transcriptionally active euchromatin (Squinto et al., 1989). Cumulatively, these results suggest that the rapid and transient induction of c-*fos* expression might serve as a biochemical determinant in the commitment of HL-60 cells along the monocyte/macrophage lineage at the expense of progression along the granulocytic lineage. Here we have chosen to test this hypothesis by selectively inhibiting the rapid activation of c-*fos* expression by PMA and TNF-α by using antisense DNA.

MATERIALS AND METHODS

Synthesis and Purification of Oligomers

Unmodified, 21-base oligodeoxynucleotides were synthesized on an automated solid-phase synthesizer (Applied Biosystems, Inc.) using standard phosphoramidite chemistry and were purified by either high-performance liquid chromatography or by polyacrylamide gel electrophoresis (Gait, 1984). All oligomers used were lyophilized and resuspended in

phosphate-buffered saline (pH 7.4). The antisense DNAs were targeted against the translation initiation region of the human c-*fos* protooncogene.

Determination of Oligomer Stability

Gel purified oligomers were tagged at their 5'-termini with fluorescein isothiocyanate (FITC) using a commercially available kit (5'-3', Inc). FITC-tagged c-*fos* oligomers (1 μM) were incubated with logarithmically growing HL-60 cells cultured in RPMI 1640 medium supplemented with 20% fetal bovine serum (FBS) and 2 mM glutamine. Cells were seeded into 175-cm^2 tissue culture flasks at a density of 2×10^5 cells per ml in 100 ml. The percentage of FITC-staining HL-60 cells and the fluorescence intensity of individual HL-60 cells (between 1 and 72 hr after incubation with FITC-oligomers) was determined by analysis on a Becton-Dickinson FACStar flow cytometer.

Immunoblot Analysis of c-*fos* in Oligomer-Treated HL-60 Cells

HL-60 cells (both treated and untreated) were lysed and sonicated in buffer containing 10% sodium dodecyl sulfate (SDS) and 10% 2-mercaptoethanol and resolved on 10% SDS polyacrylamide gels. Proteins were transferred electrophoretically to nitrocellulose membranes. Membranes were treated with 5% nonfat dry milk dissolved in phosphate-buffered saline to block nonspecific protein binding sites. The membranes were then incubated first with a 1:5000 dilution of rabbit anti-*fos* polyclonal antibody (kindly provided by Dr. T. Curran) and then with a 1:2000 dilution of ^{125}I-labeled Staph protein A (kindly provided by Dr. J. Haycock). The c-*fos* protein (pp55^{fos}) and *fos*-related antigens (*jun* and *fra*) were detected by autoradiography after exposure of the membrane to Kodak XAR film at −70°C.

Determination of HL-60 Cell Proliferation and Flow Cytometric Analysis of HL-60 Cell Cycle

All studies for the determination of HL-60 cell proliferation and DNA analysis were performed with early-passage (<30) HL-60 cells and analyzed between cell concentrations of 0.1 and 5 × 10⁶ cells per ml. The HL-60 cells used in these studies had a doubling time of approximately 22 hr and a low rate of spontaneous differentiation (less than 3%) as determined by nitro blue tetrazolium (NBT) reduction, nonspecific esterase (NSE) staining, and the expression of the monocytic cell surface marker, Mo1. The effects of the antisense c-*fos* and sense c-*fos* DNAs oligomers were assessed after a single addition (1 μ*M*) to the cell culture media 1 hr prior to the single addition of the differentiation-inducing agent. Cell counts were assessed with a Coulter Counter. Cell viability was assessed by trypan blue staining.

The DNA content of the cells incubated with AS-Fos, S-Fos, or differentiation inducing agents was determined by flow cytometry using a Becton-Dickinson FACStar flow cytometer. For these procedures, cells were stained with propidium iodide. At least 10⁴ cells were counted for each analysis. The percentage of cells in each phase of the cell cycle was determined by replicate analysis.

Assessment of HL-60 Cell Differentiation

Both biochemical and functional assays were conducted to assess HL-60 cell differentiation essentially as described previously (Squinto et al., 1989).

NBT reduction. Cytoplasmic superoxide was detected by its ability to reduce soluble NBT to a blue-black precipitate, formazan. NBT reduction was determined using 0.2 ml of cell suspension.

Nonspecific esterase activity. Histochemical staining with α-naphthylacetate was performed as described by Yam et al. (1971) using 3 × 10⁵ cells removed from the HL-60 cell suspension. The cells were suspended in 1 ml of phosphate-buffered saline and incubated in α-naphthylacetate for 25 min at 37°C. The cells were then centrifuged onto microscope slides, and positive cells were scored.

Mo1 expression. The cell surface differentiation marker, Mo1, was detected by im-

munofluorescence using commercially available kits (Coulter Immunology, Inc, Hialeah, FL). Mo1 was purchased as a fluorescein-conjugated murine monoclonal antibody. At the indicated times, approximately 1 × 10⁶ HL-60 cells were used for each assay and were analyzed by flow cytometry using a Becton-Dickenson FACStar flow cytometer. Control incubations using nonspecific fluorescein-tagged mouse monoclonal antibodies exhibited no fluorescence.

RESULTS

c-*fos* Antisense Oligonucleotides Enter HL-60 Cells and Are Stable

The sequences of the sense and antisense c-*fos* DNAs (S-Fos and AS-Fos, respectively) used to assess the role of c-*fos* expression in mediating HL-60 cell differentiation are diagramed in Figure 1. As shown, these oligomers extend one codon upstream of the initiating methionine for human c-*fos* and correspond to either the sense (S-Fos) or antisense (AS-Fos) strands encoding the first 6 amino acids of the human c-*fos* protooncogene protein. The 5′-ends of these oligomers were specifically labeled with fluorescein isothiocyanate (FITC) to determine whether they were taken up by HL-60 cells and to assess their relative stability in these cells. Fluorescent-tagged HL-60 cells were analyzed on a flow cytometer as described in Materials and Methods. Table 1 indicates that approximately 90% of the HL-60 cells incorporated 5′-FITC-*fos* oligomers and that fluorescence intensity was relatively unchanged for approximately 60 hr. At 72 hr, less than 70% of the HL-60 cells were positive (Table 1) for either FITC-AS-Fos or FITC-S-Fos.

c-*fos* Antisense DNA Inhibits PMA Induction of c-*fos* Protein in HL-60 Cells

In order for any antisense approach to be successful, complete inhibition of protein expression for a specific gene of interest must be achieved. The immunoblot shown in Figure 2 demonstrates that treatment of exponentially dividing HL-60 cells with the phor-

		Met	Met	Phe	Ser	Gly	Phe	
HUMAN FOS CODING SEQUENCE	3'	TGC	TAC	TAC	AAG	AGC	CCG	AAG	5'
HUMAN FOS mRNA SEQUENCE	5'	ACG	AUG	AUG	UUC	UCG	GGC	UUC	3'
ANTISENSE FOS OLIGONUCLEOTIDE	3'	TGC	TAC	TAC	AAG	AGC	CCG	AAG	5'
SENSE FOS OLIGONUCLEOTIDE	5'	ACG	ATG	ATG	TTC	TCG	GGC	TTC	3'

Fig. 1. Antisense and sense DNA sequences and their relationship and position relative to the cod- ing sequence of the human c-*fos* protooncogene and its cognate mRNA transcript.

bol ester PMA (50 ng/ml) for 2 hr increases expression of the c-*fos*, and c-*jun* proteins as well as some as yet unidentified *fos*-related antigen (fra) (lane 2) relative to untreated control HL-60 cells (lane 1). The antibody used in these studies was a polyclonal anti- body directed against an internal peptide of the c-*fos* protein, the M peptide (residues 127–152 of the human c-*fos* protein). This polyclonal antibody has been reported to de- tect several proteins that share similar c-*fos* epitopes (Franza et al., 1988). Pretreatment of HL-60 cells for 1 hr with the AS-Fos oli- gomer (1 μM) completely blocked the PMA induction of c-*fos* related proteins (lane 4). Treatment of the HL-60 cells with AS-Fos oli- gomer alone, in the absence of any inducing agent, had no obvious effect on c-*fos* expres- sion (lane 3). Since the gene sequences for c-*fos* and c-*jun* are relatively disparate, these results suggest that expression of c-*fos* in HL- 60 cells may also be correlated with expres- sion of both the c-*jun* and fra genes.

Antisense c-*fos* DNA Affects HL-60 Cell Proliferation and Cell Cycle Phase

As seen in Figure 3, a single addition of AS- Fos oligomer (1 μM), completely arrested logarithmic HL-60 cell growth within 12 hr both in the absence or presence of TNF-α (500 U/ml). TNF-α alone had no apparent effect on HL-60 cell proliferation. Cell via- bility (approximately 80% as determined by trypan blue exclusion) was maintained in both control cultures and in all treated cultures during the 96-hr period.

Because of the profound effects of AS-Fos on HL-60 cell proliferation, we performed experiments using propidium iodide (PI) staining for DNA content and flow cytometry to assess the effects of AS-Fos and S-Fos oli- gomers on the HL-60 cell cycle. Figure 4a demonstrates the cell cycle profile of un- treated exponentially growing HL-60 cells stained with PI and analyzed on a Becton Dickinson FACStar flow cytometer. We cal-

TABLE 1. Incorporation and Stability of FITC-Anti-*fos* DNA (AS-Fos) in HL-60 Cells[a]

	% Fluorescent HL-60 cells				
	Time of incubation (hr)				
Treatment	12	24	48	60	72
Control (untreated)	3	2	4	3	5
5'-FITC-AS-Fos	93	91	88	87	65
5'-FITC-S-Fos	98	86	94	90	56

[a]Exponentially proliferating HL-60 cells were either untreated (control) or incubated with 5'-FITC-AS-Fos (1 μM) or 5'-S-Fos (1 μM) for the time periods indicated. The percentage of fluorescent HL-60 cells was quantitated by flow cytometry as described in Materials and Methods.

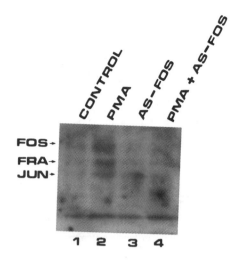

FOS→
FRA→
JUN→

1   2   3   4

Fig. 2. Autoradiograph of immunoblot demonstrating the effects of PMA, AS-Fos, and both on Fos expression in HL-60 cells. Exponentially proliferating HL-60 cells were either untreated (lane 1) or treated with PMA (50 ng/ml) or AS-Fos (1 μM) for 2 hr (lanes 2 and 3, respectively) or preincubated with AS-Fos (1 μM) for 1 hr followed by PMA (50 ng/ml) treatment for 2 hr (lane 4) and cell lysates were prepared for gel electrophoresis as described in Materials and Methods. Proteins were transferred to nitrocellulose membranes which were probed with a polyclonal antibody to c-*fos* followed by ^{125}I-labeled protein A. FOS, pp55^{c-fos}; JUN, p39^{jun}; FRA, Fos-related antigen (p45).

culated that under control conditions at 48 hr approximately 51% of the cells were in the G_1/G_0 phase, 35% were in the S phase, and 12% were in the $G_2 + M$ phase (Table 2). Treatment of the cells with TNF-α for 48 hr had little effect on the cell cycle profile of the HL-60 cells (Fig. 4b and Table 2). PMA treatment of HL-60 cells (known to rapidly induce end-stage monocyte/macrophage differentiation) for 48 hr resulted in a significant proliferation arrest with most of the cells accumulating in the G_1/G_0 phase of the cell cycle (Fig. 4c and Table 2). Figure 4d and Table 2 demonstrate that treatment of proliferating HL-60 cells for 48 hr with AS-Fos (1 μM) had a profound effect on the cell cycle profile with most of the cells apparently accumulating at the G_1/S border. In fact, [^{3}H]thymidine-labeling studies confirmed that

most of the AS-Fos-treated cells had not yet entered S phase (not shown). S-Fos treatment had no effect on the HL-60 cell cycle (inset to Fig. 4d).

Antisense c-*fos* DNA Partially Inhibits Both TNF-α and PMA-Induced Terminal Differentiation But Not RA-Induced Differentiation of HL-60 Cells

HL-60 cells can be stimulated to enter a differentiation pathway by a variety of compounds including phorbol esters (i.e., PMA), polar planar compounds (i.e., DMSO), 1,25-dihyroxyvitamin D_3, dibutyryl-cyclic AMP (db-cAMP), interferon-γ (IFN-γ), retinoic acid (RA), interleukin-2 (IL-2), and granulocyte–macrophage colony-stimulating factor (GM-CSF), among others (reviewed in Collins, 1987). We have recently demonstrated that TNF-α induces differentiation of HL-60 cells to end-stage macrophage-like cells (Squinto et al., 1989).

TNF-α is a macrophage-derived cytokine elicited during cellular responses to various microbial infections. TNF-α exerts direct cytotoxicity toward some tumor cells in vitro and produces hemorrhagic tumor necrosis in vivo. In HL-60 cells, TNF-α induces differentiation along the monocyte/macrophage lineage and these effects are preceded by an approximate 5-fold increase in c-*fos* mRNA levels followed immediately by a rapid and transient increase in the nuclear protein product of the c-*fos* protooncogene, pp55^{c-fos} (Squinto et al., 1989). Our findings suggest that pp55^{c-fos} is involved in the signal transduction system initiated by TNF-α during the induction of HL-60 cell differentiation. However, our earlier findings did not establish a direct functional correlation between c-*fos* induction by TNF-α and monocyte/macrophage differentiation of HL-60 cells.

Here, we have examined the biochemical changes induced in HL-60 cells under the influence of TNF-α and have assessed these changes under conditions of c-*fos* inhibition by antisense DNA. Induction of monocyte/macrophage differentiation of HL-60 cells has been shown to be associated with an increase of nonspecific esterase (NSE) activity (Yam

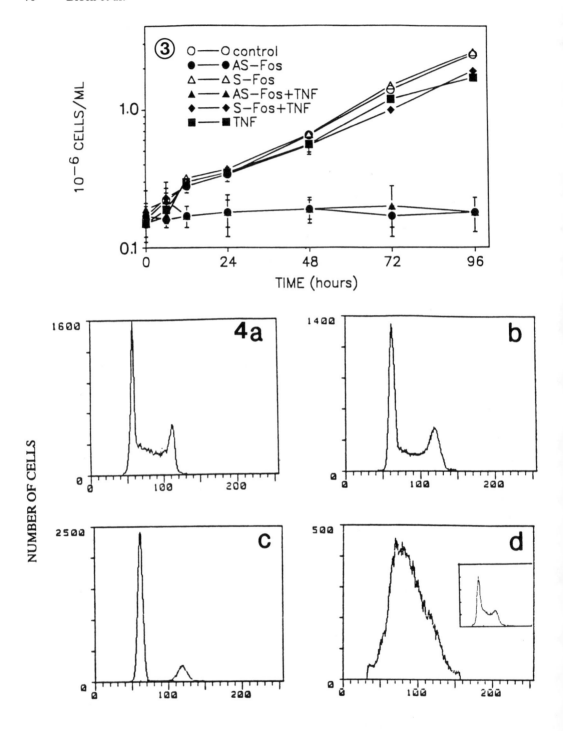

FLUORESCENCE INTENSITY

(Channel Number)

TABLE 2. Flow Cytometric Analysis of DNA Content of HL-60 Cells[a]

Treatment	G_0/G_1 (%)	S (%)	G_2/M (%)
Control	51	35	12
TNF-α (500U/ml)	49	36	13
PMA (50 ng/ml)	78	18	4
AS-Fos	64	27	6
S-Fos	52	33	11

[a]Exponentially proliferating HL-60 cells were incubated for 48 hr with either TNF-α, PMA, AS-Fos, or S-Fos. The cells were stained with propidium iodide and analyzed for their cell cycle distribution by flow cytometry as described in Materials and Methods.

et al., 1971), the increased ability to undergo a stimulus-induced respiratory burst as measured by the reduction of nitroblue tetrazolium (NBT-positive) (Collins et al., 1979), and the expression of the monocyte-specific cell surface receptor Mo1 (Todd et al., 1981). The percentage of control PBS-treated HL-60 cells expressing both functional and phenotypic markers of monocyte/macrophage differentiation was consistently low (between 0 and 6%; Figs. 2, 3, and 4). Figure 5A indicates that both TNF-α and PMA increase the number of NSE-positive HL-60 gradually between 6 and 120 hr. After 48 hr, approxi-

mately 60 to 80% of the HL-60 cell population is NSE-positive. The increase in NSE-positive cells treated with either TNF-α or PMA was inhibited approximately 2-fold (30 to 40% NSE-positive) at 48 hr by preincubating the cells for 1 hr with 1 μM AS-Fos (Fig. 5B). S-Fos (1 μM) had no inhibitory effect on either TNF-α- or PMA-induced NSE staining (not shown).

As observed with NSE staining, TNF-α, PMA, and RA all increased the number of NBT-positive HL-60 cells between 6 and 120 hr after treatment. At 48 hr, PMA treatment resulted in approximately 90% NBT positive cells while TNF-α and RA each resulted in approximately 40% NBT-positive cells (Fig. 6A). An approximate 2-fold inhibition of NBT-positive cells could also be achieved by preincubating HL-60 cells with AS-Fos (1 μM) 1 hr prior to the addition of either TNF-α or PMA (Fig. 6B). AS-Fos preincubation, however, had no effect on the granulocytic differentiation-promoting activity of RA (compare Fig. 6A with 6B).

Figure 7A demonstrates that PMA significantly increases staining for the monocytic cell surface marker Mo1 within 12 hr (30% Mo1-positive cells) and peaks at about 96 hr where almost 100% of the HL-60 cell population is Mo1-positive. TNF-α also increases Mo1 staining but to a lesser degree: only 20% of the HL-60 cell population is Mo1-positive by 48 hr and the maximum effect at 120 hr was 45% (Fig. 7A). AS-Fos (1 μM) completely inhibited the TNF-α induction of Mo1 staining within 12 hr but only partially inhibited (approximately 50%) the induction

◀ **Fig. 3.** Assessment of HL-60 cell proliferation under the influence of AS-Fos, S-Fos, TNF-α, or a combination of oligomer and TNF-α. Exponentially proliferating HL-60 cells were seeded at a density of 1×10^5 cells per ml and maintained in RPMI 1640 media supplemented with 20% heat-inactivated FBS. The HL-60 cells were either untreated (control) or treated with either TNF-α (500 U/ml), AS-Fos, or S-Fos (1 μM) or preincubated for 1 hr with AS-Fos or S-Fos (1 μM) prior to TNF-α treatment. Cell number was determined with a Coulter Counter.

Fig. 4. Flow cytometric cell cycle analysis of HL-60 cells. Exponentially proliferating HL-60 cells were either untreated (**a**) or treated for 48 hr with either TNF-α (500 U/ml) (**b**), PMA (50 ng/ml) (**c**), or AS-Fos (1 μM) (**d**). The inset to (**d**) represents HL-60 cells treated for 48 hr with 1 μM S-Fos. Cells were stained with propidium iodide as described in Materials and Methods and analyzed on a Becton-Dickinson FACStar flow cytometer. The x axis indicates fluorescence intensity and the y axis indicates cell number.

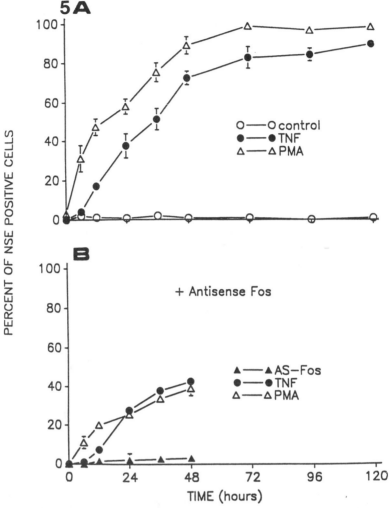

Fig. 5. Nonspecific esterase (NSE) activity of HL-60 cells. In (**A**) the NSE activity of HL-60 cells was measured in control (untreated) cells and cells treated with either TNF-α (500 U/ml) or PMA (50 ng/ml) for between 6 and 120 hr. In (**B**) cells were preincubated for 1 hr with 1 μM AS-Fos prior to the addition of differentiation inducing agents. Histochemical staining with α-naphthylacetate esterase was performed using 3 × 10⁵ cells for each determination as described in Materials and Methods. Data are presented as the means ± SEM for three independent determinations.

of Mo1 staining by PMA (compare Fig. 7A with 7B).

For all three of these differentiation markers, we found that increasing the concentration of AS-Fos from 1 μM to as much as 100 μM had no additional inhibitory effect of TNF-α- or PMA-induced cell differentiation (not shown). However, cell viability was greatly decreased in the presence of 100 μM AS-Fos.

DISCUSSION

Chemical agents capable of inducing monocyte/macrophage differentiation (i.e., PMA, INF-γ, and TNF-α) are also capable of rapidly and transiently inducing the levels of c-*fos* mRNA. Induction of granulocytic differentiation by some chemical agents (i.e., retinoic acid and DMSO) is not temporarily correlated with the rapid and transient increases in c-*fos* expression during monocytic differentiation (Collins, 1987). This had led to the hypothesis that c-*fos* expression is lineage-specific. Cumulatively, these results suggest that the rapid and transient induction of c-*fos* expression might serve as a biochemical determinant in the commitment of HL-60 cells along the monocytic/macrophage lineage at the expense of progression along the granulocytic lineage.

In this study, we employed oligomers to specifically inhibit expression of the c-*fos* protooncogene by using the antisense DNA method. The use of antisense methods to bypass cell surface events and to selectively inhibit the expression of nuclear proteins represents a novel and useful tool for the clarification of the direct role of protooncogene expression and cellular differentiation.

Incubation of HL-60 cells with the c-*fos* antisense DNA resulted in a decrease of Fos expression, a rapid and significant decrease in cell proliferation, cell cycle arrest in G_0/G_1, and the partial loss of the TNF-α and PMA-induced differentiated phenotype. However, pretreatment of HL-60 cells with AS-Fos had no effect on the granulocytic differentiation program stimulated by retinoic acid. Because the antisense DNA was stably incorporated into approximately 90% of the total HL-60 cell population for approximately 48 hr and complete inhibition of c-*fos* expression was achieved in HL-60 cell cultures, the partial inhibition (approximately 50%) of the differentiated phenotype is somewhat puzzling. While it appears from our findings that the low level expression of c-*fos* in HL-60 cells is absolutely required for the maintenance of cell proliferation, c-*fos* expression may be only partially related to TNF-α and PMA-induced cell differentiation along the monocytic pathway. Previous studies using antisense DNA directed against the c-*myc* protooncongene have demonstrated the requirement for c-*myc* expression in HL-60 cells for the maintenance of both cell proliferation and the undifferentiated phenotype (Holt et al., 1988; Wickstrom et al., 1988).

It is possible that c-*fos* expression is an early nuclear signaling event in a cascade of events that are cumulatively important for monocyte-macrophage differentiation of HL-60 cells. In this model, the function of the c-*fos* gene alone is minimized and the coordinate regulation of multiple differentiation-related genes is emphasized. This model is attractive in that Fos may function as a transcriptional activator of other cellular genes when in a heteroduplex with one of several other immediate-early gene products such as c-*jun* (Rauscher et al., 1989; Sassone-Corsi et al., 1988; Ransone et al., 1989). The Fos-associated nucleoprotein complex has been shown to bind with sequence specificity to AP-1 transcriptional regulatory sequences that may lead to the transcriptional activation or repression of these target genes (Rauscher et al., 1989; Gentz et al., 1989). Since the c-*myc* gene is known to contain an AP-1 sequence within its 5' flanking region (unpublished observation), it is possible that altered expression of c-*fos* leads to changes in the expression of c-*myc*, which is then followed by changes in the differentiated phenotype of HL-60 cells. Further studies are required to assess this hypothesis. Antisense DNA technology should provide a powerful and technically simple in vitro model to further dissect these potential relationships in HL-60 cell differentiation.

SUMMARY

Regulation of the genetic reprogramming that occurs in human promyelocytic leukemia (HL-60) cells during signal-stimulated termination of proliferation and the contiguous commitment of the cell to a differentiated phenotype has not been clearly defined. It has been suggested that the rapid and transient expression of the c-*fos* protooncogene may be a genetic determinant of the lineage-specific differentiation of HL-60 cells. This suggests a functional relationship between c-*fos* expression and monocytic differentiation. To examine this putative relationship directly, we have synthesized an antisense DNA to specifically block c-*fos*

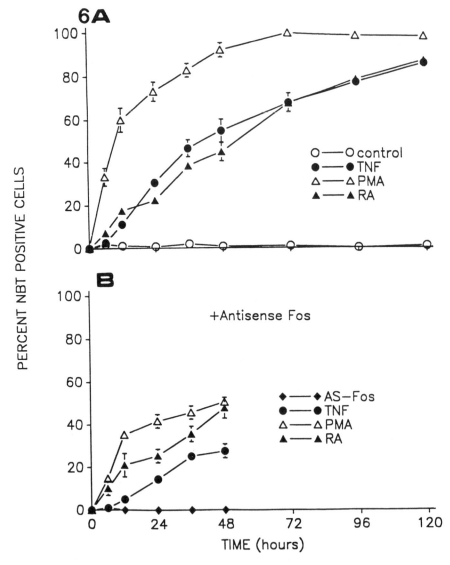

Fig. 6. Nitroblue tetrazolium (NBT) reducing activity of HL-60 cells. In (**A**) the NBT reducing activity was measured in control (untreated) cells and cells treated with either TNF-α (600 U/ml), PMA (50 ng/ml), or RA (1 μ*M*) for between 6 and 120 hr. In (**B**) the cells were first preincubated with 1 μ*M* AS-Fos prior to the addition of differentiating agents. NBT reducing activity was assessed as described in Materials and Methods. At least 250 cells were scored for each determination. Data are presented as the means ± SEM for three independent determinations.

expression in HL-60 cells. The c-*fos* antisense DNA synthesized (AS-Fos) was a 21-mer of human c-*fos* and included the translation initiation methionine codon. Incubation of exponentially proliferating HL-60 cells with AS-Fos (1 μ*M*) had profound inhibitory effects on cell proliferation within 12 hr and an altered cell cycle profile within 48 hr as determined by flow cytometric techniques. The complementary sense DNA (S-Fos) had no effect on these parameters. We and others have shown previously that the phorbol es-

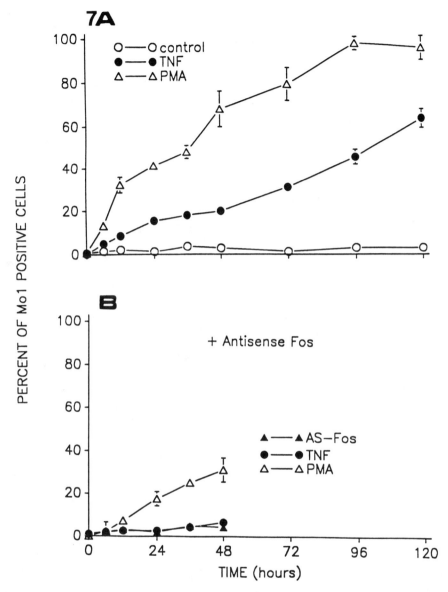

Fig. 7. Expression of the monocytic cell surface antigen Mo1 in HL-60 cells. In (**A**) the expression of Mo1 was assessed in control (untreated) cells and cells treated with either TNF-α (500 U/ml) or PMA (50 ng/ml) for between 6 and 120 hr. In (**B**) the cells were first preincubated with 1 μ*M* AS-Fos prior to the addition of differentiation-inducing agents. Mo1 expression was measured by flow cytometry as described in Materials and Methods.

ter, PMA, and human recombinant tumor necrosis factor-α (TNF-α) induce terminal differentiation of HL-60 cells to macrophage-like cells (Squinto et al., 1989). The differentiated phenotype is contiguous with an induction of c-*fos* expression at both the mRNA and protein level. c-*fos* expression is followed by cytostasis, the acquisition of an adherent phagocytic phenotype, and the expression of a specific monocytic cell surface antigen that is identified by monoclonal antibody, Mo1. Here we demonstrate that preincubation of HL-60 cells with AS-Fos for 1 hr resulted in a partial inhibition (approximately 50%) of the differentiation-inducing effects of both PMA and TNF-α while S-Fos had no such effect. However, AS-Fos had no inhibitory effect on granulocyte maturation induced by retinoic acid. Cumulatively, these results suggest that c-*fos* expression may be partially responsible for the phenotypic changes that occur in HL-60 cells in response to lineage-specific, macrophage-promoting differentiation agents and that c-*fos* expression is not linked to the granulocytic differentiation program.

REFERENCES

Collins SJ (1987): The HL-60 promyelocytic leukemia cell line: Proliferation, differentiation and cellular oncogene expression. Blood 70:1233–1244.

Collins SJ, Ruscetti FW, Gallagher RE, Gallo RC (1979): Normal functional characteristics of HL-60 after induction of differentiation by DMSO. J. Exp. Med. 149:969–975.

Franza RB Jr, Rauscher III FJ, Josephs SF, Curran T (1988): The Fos complex and Fos-related antigens recognize sequence elements that contain AP-1 binding sites. Science 239:1150–1153.

Gait MJ (ed) (1984): Oligonucleotide Synthesis. Oxford: IRL Press.

Gentz R, Rauscher III FJ, Abate C, Curran T (1989): Parallel association of Fos and Jun leucine zippers juxtaposes DNA binding domains. Science 243:1695–1699.

Grosso L, Pitot H (1985): Transcriptional regulation of c-*myc* during chemically-induced differentiation of HL-60 cell cultures. Cancer Res. 45:847–852.

Hastie ND, Held WA (1978): Analysis of mRNA populations by cDNA-mRNA hybrid-mediated inhibition of cell-free protein synthesis. Proc. Natl. Acad. Sci. U.S.A. 75:1217–1221.

Holt JT, Redner RL, Nienhuis A (1988): An oligomer complementary to c-*myc* mRNA inhibits proliferation of HL-60 promyelocytic cells and induces differentiation. Mol. Cell. Biol. 8:963–973.

Kawasaki ES (1985): Quantitative hybridization-arrest of mRNA in *Xenopus* oocytes using single-stranded complementary DNA or oligonucleotide probes. Nucl. Acids Res. 13:4991–5004.

Marcus-Sekura CJ, Woerner AM, Shinozuka K, Zon G, Quinnan GV (1987): Comparative inhibition of chloramphenicol acetyltransferase gene expression by antisense oligonucleotide analogues having alkyl phosphotriester, methylphosphonate and phosphorothionate linkage. Nucl. Acids Res. 15:5749–5760.

Mitchell R, Zokas L, Schrieber RD, Verma IM (1984): Rapid induction of the expression of proto-oncogene Fos during human monocytic differentiation. Cell 36:51–60.

Muller R, Curran T, Muller D, Guilbert L (1985): Induction of c-*fos* during myelomonocytic differentiation and macrophage proliferation. Nature (London) 314:546–548.

Ransone LJ, Visvader J, Sassone-Corsi P, Verma IM (1989): Fos-Jun interaction: Mutational analysis of the leucine zipper domain of both proteins. Genes Dev. 3:770–781.

Rauscher III FJ, Sambucetti T, Curran T, Distel RJ, Spiegelman B (1989): A common DNA binding site for Fos protein complexes and transcription factor AP-1. Cell 52:471–480.

Sassone-Corsi P, Ransone LJ, Lamph WW, Verma IM (1988): Direct interaction between *fos* and *jun* nuclear oncoproteins: Role of the leucine zipper domain. Nature (London) 336:692–695.

Squinto SP, Doucet JP, Block AL, Morrow SL, Davenport WD Jr (1989): Induction of macrophage-like differentiation of HL-60 leukemia cells by tumor necrosis factor-alpha: Potential role of *fos* expression. Mol. Endocrinol. 3:409–419.

Todd R, Griffin J, Ritz J, Nadlur L, Abrahms T, Schlossman S (1981): Expression of normal monocyte-macrophage differentiation antigens on HL-60 promyelocytes undergoing differentiation induced by leukocyte-conditioned medium or phorbol ester. Leukemia Res. 5:491–502.

van der Krol AR, Mol JNM, Stuitje AR (1988): Modulation of eukaryotic gene expression by complementary RNA or DNA sequences. BioTechniques 6:958–976.

Walder RY, and Walder JA (1988): Role of RNase H in hybrid-arrested translation by antisense oligonucleotides. Proc. Natl. Acad. Sci. U.S.A. 85:5011–5015.

Wickstrom EL, Bacon TE, Gonzalez A, Freeman DL, Lyman G, Wickstrom E (1988): Human promyelocytic leukemia HL-60 cell proliferation and c-*myc* protein expression are inhibited by an antisense pentadecandeoxynucleotide targeted against c-*myc* mRNA. Proc. Natl. Acad. Sci. U.S.A. 85:1028–1042.

Yam L, Li C, Crosby W (1971): Cytochemical identification of monocytes and granulocytes. Am. J. Clin. Pathol. 55:283–293.

Yokoyama K, Imamoto F (1987): Transcriptional control of the endogenous MYC protooncogene by antisense RNA. Proc. Natl. Acad. Sci. U.S.A. 84:7363–7367.

Antisense *fos* and *jun* RNA

Dan Mercola

Department of Pathology, University of California at San Diego, La Jolla, California 92093 and the Veterans Administration Medical Center, San Diego, California 92161

INTRODUCTION

The use of antisense *fos* RNA and, to a lesser extent, antisense *jun* RNA has contributed to our understanding of the roles of these gene products in cell cycle regulation, differentiation, transcriptional regulation, and, especially, transformation. In addition, recently the c-*fos* promoter itself has been found to be a target of negative regulation by the c-*fos* product and has been studied by use of antisense techniques; the topic is at present the subject of controversy. Preliminary studies of the differential roles of the various *fos* and *jun* family members and the use of *fos* ribozymes have been started. Progress in the application of antisense RNA and oligonucleotides to these topics and implications for diagnostic and therapeutic approaches are considered.

Here the convention of italicized three-letter acronyms for genes is followed; the genomic protein product is referred to by capitalized acronyms.

Fos and Jun have been reviewed recently (Vogt and Bos, 1989; Ovitt and Rüther, 1989; Morgan and Curran, 1989; Karin, 1990).

THE FOS AND JUN FAMILIES

There are five recognized cellular members of the Fos superfamily: c-*fos* (van Beveren et al., 1983), *fos*-B (Zerial et al., 1989), *fos*-D, Fos-related antigen I (*fra*-I) (Cohen and Curran, 1988), and Fos-related antigen II (*fra*-II) (Nishina et al., 1989). The corresponding gene products are variable in size,

275–367 amino acids, but exhibit overall homology and contain highly conserved regions with amino acid sequences indicative of particular structural motifs. The central ca. 85 amino acid portion contains a "leucine zipper" defined by a helix with a regular repeat of five leucine residues forming a hydrophobic surface which is known (Kouzarides and Ziff, 1988; Sellers and Struhl, 1989) to mediate interactions with a similar motif of Jun (Fig. 1). While Jun is capable of forming homodimers via interactions of this surface, none of the Fos family members forms homodimers but all form heterodimers with Jun (Chiu et al., 1988; Franza et al., 1988; Halazonetis et al., 1988; Kousarides and Ziff, 1988; Nakabeppu et al., 1988; Rauscher et al., 1988; Sassone-Corsi et al., 1988; Zerial et al., 1989; Smeal et al., 1989). Thus the exact composition of the Fos–Jun complex, first purified from HeLa cells as activator protein-1 (AP-1)—a transactivation factor that mediates gene induction by phorbol acetate (TPA)—likely reflects the predominant Fos member(s) present in a particular cell type (Chiu et al., 1988; Rauscher et al., 1988; Karin, 1990). So far no gross differences in the DNA-binding properties of the various Fos–Jun complexes have been reported (Karin, 1990). Recognition of the general composition of AP-1 suggests that Fos or certain family members potentially mediate transactivation or repression of genes containing AP-1 or Fos–Jun complex binding sites such as α_1(III)collagen, collagenase, major histocompatibility complex (MHC) class I, or, for repression, heat shock protein 70.

Prospects for Antisense Nucleic Acid Therapy of Cancer and AIDS, pages 83–114
©1991 Wiley-Liss, Inc.

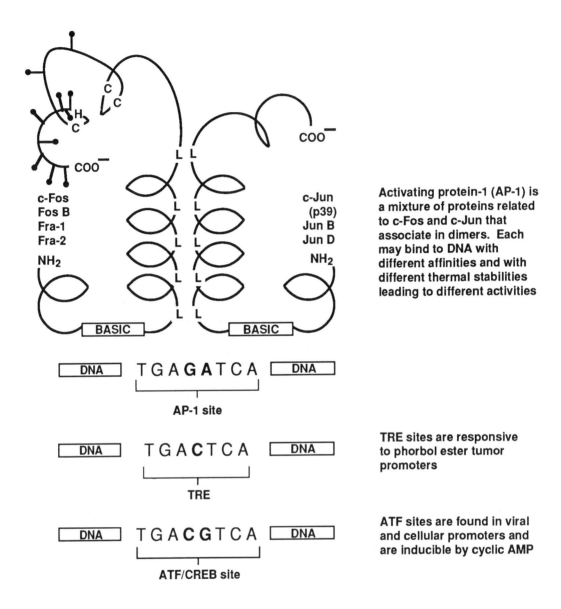

Activating protein-1 (AP-1) is a mixture of proteins related to c-Fos and c-Jun that associate in dimers. Each may bind to DNA with different affinities and with different thermal stabilities leading to different activities

c-Fos
Fos B
Fra-1
Fra-2

c-Jun
(p39)
Jun B
Jun D

NH₂

TGA**GA**TCA

AP-1 site

TRE sites are responsive to phorbol ester tumor promoters

TGA**C**TCA

TRE

ATF sites are found in viral and cellular promoters and are inducible by cyclic AMP

TGA**C**GTCA

ATF/CREB site

TRE/AP-1 containing genes: TGFß, PDGF-B
collagenase, stromelysin,
adipsin, RARß, c-*fos*, Egr-1

Fig. 1. Diagrammatic representation of structure of the Fos–Jun heterodimer and interaction with an "AP-1" site. The C-terminal portion of the Fos family proteins is rich in serine residues (—●) that undergo multiple posttranslational phosphorylations that may be important in regulation (Miller et al., 1984; Lee et al., 1988; Guis et al., 1990) and has weak homology to a "zinc finger" (Mercola et al., 1986; Mölders et al., 1987). C, H, and L are single letter amino acid code showing the relative positions of the indicated amino acids.

On the N-terminal side of the leucine zipper is a second highly homologous region of ca. 45 amino acids containing numerous basic residues known to be important for DNA binding. The Fos family members have little or no DNA-binding affinity; however, the DNA-binding affinity of the Jun family members is greatly increased in the presence of any of the Fos family members (Chiu et al., 1988; Nakabeppu et al., 1988; Halazonetis et al., 1988; Zerial et al., 1989). Thus heterodimer formation is important for high affinity DNA binding consistent with the palindromic nucleotide consensus sequence (Fig. 1).

A major characteristic of the Fos family members is very rapid and transient increase in transcription in response to a large number of stimuli. Typically, peak transcript levels occur in 15–30 min and return to basal levels in 1–2 hr. The stimulatory effects can be understood in terms of the complex array of positive regulatory elements of the c-*fos* promoter (Fig. 3; for reviews see Ovitt and Rüther, 1989; Vogt and Bos, 1989). The rapid decline is due to at least three mechanisms: a short mRNA half-life of 15–30 min (Miller et al., 1984; Meijlink et al., 1985; Treisman, 1985; Rahmsdorf et al., 1987; Wilson and Treisman 1988; Bonnieu et al., 1988; Shyu et al., 1989), possible control at the level of transcription by a short-lived suppressor protein(s) (Miller et al., 1984; Greenberg et al., 1986; Lee et

al., 1988), and control of the half-life of Fos itself possibly through phosphorylations of serine residues (Fig. 1) of the C-terminus (Miller et al., 1984; Lee et al., 1988). Fos provides one example of a suppressor protein effecting negative autoregulation (Sassone-Corsi et al., 1988; Schönthal et al., 1989; see below).

Rapid, transient responses have been recognized in a large number of early growth response or primary response genes such Early Growth Response gene-1 (*Egr*-1) as discussed below (Figure 2).

There are at present three recognized cellular members of the Jun superfamily: c-*jun* (Bohmann et al., 1987; Angel et al., 1988), *jun*-B (Ryder et al., 1988), and *jun*-D (Ryder et al., 1989; Hirai et al., 1989). They contain broadly homologous regions necessary for dimer formation and DNA binding as described above. In addition the DNA-binding domain is homologous to the DNA-binding domain of the yeast transcription factor GCN4 and can functionally substitute for the domain of GCN4 in yeast (Struhl, 1988; Angel et al., 1989). Jun-B is a much weaker transactivator than c-Jun and indeed elevated concentrations of Jun-B antagonize activation by c-Jun. The effect is traceable to the N-terminal half of the molecule (Chiu et al., 1989). These and other observations (Schütte et al., 1989) strongly suggest that Jun-B is a

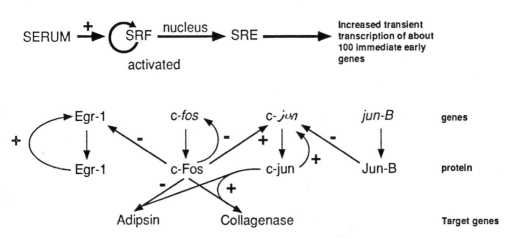

Fig. 2. Regulatory interrelations among the early growth response gene products Fos, Jun, and Egr-1.

negative regulator of c-Jun synthesis (Fig. 2) and may be responsible for the transient nature of the expression of c-*jun* (Karin, 1990). Thus the possibility exists of very different functions within the superfamilies (Fig. 2).

The antisense studies reviewed here have been carried out before and during the emergence of additional family members and therefore were not designed to discriminate among them. However, this is unlikely to be a complication. Successful application of the antisense method requires a high degree of complementarity both in extent of base pairing and in the length of contiguous pairs (Izant and Weintraub, 1985; Smith et al. 1986; Agris et al., 1986; McGarry and Lindquist, 1986; Chang et al. 1989).Cross-hybridization is unlikely with <70% identity. The sequences of the Fos and Jun family members show homology >70% only in the regions of putative DNA binding and leucine zipper formation. Virtually all sequences used for antisense studies considered here are derived from 5' noncoding and first exon sequences that exhibit <45% homology. It is very unlikely that the antisense results involve significant cross-hybridization; however, this remains unproven.

c-*fos* IN CELL-CYCLE REGULATION
Serum Regulation of Fibroblasts

The potential transforming properties of protooncogenes including c-*fos* suggest roles in the regulation of proliferation (Bishop, 1985). This conclusion, together with the rapid response of c-*fos* to a variety of mitogenic agents such as platelet-derived growth factor (PDGF) (Greenberg and Ziff, 1984; Kruijer et al., 1984; Müller et al., 1984) led to the hypothesis that c-*fos* was important in the recruitment of quiescent fibroblasts into the cell cycle (Greenberg and Ziff, 1984; Kruijer et al., 1984; Müller et al., 1984). This view is supported by studies that show that the c-*fos* promoter (Fig. 3) contains several regions such as the serum response element (SRE) and *sis*-conditioned-medium element (SCM) that mediate serum and PDGF-like factor stimulation (Triesman, 1985; Hayes et al., 1987; Greenberg et al., 1987).

The hypothesis was studied in detail for NIH-3T3 cells using antisense techniques (Nishikura and Murray, 1987). Antisense c-*fos* (mouse) RNA was expressed via a plasmid containing nearly all of the first exon of c-*fos* together with 140 bp of 5'-noncoding DNA of the c-*fos* promoter region (Fig. 3). This sequence was placed in antisense orientation between two dexamethasone-inducible mouse mammary tumor virus long terminal repeats (MMTV LTRs). Multiple copies were introduced together with pSV2*neo*, as a source of neomycin drug resistance, by the calcium phosphate–DNA transfection procedure and multiple clones were selected by addition of the neomycin derivative G418. Resistant clones appeared with a low frequency of 10^{-6} — 10-fold lower than for cells transfected with pSV2*neo* alone. A low frequency has been observed in other fibroblast-based systems (see "c-*fos* in transformation") and may be due to significant basal or "leaky" expression of antisense c-*fos* RNA, which interferes with an essential activity for growth. Thus the surviving clones may represent variants with lower basal expression of antisense RNA or possibly variants with alternate mechanisms of growth regulation.

Southern analysis of five clones revealed full-length insertion but with copy number varying from 1 to 40–50. One clone exhibited significant dexamethasone-dependent phenotypic changes and 6-fold decrease in steady-stage c-*fos* transcript levels. Consistent with this result, c-*fos* protein as judged by nuclear direct immunofluorescence studies was reduced in parallel with the decreased c-*fos* transcript levels. Antisense c-*fos* RNA production was confirmed. However, the levels were approximately 1% of sense levels and were not observed to be significantly increased by the addition of dexamethasone. Indeed for a clone with the single copy insertion and a clone with an aberrantly short antisense transcript, dexamethasone addition correlated with decreased steady-state antisense transcript levels. Thus there was poor correlation between

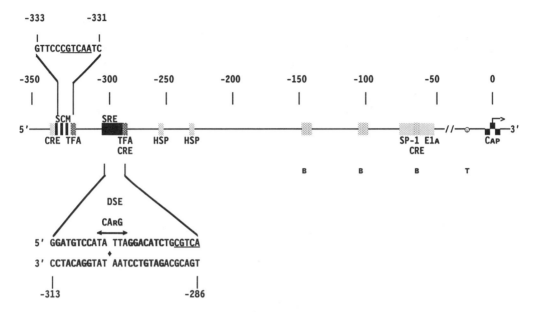

Fig. 3. Summary of the regulatory elements of the c-*fos* (mouse) promoter. TFA, AP-1-like binding site, underlined in expanded sequences; CRE, cyclic-like responsive element; SCM, *sis*-conditioned medium responsive element; SRE, serum response element or dyad symmetry element (DSE); E1a, adenovirus 2 E1a promoter homologous zone; HSP, heat-shock 70 promoter homology regions; SP-1, binding site for SP-1 transcription factor; B, sequences identified as important for regulation of basal expression; T, location of TATA box; ♦, local 2-fold axis of symmetry relating palindromic sequences shown in bold.

copy number and dexamethasone stimulation. Antisense RNA even when polyadenylated has been observed to be unstable in mammalian cells (Izant and Weintraub, 1985). Recently it has been reported that RNA duplex degrading ("unwinding") activity is widespread in mammalian cells (Bass and Weintraub, 1988; Wagner and Nishikura, 1988; Wagner, 1989). It was concluded that the small amount of antisense RNA detected did not accurately reflect transcription (Nishikura and Murray, 1987). Duplex formation and fate were not described.

One clone of antisense regulated cells was used to examine the role of c-*fos* in the PDGF stimulated recruitment of quiescent cells into the cell cycle. DNA synthesis of serum-starved quiescent fibroblasts was monitored by autoradiography of [³H]thymidine-labeled cells 24 hr after addition of purified PDGF in the presence or absence of dexamethasone. Antisense-regulated cells but not NIH-3T3 cells

or control cells showed a dramatic decrease in the number of cells undergoing DNA replication when dexamethasone was present. The findings were supported by the results of a cell cycle distribution analysis. One additional clone with high copy number exhibited a substantial dexamethasone-dependent inhibition of PDGF stimulation. It was concluded that the large increase in c-*fos* that follows growth factor stimulation of quiescent cells is indeed a requirement for, not merely an accompaniment of, renewed growth.

For asynchronous populations of cells in the log phase of growth, induction of antisense c-*fos* expression was observed to have little effect, i.e., steady-state growth was normal under conditions in which enough antisense c-*fos* RNA is being expressed to inhibit the transition from quiescence to renewed growth. Therefore a distinction was made between the role of c-*fos* in the regulation of growth renewal versus that for cycling cells. Indeed, only very low levels of c-*fos* transcripts are observed throughout the cell cycle for proliferating fibroblasts (Bravo et al., 1986).

Holt et al. (1986) examined the role of c-*fos* in cycling cells also using the MMTV LTR as a mechanism of inducible antisense RNA expression and noted a significant effect on cycling cells. Mouse c-*fos* DNA oriented in sense or antisense orientation was derived from a 301-bp fragment of mouse c-*fos* DNA that included 5'-noncoding sequences from just upstream of the cap site through most of the first exon (Fig. 4). Similar plasmids

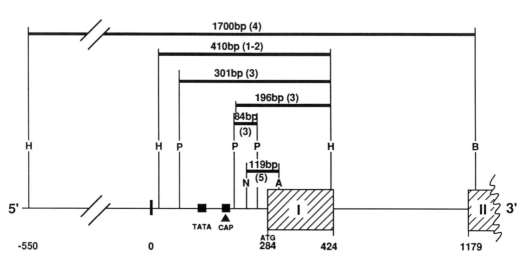

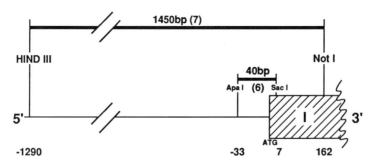

Fig. 4. Examples of sequences used for construction of plasmids expressing antisense *fos* (**A**) and *jun* (**B**) RNA. A, *Ava*I; B, *Bgl*II; H, *Hinc*II; N, *Nae*I; P, *Pst*I. References: 1. Nishikura and Murray (1987); 2. Levi and Ozato (1988); 3. Holt et al. (1986); 4. Mercola et al. (1988); 5. Schönthal et al. (1988a); 6. Schönthal et al. (1988b); 7. R. Chiu (personal communication; e.g., Kim et al., 1990). Not to scale.

based on a 196-bp fragment of the human c-*fos* sequence was also prepared. Finally a plasmid containing only an 84-bp 5′-noncoding fragment located between the cap site and the translation start site from mouse c-*fos* was also prepared. The *fos* sequences were inserted between the mouse mammary tumor virus (MMTV) promoter and a gene encoding β-gobin such that detection of β-globin mRNA-containing transcript of the appropriate size for a fusion transcript was taken as an indicator of collinear antisense RNA production. Clones containing c-*fos* fragments in antisense orientation exhibited markedly decreased cloning efficiency. The effect was most marked for the construct with only noncoding c-*fos* DNA, 4% of a control nonfunctional plasmid. It was estimated that the surviving clones expressed up to 3000 copies of hybrid RNA per cell whereas the amount of c-*fos* target mRNA has been estimated as <5 copies per cell (Kruijer et al., 1984).

Proliferation studies of the resulting clones showed that the doubling times were reversibly increased 2- to 3-fold in the presence of dexamethasone. Again no correlation could be made between the amount of antisense c-*fos* RNA produced and the extent of growth inhibition—in fact maximum inhibition of growth was observed with clonal lines expressing 250–300 copies of antisense containing RNA per cell whereas a clone expressing 10-fold higher copy number exhibited an intermediate phenotype. No inhibition was observed for control clones expressing a similar copy number of sense-oriented transcripts. It was concluded that antisense *fos* RNA sequences inhibited clone formation and the growth of cycling cells in all clones that were isolated.

Several observations support a role for c-*fos* in cycling cells. Riabowal et al (1988) demonstrated that microinjection of *fos*-specific antibodies up to 6 to 8 hr after serum stimulation of quiescent rat embryo fibroblasts efficiently blocked DNA synthesis. Injection at later times had little effect, suggesting that c-*fos* function is required early in G_1. NIH-3T3 cells transformed either by

v-*sis*, the homolog of c-*sis* that encodes the B-chain of PDGF, or transformed by activated EJ c-Ha-*ras* product exhibit rapid proliferation and loss of contact inhibition and are tumorigenic (Mercola et al., 1988; Ledwith et al., 1990). Constitutive expression of antisense c-*fos* RNA in v-*sis*-transformed NIH-3T3 cells leads to an 8-fold decrease in c-*fos* transcript levels compared to control cells and this effect correlates with a 2-fold decrease in log-phase growth rate (Mercola et al., 1987, 1988). As for PDGF, there is growing evidence that transformation by activated *ras* genes may work in part through activation of c-*fos*. Conditional expression of antisense c-*fos* RNA (using the antisense constructions of Holt et al., 1986) in EJ c-Ha-*ras*-transformed NIH-3T3 cells also led to a slower growth rate of exponentially growing cells (Ledwith et al., 1990). It may be therefore that in both these studies antisense c-*fos* RNA appears less efficient than antibody-mediated neutralization of c-*fos* protein where complete inhibition of growth was observed. The differences in efficiency may be less important in the regulation of steps requiring very low levels of Fos. However the transient accumulation of transcript levels required for the G_0 to G_1 transition is likely incomplete in antisense-regulated cells leading to decreased recruitment, a smaller growth fraction, and a decreased but finite growth rate.

Summary

Expression of antisense *fos* RNA in NIH-3T3 cells and derivatives leads to decreased steady-state c-*fos* transcript levels and prevents PDGF-stimulated accumulation of *fos* protein. Recruitment of quiescent cells into the cell cycle is restricted, strongly supporting an essential role of c-*fos* in the G_0 to G_1 transition. The doubling time of cycling populations is increased under conditions of antisense *fos* RNA expression and the combined evidence suggests that this is largely due to a role of c-*fos* at a second step in the cell cycle. These effects likely explain the markedly reduced cloning efficiency of cells bearing antisense *fos* RNA-expressing plasmids and conversely favor selection of clones with alternate mechanisms of growth. There is poor correlation between the insertion copy number and the level of antisense RNA expression and between expression and phenotypic effect. Among the many factors involved may be suppressive effects at the site of insertion and idiosyncratic growth properties resulting from selection at reduced c-*fos* transcript levels, respectively. Thus there is considerable variation in the behavior of clones. A third potential role of c-*fos* on growth control is suggested by the possible inverse relation between c-*fos* transcript levels in transformed fibroblasts and sensitivity to density-dependent growth arrest discussed under "Transformation," below. Numerous mechanistic questions remain about the role of duplex formation in sequestering target RNA, promoting breakdown, inhibiting nuclear transport, or other steps necessary for interfering in mRNA function. The fate of target RNA, optimum concentration of antisense RNA, optimum site on target RNA, optimum size of antisense RNA, and related questions are incompletely understood.

F9 Embryonal Carcinoma (EC) Cells

EC cells are multipotent mouse cells derived from tumors of mouse embryonic tissue ectopically placed beneath the kidney capsule (for a review see Silver et al., 1983). Although termed "carcinoma," these cells can be stimulated to differentiate to a wide variety of ectodermal, endodermal, and mesenchymal tissues including aggregates of synchronously beating heart-like cells, skeletal muscle, and nerve-like derivatives. The best studied line, F9, is more restricted in its differentiation potential and is widely used as a model of differentiation.

The role of c-*fos* in growth regulation of F9 EC cells may be distinct from that of fibroblasts. Edwards et al. (1988) and Levi and Ozato (1988) introduced plasmids designed to continually express antisense c-*fos* RNA. In the construction of Levi and Ozato a 410-bp fragment of mouse c-*fos* containing the first exon and 5′ sequences similar to those used for the fibroblast studies (above) was

inserted in antisense or sense orientation with respect to Rous sarcoma virus enhancer (RSV-LTR) (Fig. 4). Full-length plasmid insertion was confirmed by Southern analysis and full-length antisense RNA production was confirmed by Northern analysis. The efficacy of the constructions was tested by phorbol ester (TPA) and interferon challenges, which induce large and transient bursts of c-*fos* mRNA in F9 cells. C-*fos* expression was nearly eliminated in the antisense RNA expressing cells—the two examined clones exhibiting 35-fold and 20-fold lower levels of endogenous transcript than control cells. Thus, as with other studies (Nishikura and Murray, 1987; Mercola et al., 1987, 1988; Ledwith et al., 1990), antisense *fos* RNA expression is associated with decreased or abolished c-*fos* mRNA. However cloning efficiency of G418 resistant cells following transfection with these continuously expressing plasmids was not altered from that observed with sense control plasmids of $5-8 \times 10^{-5}$. Similarly, subsequent growth rates, although not quantitated, were described as comparable to that of untransfected cells.

Edwards et al. (1988) employed a similar fragment of c-*fos* mouse DNA as a source of antisense RNA in an SV40 early promoter-based vector. However, the vector was bifunctional, with the *fos* sequences inserted 3′ to the phosphotransferase gene of pNEO (Pharmacia). Following calcium phosphate-mediated transfection and selection (G418), full-length insertion was confirmed for several clones by Southern analysis and collinear transcription of both *neo* and antisense *fos* RNA was confirmed by sequential probing. C-*fos* transcript levels were shown to be substantially reduced—again illustrating that target RNA is largely destroyed in the presence of antisense c-*fos* RNA. Decreased Fos protein expression was also confirmed.

Cloning efficiency using the antisense *fos* expressing construction was high, 20×10^{-5}, although about 100-fold less than that for transfections using pSV2*neo* alone. Direct proliferation studies showed no differences in growth rate or final cell density for multiple clones of antisense expressing cells when compared to control clones (S. Edwards, personal communication).

These results contrast sharply with the very low cloning efficiencies of fibroblasts growing in the presence of antisense c-*fos* expressing constructions even when using the same promoter and antisense directing sequences in fibroblasts as used for the F9 cells (cf. Mercola et al., 1988 and Edwards et al., 1988). These results suggest that these cells utilize a growth mechanism that may be independent of c-*fos* or require comparatively less c-*fos* protein for cell cycle regulation. The conclusion is strongly supported by recent findings that, unlike most cell types, undifferentiated F9 stem cells do not contain Fos–Jun complex binding activity and basal c-*fos* expression is not detected (Angel et al., 1988; Chiu et al., 1988; Angel et al., 1989; Yang-Fen et al., 1990). Similarly c-*jun* and *jun*-B expression is very low or undetectable while that of *jun*-D is detectable but considerably lower than for differentiated cells (Yang-Fen et al., 1990). Thus the requirement for c-*fos* expression in growth may be cell type specific.

ANTISENSE *fos* AND *jun* RNA IN THE ANALYSIS OF THE AUTOREGULATION OF c-*fos*

Evidence for Autoregulation by Direct Binding by a Fos/Jun Complex

Little is known of the targets of *fos* protein action involved in cell cycle regulation; however, recent studies have revealed that one relevant target of c-*fos* transcription is its own promoter and that the interaction causes transcriptional repression (Sassone-Corsi and Verma, 1987). Subsequent studies, some based on antisense techniques, have led to contrasting results and the development of opposing mechanisms. At issue are whether additional factor(s) such as Jun cooperate with Fos, the identity of the DNA target sites, and which if any of the known interaction sites of the Fos protein are involved.

The autoregulatory mechanism was suggested by in vivo competition experiments that demonstrated that when DNA fragments

of the c-*fos* promoter were cotransfected with a c-*fos* promoter–reporter chimeric gene based on the bacterial chloramphenicol acetyltransferase (CAT) gene as the reporter, a preexisting repression of promoter activity was effectively relieved (Sassone-Corsi and Verma, 1987). Additional cotransfection studies demonstrated that Fos expression vectors could suppress Fos–CAT synthesis (Sassone-Corsi et al., 1988). Direct measurement of endogenous c-*fos* transcript levels in HeLa cells showed that expression of mouse or human Fos protein suppressed the endogenous c-*fos* gene. However, repression by the v-*fos* product, which has a 3′ deletion leading to 49 C-terminal amino acids completely unrelated to those of Fos, was of low efficiency and a mutant of c-*fos* with a frame-shift 17 codons upstream of the viral deletion site did not repress the c-*fos* promoter at all. Promoter deletion experiments suggested that the target of repression resided in a region containing the 14-bp serum response element (SRE; Fig. 3). The SRE binds two proteins, the serum response factor (SRF) and p62, and possibly a third protein (Ryan et al., 1989). It was suggested that a complex of these proteins interacts with a Fos–Jun complex bound at the adjacent AP-1 site (Fig. 3). Thus a model of direct interaction of Fos with its own promoter was suggested.

Other studies using a combination of antisense techniques and transcription assays have confirmed and extended these observations (Büscher et al., 1988; Schönthal et al., 1988a, 1989). The basic paradigm for these experiments was to cotransfect NIH-3T3 cells with a Fos–CAT reporter gene and expression vectors for Fos or c-*fos* antisense RNA. In most studies CAT assays are carried out 10–24 hr after serum stimulation. Therefore in the simplest case the results represent the net effect of induction of Fos–CAT and repression by coinduced endogenous Fos. For example when quiescent NIH-3T3 cells were cotransfected with the Fos–CAT gene and a plasmid that expressed the human c-*fos* product, CAT synthesis was repressed to about one-third in comparison to cotransfection with a control plasmid (Schönthal et al., 1988a).

However cotransfection of the reporter gene with a plasmid bearing an antisense-oriented 119 bp fragment of mouse c-*fos* taken from the 5'-noncoding sequences and the first three codons of exon I under the control of an SV40 promoter (pSV*sof*), led to a 4-fold increase in CAT synthesis. These results suggest, first, that the *fos* promoter is specifically repressed in quiescent NIH-3T3 cells, presumably due in part to endogenous c-*fos* activity leading to the establishment of a low basal level of c-*fos* expression. Second, the *fos* promoter may be repressed further by exogenous Fos.

These results have been extended (Schönthal et al., 1989) to an analysis of the role of the Fos–Jun complex in growth factor signal transduction under conditions of mitogenesis. The results are based on an analysis of the consequences of serum stimulation of quiescent NIH-3T3 cells previously rendered quiescent, i.e., in "G$_0$" of the cell cycle, by maintenance of near-confluent monolayers in low (0.5–1.0%) serum for 1–2 days. Serum stimulation (addition to 20% fetal calf serum) of quiescent NIH-3T3 cells cotransfected with Fos–CAT and a control plasmid led to a 5-fold increase in CAT synthesis to saturating levels of the reporter system thus demonstrating the expected behavior of an intact c-*fos* promoter. However, when a continuously productive *fos* expression vector pSV*fos* was used in place of the control plasmid, basal CAT activity was reduced about 3-fold. Serum stimulation led to only about one-fifth the previous saturated rate—consistent with suppression of the reporter construct by the pSV*fos* product. When pSV*fos* was replaced by pSV*sof*, basal CAT activity rose over 10-fold. In this case serum stimulation again led to the saturated CAT synthesis rate. That the Fos–CAT results also reflect regulation of the endogenous c-*fos* promoter was confirmed by measuring c-*fos* transcript levels using an S1 nuclease protection analysis for cells stably transfected with MMTV-LTR-*fos* expression vector. Serum induction of c-*fos* transcription was markedly reduced in the presence of dexamethasone. Conversely, for cells stably transfected with an MMTV-LTR-

sof, c-*fos* RNA expression is derepressed upon dexamethasone treatment. It was concluded that the low basal expression of c-*fos* and the rapid and effective turn-off of serum-induced c-*fos* transcription are due to autoregulation.

The participation of Jun was analyzed in analogous experiments (Schönthal et al., 1989; König et al., 1989). In this case cotransfection of quiescent NIH-3T3 cells with the Fos–CAT construct and a vector that continuously expressed Jun (pSV*jun*) again led to a 3-fold decrease in basal CAT synthesis. Serum stimulation led to increased CAT synthesis, which was, however, approximately 5-fold less than that seen with a control plasmid. Fos and Jun appeared to cooperatively repress the promoter since the use of both pSV*fos* and pSV*jun* in cotransfection studies led to over a 7-fold decrease in basal CAT synthesis (König et al., 1989).

In order to determine the effects of antisense *jun* RNA on this system a 40-bp synthetic oligonucleotide of the c-*jun* sequence starting 33 bp upstream from the translation start site was inserted in antisense orientation downstream of the SV40 promoter (Schönthal et al., 1988a). Cotransfection with this vector in place of pSV*jun* led to a 3-fold increase in basal transcription over the control value (Schönthal et al., 1989). As with antisense *fos* RNA expression, serum stimulation led to little or no additional effect, presumably owing to saturation of the reporter system. The results were interpreted to mean that in NIH-3T3 cells, the Fos product may exert its effect through formation of a Fos–Jun complex as has been shown previously for Fos-mediated transaction of AP-1 responsive genes (Chiu et al., 1988).

The above studies provide evidence that Fos and Jun function to repress the endogenous promoter following stimulation of mitogenesis. Deletion studies show that the major site of repression is the SRE (Lucibello et al., 1989; König et al., 1989; Shaw et al., 1989; Rivera et al., 1990; Guis et al., 1990) rather than the AP-1 sites (Fig. 3). Thus the SRE mediates both positive and negative effects. A number of examples of a single DNA regulatory site functioning in both positive and negative regulation are known and their common features have been reviewed recently (Karin, 1990).

The AP-1 site adjacent to the SRE (Fig. 3) appears to mediate basal repression by the Fos–Jun complex. In order to demonstrate the physiological significance of basal regulation, again advantage was taken of pSV*sof* (Schönthal et al., 1988a). If the basal rate of transcription from the *fos* promoter was physiologically determined by the endogenous level of Fos and Jun proteins acting on or via the adjacent binding site, then antisense *fos* RNA should derepress the promoter. Cotransfection studies of pSV*sof* or a control plasmid together with a modified Fos–CAT with or without the AP-1 sequence were used to demonstrate derepression (König et al., 1989). The results were obtained under physiological conditions by carrying out nuclear run-on assays of NIH-3T3 cells that had been stably transfected with each Fos–CAT construct chimeric gene. Constructions with the DSE alone proved able to confer serum sensitivity. However, in the absence of the adjacent AP-1 site, the CAT transcription rate did not return completely to basal levels but retained 10–30% of the maximal rate (see also Shaw et al., 1989; Lucibello et al., 1989). Thus one determinant of the transient nature of the c-*fos* promoter activity was directly demonstrated. A model has been proposed by Shaw et al. (1989) suggesting that direct binding of the Fos–Jun complex at a site adjacent to the SRE can alter promoter activity by interacting with one or more of the bound proteins such as SRF and p62. Similar concepts have been proposed in several other promoters where DNA sequences immediately adjacent to regulatory elements mediate activating and suppressive effects such as for the interferon gene (Goodbourn and Maniatis, 1988) and the proliferin gene (Diamond et al., 1990).

In the above studies antisense expressing plasmids have been used to dissect a complex regulatory scheme. However, it should be noted that the mechanisms remain incompletely understood. Fos and Jun may affect repression by direct binding to the SRE–protein complex or CArg box (Fig. 3) (König et al., 1989; Lucibello et al., 1989; Guis et al., 1990); however, as yet complexes have not been detected (Rivera et al., 1990). An indirect mechanism is suggested by studies of the Egr-1 promoter, which contains multiple SRE-like sequences and CArG boxes. Repression by Fos appears independent of Jun since alteration of the amino acids essential for heterodimer formation have no effect. Alternatively, deletions in the C-terminal region led to complete loss of repression. The smallest deletion was the C-terminal 27 amino acids, which eliminated repression. These experiments have been confirmed and extended recently by a converse experiment in which an expression vector encoding a nuclear localization signal linked with the coding sequence of only the C-terminal 15 amino acids of Fos sequence was observed to confer repression (V. Sukhatme, personal communication). The sum of results favor an indirect mechanism possibly involving a Fos-induced repressor (Guis et al., 1990; V. Sukhatme, personal communication; cf. Rivera et al., 1990). It would be of interest to learn the effects of antisense *fos* and *jun* RNA expression in this system since, unlike the c-*fos* promoter, it would be expected that antisense *jun* RNA would have little effect while again antisense *fos* RNA would be expected to block repression.

Summary

Antisense *fos* and *jun* RNA has been employed as a complementary approach to the use of expression vectors in the analysis of gene regulation. In these studies antisense-induced loss of function was loss of a negative regulatory capacity and therefore led to a positive effect, namely increased promoter activity and gene expression. The observation of specific gene expression helps to allay worries that loss of function was simply the result of nonspecific inhibition of an essential cellular function by antisense RNA. However, the conclusion that an antisense RNA-dependent mechanism accounts for the results rests with the biochemical consistency of the results. Antisense RNA production was not confirmed, destruction or sequestration of target RNA or translation arrest of target product synthesis was not assessed, and controls utilizing sense-oriented DNA fragments or other specificity or toxicity checks were not described. All but one (Fig. 4) of the antisense *fos* RNA constructs described included the Kozak-like sequence, which for murine c-*fos* (5'-CGACCAUGA-3') is similar to the consensus sequence [5'-CCAC-CAUG(G)-3'] associated with efficient translation of eukaryotic mRNA (Kozak, 1986) and so could potentially inhibit the synthesis of other proteins—perhaps repressors. It is of interest to note that a construct without complementary Kozak-like sequences exhibited prominent growth inhibitory properties (Holt et al., 1986; Ledwith et al., 1990).

The relative speed of transient transfection experiments has been made possible by combining a sensitive reporter gene downstream of sequences with positive or negative promoter regulatory properties, thus suggesting a method of some generality for analysis of gene regulation. A validation of the antisense mechanism occurring in transiently transfected cells will be necessary.

c-*fos* IN DEVELOPMENT

Several lines of evidence suggest that the Fos gene plays a role in differentiation and development. These include high expression in certain embryonic tissues (Adamson et al., 1983), the correlation of rapid transient expression in many cell types upon addition of agents that promote differentiation (Müller, 1983; Gonda and Metcalf, 1984; Greenberg et al., 1985; Curran and Morgan, 1985), and sustained expression during terminal differentiation of F9 EC cells (Edwards and Adamson, 1986). Consistent with these observations, transfection of F9 EC cells with Fos expression vectors leads to clonal lines that exhibit increased Fos protein levels and the appearance of Troma-1 and Troma-3—markers that accompany differentiation of F9 EC cells to parietal endoderm-like cells (Müller and Wagner, 1984; Rüther et al., 1985). How-

ever, there was poor correlation between the degree of expression of *fos* in a given clone and the expression of the markers. In addition, less than 10% of the cells of a given clonal line became differentiated by the criteria of Troma-1 and Troma-3 expression (Rüther et al., 1985). Further, the role of c-*fos* in differentiation is largely based on correlation data and has been challenged (Müller et al., 1985; Mitchell et al., 1986; Calabretta, 1987).

Two groups have employed antisense *fos* RNA techniques. Levi and Ozato (1988) and Edwards et al. (1988) examined the effects of the inducers retinoic acid (RA) and cyclic AMP (cAMP) on antisense-regulated F9 EC cells. When these cells are exposed to either RA or RA together with cAMP they are induced to differentiate over the following 4–8 days to cells with morphological and biochemical similarities to visceral and parietal endoderm, respectively. In both cases clonal populations of cells that constitutively express antisense c-*fos* RNA were examined and endogenous c-*fos* transcript levels where shown to be reduced or undetectable. Four days following stimulation of differentiation in either direction, Levi and Ozato (1988) observed morphological changes of antisense-regulated cells growing as attached cells in tissue culture. Further, the marker stage-specific embryonic antigen 1 decreased while the marker mouse histocompatibility complex (MHC) class I antigen increased, leading to the conclusion that, despite the profound inhibition of c-*fos* gene expression, clones expressing c-*fos* antisense RNA underwent differentiation after RA treatment. As a result of this observation, the expression of c-*fos* was examined in unregulated F9 EC cells and, similar to Mason et al. (1985) and Yang-Fen et al. (1990), no increase in c-*fos* transcript levels was observed to accompany differentiation during the first 2 days in contrast to previous results (Müller, 1983). However, timing and direction of differentiation may be important variables in understanding the role of c-*fos*.

Edwards et al. (1988) examined RA or RA-cAMP-induced differentiation using a variety of markers specific for either parietal or visceral endoderm. In addition the F9 EC cells were grown as unattached and aggregated "embryoid bodies" that exhibit a striking morphological criterion in the form of a well-developed epithelial-like layer that forms in known stages over the course of 6 days. Endogenous c-*fos* expression in this system was in agreement with that of Lockett and Sleigh (1987), who showed that either RA alone or in combination with dibutyryl-cAMP induced high levels of c-*fos* mRNA levels in F9 EC cells that was maximal only after 6 days. Parallel experiments were carried out with six antisense *fos* RNA-regulated clonal lines bearing the bifunctional plasmid pSV*neosof* (cf. c-*fos* in "Cell-Cycle Regulation: F9 EC Cells," above). Two clones expressed reduced or absent levels of c-*fos* transcript and protein and were devoid of expression of laminin, type IV collagen, and ɔroteoglycan-19 RNA transcripts after 4 days of induction with RA and dibutyryl-cAMP. Induction of Troma-1 and Troma-3 was also inhibited while control clones (pSV2*neo* only) behaved normally—in all suggesting inhibition of the formation of parietal endoderm. However, the antisense-regulated cells were not inhibited in the differentiation pathway to visceral endoderm since the α-fetoprotein gene was activated normally. The differential affect argues against an inhibition of parietal endoderm formation based on a nonspecific toxic effect associated with antisense c-*fos* RNA production or duplex formation. It was concluded that c-*fos* antisense expression effectively inhibits some aspects of differentiation in F9 cells and, moreover, the influence of c-*fos* on differentiation is specific to the parietal endoderm pathway.

The role of c-*fos* expression in the differentiation of rat phaeochromocytoma (PC12) cells has also been examined (Kindy and Verma, 1988). Addition of nerve growth factor (NGF) to these cells in vitro leads to a large but transient increase in expression of c-*fos*, which is maximal in 30–60 min and returns to basal levels in about 2 hr. This is followed by the appearance of neurite-like outgrowths in several days. Addition of a synthetic 14-bp antisense DNA at 10μ*M* 4 hr prior to stimulation with inducer led to 80–100% inhibition of c-*fos* protein expression 50 min after addition of NGF. However, neuronal differentiation followed normally and it was concluded that NGF-stimulated c-*fos* expression was not necessary for NGF-stimulated differentiation of PC12 cells.

Summary

The combined results show that the use of antisense *fos* RNA has been effective in the analysis of differentiation. Many differentiation-inducing agents stimulate transient expression of c-*fos*; however, this appears to be unrelated to differentiation in many cases (e.g., Müller et al., 1985; Mitchell et al., 1986; Calabretta, 1987; Kindy and Verma, 1988). Elevated and continuous c-*fos* expression is required for the differentiation of F9 EC cells and this is specific to the parietal endoderm pathway.

THE FOS AND JUN FAMILIES IN TRANSFORMATION

Overexpression of c-*fos* causes transformation in susceptible cell types (Miller et al., 1984; Raymond et al., 1989) including primary fibroblast (Iba et al., 1988) and causes tumors in transgenic mice (Rüther et al., 1987, 1988, 1989). Antisense studies have implicated the continuous activation of c-*fos* in the mechanism of action of a wide variety of other protooncogenes including c-*sis*, c-Ha-*ras*, c-*mos*, and v-*src*. C-*fos* expression may be essential for the production of transformation-related products such as transin and stromelysin and c-*fos* amplification and increased expression have been associated with drug resistance of human colon carcinoma (Kashani-Sabet et al., 1990). C-*fos* is overexpressed in the majority of human osteosarcomas (Wu et al., 1990). Application of antisense has been useful in elucidating these relationships.

Sis

Overexpression of either v-*sis* (Favera et al., 1981; Devare et al., 1983; Huang et al.,

1984; Leal et al., 1985) or human c-*sis* (Clarke et al., 1984; Westin et al., 1984; Johnsson et al., 1985; Kelly et al., 1985; Stevens et al., 1988), which encode homologs of the B-chain of platelet-derived growth factor (PDGF), in established or primary fibroblast cell lines leads to complete transformation including tumorigenesis. Similar, albeit less dramatic results, hold for the A-chain of PDGF (Beckmann et al., 1988). Thus the observations of continuous or elevated expression of one or both of these products in a large number of human tumor types (Table 1) is provocative and suggests that one or more of the three possible isoforms of PDGF—consisting of AA, BB, or AB chain combinations—may play an important role in tumorigenesis (see Table 1 for references). Many of these tumor types have been shown to express one or both of the two cognate receptors of PDGF (Table 1). Thus autocrine stimulation of these receptors by PDGF molecules has been postulated as a mechanism of initiation and/or growth of these tumors (e.g., Garret et al., 1984; Graves et al. 1984; Betsholtz et al. 1984; Matsui et al. 1989). In view of the essential role of c-*fos* in PDGF-stimulated growth, it appears likely that c-*fos* activation may mediate *sis*-dependent transformation.

This hypothesis was tested by preparing stable clonal lines of v-*sis* transformed NIH-3T3 cells (SSV-3T3) bearing plasmids designed to continuously express antisense *fos* RNA (Mercola et al., 1987, 1988). The fragment consisted of a 1.7-kb portion of genomic mouse c-*fos* from the 5'-noncoding region and extending approximately half-way through the second exon, i.e., overlapping the antisense *fos* constructs described above. Control cells transfected with pSV2*neo* alone or with a plasmid containing a sense-oriented fragment continuously overexpressed c-*fos* by at least 8-fold—presumably due to continuous stimulation by the v-*sis* product. However, the antisense-regulated cells exhibited c-*fos* transcript and protein levels less than those of NIH-3T3 cells. The antisense-regulated cells exhibited several phenotypic criteria of reversal of transformation such as a flatter and larger cell morphology, de-

creased growth rate, and restoration of contact inhibition as judged by the reemergence of a saturation density with over 90% of the cells in G_0/G_1 of the cell cycle (Fig. 5). It was concluded that the overexpressed c-*fos* indeed functioned in the *sis*-dependent transformed phenotype to effect more rapid growth and loss of contact inhibition.

However, overexpression of c-*fos* does not appear to be sufficient for transformation. The antisense-regulated cells also exhibited anchorage-independent growth and tumorigenicity—albeit with a slow growth rate (Mercola et al., 1988). Thus either the SSV-3T3 parental cell line or transfected clonal lines had acquired additional transforming alterations, or there exist additional *sis*-dependent transforming pathways.

The confounding possibility of additional acquired genetic changes also arises when contemplating the significance of the *sis*-autocrine system apparent in the numerous and aneuploid human tumor lines (viz. Table 1). To distinguish among these possibilities we studied reversible transformation of NIH-3T3 cells bearing an inducible metallothionein–v-*sis* construct (Carpenter et al., 1989, 1991). These cells exhibited zinc-dependent and reversible transformation as judged by morphology, growth rate, loss of contact inhibition, and anchorage-independent growth (Fig. 6). Thus the *sis*-dependent pathway alone is sufficient for transformation in vitro. However, studies in vivo revealed two populations of tumor cells (Fig. 7). Cells subcultured from 10 of 18 tumors, all <0.1 g and ≤21 days in age, reverted to a normal phenotype by four criteria but could be induced to transform by the addition of zinc. Thus these tumor cells were also reversibly transformed, arguing that activation of the v-*sis* autocrine loop alone is sufficient for initiation of tumors. Cells from the remaining and larger, 0.5 ± 0.7 g, tumors did not revert by any criteria and therefore had become irreversibly transformed in vivo (Fig. 7). Southern analysis of tumor DNA suggested that there was no gross change in the pMT*sis* insert and Northern analysis of RNA from certain irreversibly transformed tumors demonstrated

TABLE 1. Expression of PDGF-Like Gene Products and Receptors in Selected Human Tumor Cell Lines and Tissues[a]

Cells and tissues		Receptor status	PDGF-like gene product[b]
HT1080	Fibrosarcoma	ND	PDGF-A and -B (1–2)
B5-GT	Giant cell sarcoma	ND	PDGF-A only (4)
SKIMS	Leiomyosarcoma	α only (3)	PDGF-A only (4)
G402	Leiomyosarcoma	α only (3)	PDGF-A only (4)
RD	Rhabdomyosarcoma	ND	PDGF-A only (5)
A1186	Rhabdomyosarcoma	α > β (3)	
A204	Rhabdomyosarcoma	α only (3)	
A875	Melanoma	Not detected (3)	
WM266-4	Melanoma	ND	PDGF-A, short > long (6)
SW691	Melanoma	ND	PDGF-A only (7)
U2-OS	Osteosarcoma	PDGFR (8, 9)	PDGF-A > -B (4,8,10); PDGF-A, short ≥ long (6,11); PDGF-B > -A (9, see also 33)
Saos II	Osteosarcoma	ND	c-*sis*/PDGF-B only (4)
Saos	Osteosarcoma	ND	PDGF-B (5)
OS 1	Osteosarcoma	ND	PDGF-B not detected (5)
OS 2	Osteosarcoma	ND	PDGF-B not detected (5)
OS 3	Osteosarcoma	ND	PDGF-B (5)
TE	Osteosarcoma	ND	PDGF-B (5)
MG 63	Osteosarcoma	PDGFR present, not mitogenic (14)	Not detectable (15)
U1810	Osteosarcoma	ND	PDGF-A, short (16)
HOS	Osteosarcoma	ND	PDGF-A, short (16)
8842	Osteosarcoma	ND	PDGF-A, short only (11)
B5GT	Osteosarcoma	ND	PDGF-A not detected (16)
T-2	Teratocarcinoma (clone 13)	ND	PDGF-A, short > long (6)
EJ	Bladder carcinoma	ND	c-*sis*/PDGF-B (18)
MCR-7	Breast carcinoma MDA-MB-231	ND	PDGF-A and -B (19)
T47D	Breast carcinoma	ND	PDGF-like factor (32)
	Breast carcinoma MDA-MB-157	ND	PDGF-like factor (32)
MCF-7	Breast carcinoma	Not detectable (21)	PDGF-A and -B (21)
BT-20	Breast carcinoma	Not detectable (21)	PDGF-A and -B (21)
ZR 75-1	Breast carcinoma	ND	PDGF-A and -B (21)
ARM	Ovarian carcinoma	Not detectable (21)	PDGF-A only (21)
DUN	Ovarian carcinoma	Not detectable (21)	PDGF-A and -B (21)
MAC	Ovarian carcinoma	ND	PDGF-A and -B (21)
SAM	Ovarian carcinoma	Not detectable (21)	PDGF-A and -B (21)

TABLE 1. *Continued*

Cells and tissues		Receptor status	PDGF-like gene product
A549	Lung carcinoma	Not detectable (21)	PDGF-A and -B (21)
CALU-1	Lung carcinoma	ND	PDGF-A and -B (21)
COLO201	Colon carcinoma	ND	PDGF-A and -B (21)
COLO205	Colon carcinoma	Not detectable (21)	PDGF-A and -B (21)
COLO357	Colon carcinoma	ND	PDGF-A>>-B (34)
WiDR	Colon carcinoma	ND	PDGF-B>>-A (34)
SW850	Pancreatic adenocarcinoma	ND	Not detected (34)
QGP1	Pancreatic adenocarcinoma	ND	PDGF-A>> -B (34)
Panc 89	Pancreatic adenocarcinoma	ND	PDGF-B>> -A (34)
Panc Tu1	Pancreatic adenocarcinoma	ND	PDGF-B>> -A (34)
Panc Tu2	Pancreatic adenocarcinoma	ND	PDGF-B>> -A (34)
ASPC 1	Pancreatic adenocarcinoma	ND	PDGF-A and -B (34)
BXPC 3	Pancreatic adenocarcinoma	ND	PDGF-A and -B (34)
HPAF	Pancreatic adenocarcinoma	ND	Not detected (34)
A590	Pancreatic adenocarcinoma	ND	PDGF-A>> -B (34)
PT 45P1	Pancreatic adenocarcinoma	ND	PDGF-A>> -B (34)
A-818-1	Pancreatic adenocarcinoma	ND	PDGF-A (34)
A-818-4	Pancreatic adenocarcinoma	ND	PDGF-A (34)
A-818-7	Pancreatic adenocarcinoma	ND	PDGF-A> -B (34)
KATO III	Gastric carcinoma	Not detectable (21)	PDGF-A and -B (21)
A431	Cervical squamous carcinoma	ND	PDGF-A, short > long (6) PDGF-B > -A (34)
HL-60	Myelocytic leukemia	ND	PDGF-A (18, cf. ref. 22)[c]
U937	Promyelocytic leukemia	ND	PDGF-A (22)[c]
MOLT-4	T cell leukemia	ND	PDGF-A and -B (22)
K562	Erythroleukemia	ND	PDGF-A and -B (22)
A1690	Astrocytoma	β > α (28)	c-*sis*/PDGF-B (11,12,18,23)
A 172	Glioblastoma	β only (28)	PDGF-A, short only (11)
A2781	Glioblastoma	ND	c-*sis*/PDGF-B (18)
A1207	Glioblastoma	α only (28)	

	Relative receptor mRNA[d]	Relative mRNA[d]	
		PDGF-A	PDGF-B
Malignant gliomas (9)			
U-87 MG	+	+++	++
U-105 MG	+++	+++	-
U-118 MG	+++++	+++	(+)
U-138 MG	+++++	+	-
U-178 MG	+++++	+++++	++
U-251 MGO	-	(+)	+++
U-251 MGAg Cl. 1	-	+	+++
U-251 MGsp	+(+)	+	+++
U-343 MG	-	+++	-
U-343 MGa Cl. 2:6	ND	+++	++++
U-343 MGa Cl. 2:6	ND	PDGF-A-like (29)	
U-343 MGa Cls. 26L and 5H	ND	PDGF-A > -B (30)	
U-343 MGa Cl. 12.6 (D-1)	+(+)	PDGF-A >> -B (4)	
U-372 MG	ND	+	++++
U-373 MG	++(+)	+++	-
U-399 MG	++++	+++	+
U-410 MG	+++	+	+
U-489 MG	+	+	+
U-539 MG	-	ND	ND
U-563 MG	+	+	+
U-706 S	(+)	+++	+++
U-1231 MG	++++	+++	+++
U-1240 MG	+(+)	+++	+++++
U-1242 MG		+	+++
U-1796 MG		+	+

	Receptor status	PDGF-like gene product
Primary cultures and biopsies		
Glioblastoma ($n = 3$)	β, 3 of 3 (28)	PDGF-A and -B, 3 of 3 (28)
Mesothelioma ($n = 10$)	ND	PDGF-A and -B (24)
Astrocytoma ($n = 50$)	PDGFR in 50% (31)	PDGF-A and -B in 80% (31)
Breast carcinoma ($n = 4$)	ND	PDGF-B in 3 of 4 (25)
Breast benign lesions ($n = 2$)	ND	PDGF-B by in situ hyb (25)
Fibrosarcoma ($n = 4$)	ND	PDGF-B in 3 of 4 (5)
Melanoma ($n = 1$)	ND	PDGF-B not detected (5)

TABLE 1. *Continued*

	Receptor status	PDGF-like gene product
Lymphoma, B cell (n = 1)	ND	PDGF-B (5)
WM9 Melanoma	ND	Undetectable (7)
WM115 Melanoma	ND	PDGF-A >> -B (7)
WM239A Melanoma	ND	PDGF-A >> -B (7)
WM266-4 Melanoma	ND	PDGF-A only (7)
SW691 Melanoma	ND	PDGF-A only (7)
Normal or immortalized cells		
Human diploid fibroblasts	α and β (3,26,27,35)	PDGF-A, cell-cycle dependent (13)
Murine diploid fibroblasts	α and β (3,26,27,35)	Undetectable (20)
Human umbilical vein endothelial cells		PDGF-A, short (16).

[a] Abbreviations used: PDGF-B = PDGF-II, transcript or protein product of *c-sis* gene; PDGF-A, transcript or protein product of the PDGF A-chain gene; PDGF-AB, heterodimer of PDGF A- and B-chains; PDGFR, PDGF receptor not specified as to subtype; α and β refer to PDGFR α and PDGFR β type receptor molecules; ND, not described. The PDGF A-chain is specified as "long" or "short" depending on whether a 15 residue C-terminal fragment (GRPRESGKKRKKRKRL) of exon 6 is known to be present or absent.

[b] References: 1. Eva et al. (1982); 2. Pantazis et al. (1985); 3. Matsui et al. (1989); 4. Betsholtz et al. (1986); 5. Fahrer et al. (1989); 6. Rorsman et al. (1988); 7. Westermark et al. (1986); 8. Betsholtz et al. (1984); 9. Nister et al. (1988a); 10. Heldin et al. (1986); 11. Tong et al. (1987); 12. Harsh et al. (1989); 13. Paulsson et al. (1987); 14. Womer et al. (1987); 15. Graves et al. (1984); 16. Collins et al. (1987); 17. Weima et al. (1989); 18. Igarashi et al. (1987); 19. Bronzert et al. (1987); 20. Bowen-Pope et al. (1984); 21. Sariban et al. (1988); 22. Alitalo et al. (1987); 23. Pantazis et al. (1985); 24. Versnel et al. (1988); 25. Ro et al. (1989); 26. Heldin et al. (1988); 27. Hart et al. (1988); 28. Hermansson et al. (1988); 29. Nister et al. (1988b); 30. Nister et al. (1987); 31. Maxwell et al. (1989); 32. Rozengurt et al. (1985); 33. Graves et al. (1986); 34. Kalthoff et al. (1991); 35. Bywater et al. (1988).

[c] PDGF-B expression is inducible by addition of the differentiation promoting phorbol esters to either HL-60 or U-937 cells (18,22).

[d] Arbitrary units of Nister et al. (1988a).

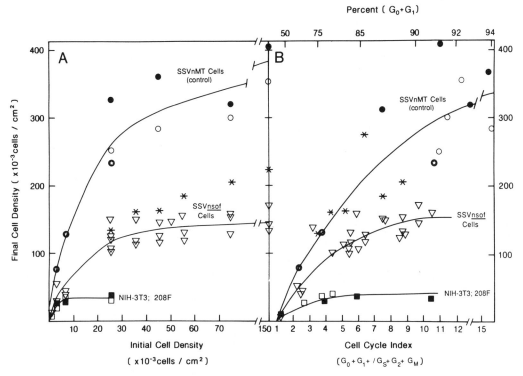

Fig. 5. Antisense *fos* regulated v-*sis*-transformed cells exhibit restoration of contact inhibition. (**A**) The Y axis gives the final saturation density obtained following plating at the density shown on the X axis and allowed to grow through log-phase growth, 4 days. NIH-3T3, parental cells; SSVnMT, control cells (v-*sis*-transformed NIH-3T3 cells transfected with pSV2*neo* and an irrelevant plasmid; solid curves are average of five clones) and SSV*nsof* antisense regulated cells (solid curves are average of six clones). Note that antisense-regulated cells do not grow above 150,000 cells/cm² even if seeded at that density 4 days prior to the measurement, unlike transformed control cells that achieve a postlog-phase density proportional to seeding. (**B**) The fraction of cells in G_0/G_1 of the cell cycle as a function of final saturation density. Note that for any density (Y axis) antisense-regulated cells exhibit a higher proportion of quiescent cells than the corresponding control cells. Reproduced from Mercola et al. (1988), with permission of the publisher.

that zinc-dependent pMT*sis* expression remained intact. The combined observations suggest that extended tumor growth is accompanied by acquisition of other irreversible change(s) that occur at high frequency in vivo. Thus an activated *sis* gene may initiate a multistep process.

For the many human tumor types that continually express c-*sis* (i.e., Table 1), it may be possible, as our data suggest, that autocrine stimulation is related to initiation and may contribute to growth, but that additional genetic changes are likely to occur and contribute to the phenotype by the time these tumors are clinically evident. Thus a multistep (Fearon and Vogelstein, 1990) model best accounts for the sum of results.

c-Ha-*ras*, v-*mos*, and v-*src*

In studies carried out in parallel to the analysis of c-*fos* autoregulation, Schönthal et al. (1988a) observed that NIH-3T3 cells cotransfected with a reporter construct containing the AP-1 binding sequence of the human collagenase promoter together with prospective activators of c-*fos* such as MMTV-LTR-c-Ha-*ras*, -*mos*, or -*src* exhibited prominent and dexamethasone-dependent CAT

A

- ZINC   + ZINC

pMT*SIS*
TRANSFORMED
NIH-3T3 CELLS

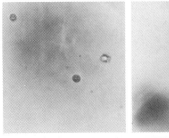

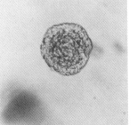
pMT*SIS*
TRANSFORMED
NIH-3T3 CELLS

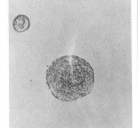
v-*SIS*
TRANSFORMED
NIH-3T3 CELLS

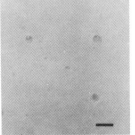
NIH-3T3 CELLS

B

Disaggregated NSV*neo*MT*sis* Colony Cells

- ZINC   + ZINC

d = 4

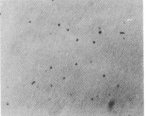

d = 28

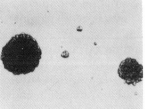

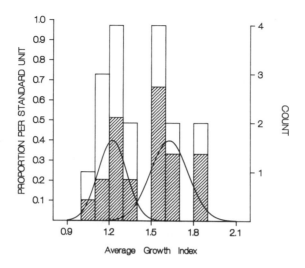

Fig. 7. The frequency distribution of the growth index of tumor cells is bimodal. Tumors of pMT*sis*-transformed cells were recovered from athymic mice and subcultured in the presence and absence of zinc. Reversibility of growth was determined by measurement of zinc induction of growth (growth index, GI)—defined as the total number of cells observed after log-phase growth (typically day 5 after seeding at 125,000 cells/cm^2) in the presence of zinc to that in the absence of zinc. The average GI of four independent clones of low passage pMT*sis*-transformed cells was 1.83 ± 0.2, whereas the average GI of control clones as well as the NIH-3T3 parental population was 1.0 ± 0.1. Approximately half of the tumor subcultures grew as fast as transformed cells in the absence of zinc, yielding an average GI of this of 1.22 ± 0.1, whereas the remaining tumor-derived cells reverted to a normal phenotype with normal morphology and growth but exhibit prominent zinc-dependent increase in growth with an average GI of 1.63 ± 0.13. The difference is highly significant, $P < 0.005$. Hatched bars, left ordinate; open bars, right ordinate.

synthesis. CAT synthesis increased up to a third that observed when Fos was overexpressed by pSV*fos* while nonproductive *fos* mutants and irrelevant plasmids had no effect. The three oncogene expression vectors separately activated a c-*fos* promoter in a dexamethasone-dependent manner up to 5-fold. To determine whether activation of the collagen promoter occurred via activation of the endogenous c-*fos* gene, stable cell lines containing each of the oncogene constructs were transfected with the collagen promoter–CAT construct with or without pSV*sof* and treated with dexamethasone. The antisense condition was associated with 1.7- to 3.2-fold less dexamethasone-stimulated activation. It was concluded that c-*fos* plays a key role in signal transduction by these agents. Further support derives from the observation that revertants of *fos*-transformed fibroblasts

◄**Fig. 6.** Anchorage-independent growth of conditional v-*sis* transformation is reversible. (**A**) NIH-3T3 cells stably transfected with pMT*sis* (NSV*neo*MT*sis* cells) exhibit zinc-dependent growth in methylcellulose, day = 12 after suspension. (**B**) Colonies recovered from methylcellulose and disaggregated by treatment with trypsin again exhibit zinc-dependent growth in soft agar illustrating that viable and mitotically active cells were recovered. However, no proliferation occurs after 4 weeks in the absence of zinc, arguing that no irreversible transforming events have occurred during *sis* expression and anchorage-independent growth, cf. **A**. Bar = 50 μm.

cannot be retransformed by the *fos*, *ras*, or *mos* oncogenes (Zarbl et al., 1987).

EJ c-Ha-*ras*-transformed NIH-3T3 cells have been studied in more detail (Ledwith et al., 1990). Using the constructs of Holt et al. (1986), stable transfected clones of the *ras* transformed cells were isolated and examined. As for SSV-3T3 cells (Mercola et al., 1988), antisense *fos* RNA expression led to a decrease in sense mRNA expression, morphological change, slower growth, and restoration of contact inhibition, and in addition exhibited inhibition of anchorage-independent growth (similar effects of antisense *fos* RNA in *ras*-transformed cells are quoted by Schönthal et al., 1988a). Although tumors of the antisense-regulated cells formed in athymic mice, in the presence of dexamethasone, a marked inhibition of tumor growth rate was observed. It was concluded that c-*fos* expression is required for *ras* transformation. The results are consistent with other studies implicating c-*fos* in the *ras* pathway (Stacey et al., 1987; Kerr et al., 1988).

Finally it is noted that, as for antisense *fos* RNA expression in *sis*-transformed cells (Mercola et al., 1988), decreased c-*fos* expression was associated not only with morphological and growth rate properties but also restored density dependence of growth arrest, suggesting the possibility that elevated c-*fos* levels antagonize the mechanism by which cell–cell contact inhibits cell division at characteristic cell densities.

Jun Family

Transactivation by Fos implies complex formation with a member of the Jun family since Fos alone has little of no DNA binding properties (Chiu et al., 1988; Halazonetis et al., 1988; Kousarides and Ziff, 1988; Nakabeppu et al., 1988; Rauscher et al., 1988; Sassone-Corsi et al., 1988; Zerial et al., 1989; Smeal et al., 1989). Thus it might be expected that certain of the Jun family members may also mediate signal transduction of the same oncogenic pathways. Antisense methods have been employed to explore the differential roles of the Jun family members in regulation of the interkeukin-2 (IL-2) gene (M. Karin, per-

sonal communication). Activators of the IL-2 gene of EL4 cells—transformed murine T cells—strongly induce expression of *jun*-B but not c-*jun* or *jun*-D and strongly induce c-*fos* and *fos*-B but not *fra*-I. Addition of synthetic conventional DNA oligonucleotides complementary to 18 residues of the 5'-coding regions led to differential effects: antisense DNA complementary to *jun*-B or *fos*-B inhibited activation of the IL-2 promoter while antisense *jun*-D or c-*fos*, which strongly inhibited expression from a reporter construct containing three copies of a consensus AP-1 site, had no effect of expression of the IL-2 construct. The results suggest that Jun-B/Fos-B heterodimers and/or Jun-B homodimers are the preferred activators in this system. The results also suggest the possibility of differentiating the roles of the family members by use of antisense methods.

As noted, previous studies have suggested that elevated levels of Jun-B may antagonize activation by the c-*jun* product, possibly due in part repression of the c-*jun* enhancer (Chiu et al., 1989; M. Karin, personal communication). Similar results have been observed in 3T3 mouse fibroblasts, SK-Br-3 human mammary carcinoma cells, and PC12 cells (Brysch and Schlingensiepen, 1991; W. Brysch and K.-H. Schlingensiepen, personal communication). Growth studies were carried out for 5 days following addition to 2 μM phosphorothioate DNA oligonucleotides compelementary to 14 nucleotides of the target sequence to freshly seeded (1 hr) cells. In all cases treatment with antisense to *jun*-B led to increased growth, which usually peaked in 1 to 2 days. Treatment with antisense to *jun*-D to the transformed cells led to a small increase in growth, which was maximal in 1 day while for 3T3 cells the stimulation was similar to that with seen with antisense to *jun*-B. Treatment with antisense to c-*jun* led to decreased growth in 3T3 and SK-Br-3 cells compared to cells treated with an oligomer with a randomized sequence. Further, addition of antisense DNA complementry to the p53 gene sequence to 3T3 or SK-Br-3 cells also led to increased growth maximal on days 1–2. The results were taken to suggest that *jun*-B and *jun*-D have characteristics of functional antagonists of c-*jun*.

Stromelysin/Transin

Transin is one of a family of metal-dependent proteases secreted by many transformed cells and may mediate metastasis via breakdown of collagen and extracellular matrix proteins (for a review see Matrisian and Hogan, 1990). Transin is induced by PDGF, epidermal growth factor (EGF), v-*fos*, H-*ras*, v-*src*, and others depending on cell type, suggesting a role for c-*fos*. McDonnell et al. (1990) examined stably transfected NIH-3T3 cells bearing the dexamethasone-inducible constructs of Holt et al. (1986). PDGF-induced transin transcript levels were nearly eliminated in the presence of dexamethasone, whereas in the absence of the steroid or with EGF, transin induction occurred normally, suggesting that both *fos*-dependent and *fos*-independent pathways for the two growth factors exist in NIH-3T3 cells. EGF-dependent induction of transin has been examined in rat-1 cells using synthetic antisense *fos* and *jun* DNA complementary to the first 18 coding nucleotides as well as the antisense RNA-expressing plasmids. EGF stimulation caused a transient increase in c-*fos* and c-*jun* transcript level, which was followed by the appearance of transin transcripts. However, when serum-deprived cells were first treated with antisense DNA (0.5 μM), immunoreactive Fos and Jun were eliminated and transin transcription was reduced by a third to a half. Sense control DNA had no effect. Similarly a 50% decrease in EGF-induced transin mRNA levels was observed for stably transfected cell lines bearing antisense *fos* RNA expressing plasmids. In support of these findings, Kerr et al. (1990) observed a novel binding site in the transin promoter capable of binding Fos induced in vivo by EGF.

Transforming Growth Factor-β1 (TGF-β1)

TGF-β1 has numerous effects, including growth-retarding effects in many cells (for a review see Moses et al., 1990), c-*fos*-dependent inhibition of EGF induction of transin (Kerr et al., 1990), and positive induction of its own promoter (Kim et al., 1990). Promoter sequence and deletion studies implicated two AP-1 binding sites as required for autoregulation, suggesting a role for the Fos–Jun complex. Footprinting studies confirmed binding at these sites by a purified Fos–Jun complex. The functional significance was studied by use of antisense *fos* and *jun* RNA expressing plasmids (Kim et al., 1990) using the antisense *jun* construct of Robert Chiu (Fig. 4) while the antisense *fos* construct utilized the fragment of Holt et al. (1986) in conjunction with a 1.6-kb mouse metallothionein promoter. In experiments similar to those for the analysis of autoregulation, A-549 human lung adenocarcinoma cells were cotransfected with a plasmid containing a TGF-β1 promoter–CAT construct together with antisense c-*fos* or c-*jun* or control plasmid DNA. In the presence of either antisense RNA expressing construct, TGF-β1-stimulated CAT synthesis was inhibited by over 80%. It was concluded that the AP-1 binding site mediated the autoregulation of TGF-β1 and that expression of both Jun and Fos components of the AP-1 complex is required.

Control of Chromosomal Aberrations

Chromosomal instability is a characteristic of tumor cells that may be important in the multistep progress of tumor progression (Fearon and Vogelstein, 1990). van den Berg et al. (this volume) have explored the role of FOS in this process and provide an additional potential effect of overexpression of c-*fos* in certain human tumors (Wu et al., 1990).

PROSPECTS AND APPLICATIONS

Antitumor therapy is an attractive potential of the antisense technique. In the case of c-*fos* and c-*jun* a number of concerns are apparent. Many of the results described here, especially for the various protooncogenes, support the emerging view that c-*fos* and c-*jun* are points of convergence of signal transduction pathways, so these genes may play fundamental roles in many normal cells (Vogt and Bos, 1989; Marx, 1989). For example, c-*fos* may be essential for normal cell division in fibroblasts. Thus the use of antisense *fos* and *jun* RNA may lead to detrimental effects in normal cells and may be a poor use of the potential specificity of the antisense approach. Further, the role of c-*fos* in transformed cells is unclear. Although overexpression of c-*fos* causes chondroosseous tumors in transgenic animals, overexpression is rarely observed in human tumors, even in tumor types that are closely related to v- and c-*fos*-promoted murine tumors (Wu et al., 1990). We examined the expression of Fos protein in 30 human osteosarcomas using a variety of specific antisera at five concentrations and measured the extent of immunoreactivity, either by microdensitometry or by means of a panel of three trained pathologists (Wu et al., 1990). Nonlinear least-squares refinement of the data according to a model for immunotitration was used to show that 61% of the cases overexpressed Fos. However, the average increase in Fos expression was 1.5 times that of normal tissues or benign lesions (Wu et al., 1990). Similarly, in the case of EJ Ha-*ras*-dependent transformation of NIH-3T3 cells, a process that is believed to required activation of c-*fos*, increased c-*fos* transcription was not observed. Indeed in the EJ Ha-*ras*-transformed cells the ability of added growth factors to induce c-*fos* transcription was reduced—as for several other genes of *ras*-transformed cells (Ledwith et al., 1990). In the case of *sis*-transformed fibroblasts, c-*fos* was elevated in several transfected and G418-resistant control clones (Mercola et al., 1988). However, S. Edwards (personal communication) examined c-*fos* expression in over 20 subclones of the parental SSV-3T3 cells and observed that, while v-*sis* transcription was high and constant, c-*fos* expression was only slightly increased in a minority of subclones. Similarly, Fahrer et al. (1989) observed that, while c-*sis* and c-*fos* were commonly expressed in a series of 12 human sarcoma cell lines or biopsies, there was little correlation. On the other hand, it may be that the relevant level of Fos depends critically

on tissue susceptibility. For example, E. Wagner and co-workers (E. Wagner, personal communication) have prepared a transgenic mice with exogenous c-*fos* expression determined by an H2 promoter and have prepared chimeric mice derived from embryonal carcinoma stem cells bearing human MT-c-*fos* constructs, all with 3' viral LTRs. The tissue distribution of Fos expression was different from that of previous studies (Rüther et al., 1989) but again only chondroosseous lesions were observed—thus emphasizing the susceptibility of certain precursor cells to the effects of Fos expression.

A final caveat derives from the lack of any demonstration that complete tumor suppression can be achieved by use of antisense *fos* or *jun* RNA. Even for the two reported cases of reduced tumor growth (Mercola et al., 1988; Ledwith et al., 1990) antisense RNA expression and other mechanistic feature of an antisense RNA-dependent effect were not assessed. Thus a demonstration of the efficacy of antisense *fos* or *jun* RNA in suppression of solid tumor growth in vivo and determination of the consequences of exposing normal tissue to efficacious treatment levels would be very valuable.

Many of the studies summarized here provide evidence for the efficacy of antisense RNA as an analytical tool. This approach may be of use in determining the role of c-*fos*, c-*jun*, and other gene products in transformed cells including human tumors via analyses of subcultures. An appealing extension of this approach is the use of antisense DNAs to study functional differences between members of the superfamilies. Indeed, as noted, antisense DNAs with sequences complementary to individual Fos and Jun family members may be useful in differentiating among the roles of family members (Morris et al., 1991; Brysch and Schlingensien, 1991). Thus it will be of considerable interest to determine whether transformed cells exhibit particular specificities that can be targeted separately from universal or essential signal transduction pathways. Recent studies have clarified the nature of the acquisition of activated oncogenes and altered suppressor genes in several carcinomas (Fearon and Vogelstein, 1990). For a given tumor type, the acquisition in time of a finite number of recurring genetic alterations, but in varying combinations depending on the particular tumor, appears to be a reasonable expectation. Thus by the use of a series of antisense RNA expressing plasmids and/or DNAs, it may be possible to evaluate the significance of each candidate gene on phenotype of tumor cells in tissue culture. Further, as the targets of suppressor gene regulation become known, it may be possible to apply antisense RNA to these targets and further evaluate—and reverse—the effects of loss of suppressor regulation. Recently one such target for the retinoblastoma (suppressor) gene has been suggested to be c-*fos* itself (Robbins et al. 1990). Such information may be of value in defining the consequences of genetic alterations, in subtyping tumors, and in defining a minimum list of target genes for a rational therapy. Thus by using a combination of antisense agents and focusing on the particular "profile" of a particular tumor type, it may be possible to use lower doses of antisense agents that are less detrimental to normal processes. Recent advances in targeting double-stranded DNA and devising forms of DNA stable to enzymatic degradation and increased efficacy against a variety of viruses further suggest a broadening range of applications (Pine, 1990; J. Cohen, personal communication). Rossi and colleagues have developed a promising new approach based on the first application of ribozyme technology to *fos* transcripts (Scanlon et al., 1989). Ribozymes (synthetic hammerhead) were constructed to specifically bind to the fos mRNA at a 24 nt stretch that contained a ribozyme-susceptible GUC sequence. Addition of $MgCl_2$ led to specific clevage. Complete digestion by the *fos* ribozyme required the presence of 3 mM $MgCl_2$. Thus ribozymes may be a useful adjunct.

There appear, therefore, to be a variety of analytical, diagnostic, and potential therapeutic approaches that may be of value in understanding and intervening in the expression of the transformed phenotype by antisense RNA methods.

SUMMARY

The use of antisense *fos* and *jun* RNA has contributed to our understanding of cell-cycle regulation, differentiation, gene regulation, and, in particular, transformation. In all effective cases, expression of antisense *fos* RNA via stable transfection of plasmids has led to a reduction of steady-state c-*fos* transcript levels implying a mechanism of action that involves the breakdown of RNA involved in RNA duplex formation. A wide range of sequences all containing 5' portions of the c-*fos* gene have been used as a source of antisense RNA production or DNA antisense oligonucleotide synthesis. Similarly the 5' coding region of c-*jun* has been used for the preparation of plasmids designed to express antisense RNA. Induction of antisense c-*fos* RNA in fibroblasts that have been stimulated to divide by platlet-derived growth factor leads to inhibition of the characteristic transient expression of c-*fos* and cell division illustrating the essential role of c-*fos* expression for the reentry of quiescent fibroblasts into the cell cycle. However at least one other cell type, mouse F9 embryonal carcinoma cells, do not require c-*fos* for growth.

The introduction of plasmids designed to express antisense *fos* or *jun* RNA into fibroblasts relieves repression of the c-*fos* promoter thus providing confirmatory evidence that the c-*fos* product acts with the c-*jun* product as a repressor of its own promoter. Similar transient transfection studies have been carried with both Fos expression vectors and cotransfected chimeric promoter–chloramphenicol transferase (CAT) constructs in order to study the role of Fos and Jun products in regulation. In a variation of this approach stably transfected fibroblast lines that conditionally express c-Ha-*ras*, c-*mos*, or c-*src* were inhibited in their ability to conditionally activate a collagen promoter-CAT construct but cotransfection of a antisense *fos* RNA expressing plasmid blocked activation thus suggesting that c-*fos* activity is required for regulation by these protooncogenes. Expression of antisense *fos* RNA in stably transfected fibroblasts lines transformed by c-Ha-*ras* or v-*sis* demonstrates that c-*fos* mediated many of the manifestations of the transformed phenotype in *ras* and *sis* transformed cells including tumor formation and growth rate. Recent preliminary studies using synthetic single-stranded oligonucleotide DNA complementary to the *fos* or *jun* family members suggest that c-*jun* expression is required for growth of a human mammary carcinoma cell but had little effect on murine fibroblasts whereas antisense complementary to *jun*-B or *jun*-D led to markedly increased growth in both cases. Differential effects of the Fos and Jun family members have been reported in the control of the interleukin-2 (IL-2) gene of thymoma and T cells. Antisense DNA complementary to *jun*-B or *fos*-B inhibited activation while antisense *jun*-D or c-*fos* had no effect on the expression of the IL-2 construct. The results suggest that for *fos* and *jun* expression the various family members may have differential roles that can be discriminated by antisense techniques. *Fos* ribozymes have been prepared and shown to cleave *fos* RNA in vitro. Confirmatory studies including a demonstration that antisense RNA and DNA inhibit target gene expressing are required. The application of antisense *fos* and *jun* RNA and DNA for therapy was discussed.

ACKNOWLEDGMENTS

We are grateful to Drs. E. Adamson, W. Brysch, R. Chiu, H. Kalthoff, M. Karin, B. Ledwith, A. Schönthal, and V. Sukhatme for making manuscripts or other information available prior to publication and to Drs. E. Adamson, D. O'Connor, and Mr. J. Westwick for thoughtful comments on this manuscript. Figures 1 and 2 are based on drawings kindly supplied by Dr. E. Adamson.

REFERENCES

Adamson ED (1986): Cell lineage-specific gene expression in development. In Rossant J, Pedersen R, (eds): Experimental Approaches to Mammalian Embryonic Development. New York: Cambridge Univ Press, pp 321–364.

Adamson ED, Müller R, Verma I (1983): Expression of c-*onc* genes c-*fos* and c-*fms* in developing mouse tissues. Cell Biol. Int. Rep. 7:557–558.

Agris C, Blake K, Miller PS, Reddy MP, Ts'o P (1986): Inhibition of vesicular stomatitis virus protein synthesis and infection by sequence-specific oligodeoxycells cycloribonucleoside methylphosphonates. Biochemistry 25:6268–6275.

Alitalo R, Andersson LC, Betsholtz C, Nilsson K, Westermark B, Heldin C-H, Alitalo K (1987): Induction of PDGF expression during megakaryoblastic and monocytic differentiation of human leukemia cell lines. EMBO J. 6:1213–1219.

Angel P, Allegretto EA, Okino S, Hattori K, Boyle WJ, Hunter T, Karin M (1988): Oncogene jun encodes a sequence specific transactivator similar to AP-1. Nature (London) 332:166–171.

Angel P, Smeal T, Meek J, Karin M (1989): Jun and v-Jun contain multiple regions that participate in transcriptional activation in an interdependent manner. New Biol. 1:35–43.

Bass BL, Weintraub H (1988): An unwinding activity that covalently modifies its double-stranded RNA substrate. Cell 55:1089–1098.

Beckmann M, Betsholtz C, Heldin C-H, Westermark B, Di Marco E, Di Fiore PP, Robbins K, Aaronson S (1988): Comparison of biological properties and transforming potential of human PDGF-A and PDGF-B chains. Science 241:1346–1349.

Betsholtz C, Westermark B, Ek B, Heldin C-H (1984): Coexpression of a PDGF-like growth factor and PDGF receptors in a human osteosarcoma cell line: Implications for autocrine receptor activation. Cell 39:447–457.

Betsholtz C, Johnsson A, Heldin C-H, Westermark B, Lind P, Urdea MS, Edy RT, Shows TB, Philpott K, Miller AL (1986): cDNA sequence and chromosomal localization of human platelet-derived growth factor A-chain and its expression in tumor cell lines. Nature (London) 320:695–699.

Bishop JM (1985): Viral oncogenes. Cell 42:23–38.

Bohmann D, Bos TJ, Admon A, Nishimura T, Vogt PK, Tjian R (1987): Human proto-oncogene c-*jun* encodes a DNA binding protein with structural and functional properties of transcription factor AP-1. Science 238:1386–1392.

Bonnieu A, Piechaczyk M, Mary L, Cuny M, Blanchard J-M, Fort P, Jeanteur P (1988): Sequence determinants of *myc* mRNA turnover: Influence of 3' and 5' non-coding regions. Oncogene Res. 3:155–166.

Bowen-Pope DF, Vogel A, Ross R (1984): Production of platelet-derived growth factor-like molecules and reduced expression of platelet-derived growth factor receptors accompany transformation by a wide spectrum of agents. Proc. Natl. Acad. Sci. U.S.A. 81:2396–2400.

Bravo R, Burckhardt T, Curran T, Müller R (1986): Expression of c-*fos* in NIH-3T3 cells is very low but inducible throughout the cell cycle. EMBO J. 5:695–700.

Bronzert D, Pantazis P, Antoniades HN, Kasid A, Davidson N, Dickson RB, Lippman M (1987): Synthesis and secretion of platelet-derived growth factor by human breast cancer cell lines. Proc. Natl. Acad. Sci. U.S.A. 84:5763–5767.

Brysch W, Schlingensiepen K-H (1991): c-*jun* and *jun*-B have opposite effects on cell proliferation—a study with antisense phosphorothioate oligodeoxynucleotides. J. Cell. Biochem. Suppl. 15D:35.

Büscher M, Rahmsdorf HJ, Litfin M, Karin M, Herrlich P (1988): Activation of the c-*fos* gene by UV and phorbol ester: Different signal transduction pathways converge on the same enhancer element. Oncogene 3:301–311.

Bywater M, Rorsman F, Bongcam-Rudkoff E, Mark G, Hammacher A, Heldin C-H, Westermark B, Betsholtz C (1988): Expression of recombinant PDGF-A and -B homodimers in rat-1 cells and human fibroblasts reveals differences in protein processing and autocrine effects. Mol. Cell. Biol. 8:2753–2762.

Calabretta B (1987): Dissociation of c-*fos* induction from macrophage differentiation in human myeloid leukemic cell lines. Mol. Cell. Biol. 7:769–774.

Carpenter P, Mercola M, Mercola D (1989): Conditional transformation by v-*sis* in NIH-3T3 cells. J. Cell. Biochem. Suppl. 13B:119.

Carpenter P, Mercola M, Mercola D (1991): A model for analysis of *sis*-dependent transformation by antisense methods. Antisense Res. Dev. (in press).

Chang EH, Yu K, Shinozuka K, Zon G, Wilson WD, Strekowska A (1989): Comparative inhibition of *ras* p21 protein synthesis with phosphorus-modified antisense oligonucleotides. Anti-Cancer Drug Design 4:221–232.

Chiu R, Boyle WJ, Meek J, Smeal T, Hunter T, Karin M (1988): The c-*fos* protein interacts with c-*jun*/AP-1 to stimulate transcription from AP-1 responsive genes. Cell 54:541–552.

Chiu R, Angel P, Karin M (1989): *Jun*-B differs in its biological properties and is a negative regulator of c-*jun*. Cell 59:979–986.

Clarke M, Westin E, Schmidt D, Joesphs S, Ratner L, Wong-Staal F, Gallo R, Reitz M Jr (1984): Transformation of NIH-3T3 cells by a human c-*sis* DNA clone. Nature (London) 308:464–466.

Cohen DR, Curran T (1988): fra-1: A serum inducible, cellular immediate-early gene that encodes a Fos-related antigen. Mol. Cell. Biol. 8:2063–2069.

Collins T, Bonthron DT, Orkin SH (1987): Alternative RNA splicing affects the function of encoded platelet-derived growth factor A chain. Nature (London) 328:621–624.

Curran T, Morgan JI (1985) Superinduction of c-*fos* by nerve growth factor in the presence of peripherally active benzodiazepines. Science 229:1265–1268.

Devare S, Reddy P, Law J, Robbins K, Aaronson S (1983): Nucleotide sequence of the simian sarcoma virus genome: Demonstration that its acquired cellular sequences encode the transforming gene product p28^{sis}. Proc. Natl. Acad. Sci. U.S.A. 80:731–735.

Diamond MI, Miner JN, Yoshinaga SK, Yamamoto KR (1990): Transcription factor interactions: Selectors of positive or negative regulation from a single DNA element. Science 249:1266–1272.

Distel RJ, Ro H-S, Rosen BS, Groves DL, Spiegelman BM (1987): Nucleoprotein complexes that regulate gene expression in adipocyte differentiation direct participation of c-*fos*. Cell 49:835–844.

Edwards SE, Adamson ED (1986): Induction of c-*fos* and AFP expression in differentiating embryonal carcinoma cells. Exp. Cell Res. 165:473–480.

Edwards SE, Rundell AYK, Adamson ED (1988): Expression of c-*fos* antisense RNA inhibits differentiation of F9 cells to parietal endoderm. Dev. Biol. 129:91–102.

Eva A, Robbins K, Andersen P, Srinivasan A, Tronick S, Reddy E, Ellmore N, Galen A, Lautenberger J, Papas T, Westin E, Wong-Staal F, Gallo R, Aaronson S (1982): Cellular genes analogous to retroviral *onc* genes are transcribed in human tumor cells. Nature (London) 295:116–119.

Fahrer C, Brachmann R, von der Helm K (1989): Expression of c-*sis* and other cellular protooncogenes in human sarcoma cell lines and biopsies. Int. J. Cancer 44:652–657.

Favera R, Gelmann E, Gallo R, Wong-Staal F (1981): A human *onc* gene homologous to the transforming gene (v-*sis*) of simian sarcoma virus. Nature (London) 292:31–33.

Fearon ER, Vogelstein B (1990): A genetic model of colorectal tumorigenesis. Cell 61:759–767.

Franza BR, Rauscher III FJ, Josephs SF, Curran T (1988): The *fos* complex and fos-related antigens recognize sequence elements that contain AP-1 binding sites. Science 239:1150–1153.

Garrett J, Coughlin SR, Niman HL, Tremble PM, Giels G, Williams LT (1984): Blockade of autocrine stimulation in simian sarcoma virus transformed cells reverses down-regulation of PDGF receptors. Proc. Natl. Acad. Sci. U.S.A. 81:7466–7470.

Gonda TJ, Metcalf D (1984): Expression of *myb*, *myc*, and *fos* proto-oncogenes during the differentiation of a murine myeloid leukaemia. Nature (London) 310:249–251.

Goodbourn S, Maniatis T (1988): Overlapping positive and negative regulatory domains off the human β-interferon gene. Proc. Natl. Acad. Sci. U.S.A. 85:1447–1451.

Graves D, Owen ASJ, Barth RK, Tempst P, Winoto A, Fors L, Hood LE (1984): Detection of c-*sis* transcripts and synthesis of PDGF-like protein by human osteosarcoma cells. Science 234:972–294.

Graves D, Owen A, Williams S, Antoniades HN (1986): Identification of processing events in the synthesis of platelet-derived growth factor-like proteins by human osteosarcoma cells. Proc. Natl. Acad. Sci. U.S.A. 83:4636–4640.

Greenberg ME, Ziff EB (1984): Stimulation of 3T3 cells induces transcription of the c-*fos* proto-oncogene. Nature (London) 311:433–438.

Greenberg ME, Greene LA, Ziff EB (1985): Nerve growth factor and epidermal growth factor induce rapid transient changes in proto-oncogene transcription in PC12 cells. J. Biol. Chem. 260:14100–14110.

Greenberg ME, Hermanoski AL, Ziff EB (1986): Effect of protein synthesis inhibitors on growth factor activation of c-*fos*, c-*myc* and actin gene transcription. Mol. Cell. Biol. 6:1050–1057.

Greenberg ME, Siegried Z, Ziff, EB (1987): Mutation of the c-*fos* gene dyad symmetry element inhibits serum inducibility of transcription *in vivo* and the nuclear regulatory factor binding *in vitro*. Mol. Cell. Biol. 7:1217–1225.

Guis D, Cao X, Rauscher III FJ, Cohen DR, Curran T, Sukhatme VP (1990): Transcriptional activation and repression by Fos are independent functions: The C terminus represses immediate-early gene expression via CArG elements. Mol. Cell. Biol. 10:4243–4255.

Halazonetis TK, Georgopoulos K, Greenberg ME, Leder P (1988): cJun dimerizes with itself and with cFos, forming complexes of different DNA binding affinities. Cell 55:917–924.

Harsh GR, Kavanaugh WM, Starksen NF, Williams LT (1989): Cyclic AMP blocks expression of the c-*sis* gene in tumor cells. Oncogene Res. 4:65–73.

Hart C, Forstrom JW, Kelly JD, Seifert RA, Smith RA, Ross R, Murray MJ, Bowen-Pope DF (1988): Two classes of PDGF receptor recognize different isoforms of PDGF. Science 240:1529–1531.

Hayes T, Kitchen A, Cochran B (1987): A rapidly inducible DNA-binding activity which binds upstream of the c-*fos* proto-oncogene. Proc. Natl. Acad. Sci. U.S.A. 84:1272–1276.

Heldin C-H, Johnsson A, Wennergren S, Wernstedt C, Betsholtz C, Westermark B (1986): A human osteosarcoma cell line secretes a growth factor structurally related to a homodimer of PDGF A-chains. Nature (London) 319:511–514.

Heldin C-H, Backstrom G, Ostman A, Hammacher A, Ronnstrand L, Rubin K, Nister M, Westermark B (1988): Binding of different dimeric forms of PDGF to human fibroblasts: Evidence for two separate receptor types. EMBO J. 7:1387–1391.

Hermansson M, Nister M, Betsholtz C, Heldin C-H, Westermark B, Funa K (1988): Endothelial cell hyperplasia in human glioblastoma: Coexpression of mRNA for platelet-derived growth factor (PDGF) B chain and PDGF receptor suggests autocrine growth stimulation. Proc. Natl. Acad. Sci. U.S.A. 85:7748–7752.

Hirai SI, Ryseck RP, Mechta F, Bravo R, Yaniv M (1989): Characterization of Jun-D: A new member of the *jun* proto-oncogene family. EMBO J. 8:1433–1439.

Holt JT, Gopal TV, Moulton AD, Neinhuis AW (1986): Inducible production of c-*fos* antisense RNA inhibits 3T3 cell proliferation. Proc. Natl. Acad. Sci. U.S.A. 83:4794–4798.

Huang JS, Huang SS, Deuel TF (1984): Transforming protein of simian sarcoma virus stimulates autocrine growth of SSV-transformed cells through PDGF cell-surface receptors. Cell 39:79–87.

Iba H, Shindo Y, Nishina H, Yoshina T (1988): Transforming potential and growth stimulating activity of the v-*fos* and c-*fos* genes carried by avian retrovirus vectors. Oncogene Res. 2:121–133.

Igarashi H, Rao C, Leal S, Robbins K, Aaronson S (1987): Detection of PDGF-2 homodimers in human tumor cells. Oncogene 1:79–85.

Izant J, Weintraub H (1985): Constitutive and conditional suppression of exogenous and endogenous genes by anti-sense RNA. Science 229:345–352.

Johnsson A, Betsholtz C, Heldin C-H, Westermark B (1985): Antibodies against platelet-derived growth factor inhibit acute transformation by simian sarcoma virus. Nature (London) 317:438–440.

Kalthoff H, Roeder C, Humburg I, Heinz-Günter T, Greten H, Schmiegel (1991): Modulation of platelet-derived growth factor A- and B-chain/c-*sis* by tumor necrosis factor and other agents in human pancreatic cancer cells. Submitted.

Karin M (1990): Too many transcription factors: Positive and negative interactions. New Biol. 2:126–131.

Kashani-Sabet M, Lu Y, Leong L, Haedicke K, Scanlon KJ (1990): Differential oncogene amplification in tumor cells from a patient treated with cisplatin and 5-fluorouracil. Eur. J. Cancer 26:383–390.

Kelly J, Raines E, Ross R, Murray M (1985): The B-chain of platelet-derived growth factor is sufficient for mitogenesis. EMBO J. 4:3399–3405.

Kerr KD, Holt JT, Matrisian LM (1988): Growth factors regulate transin gene expression by c-*fos*-dependent and c-*fos*-independent pathways. Science 242:1424–1427.

Kerr LD, Miller DB, Matrisian LM (1990): TGF-β1 inhibition of transin/stromelysin gene expression is mediated through a fos binding sequence. Cell 61:267–278.

Kim S-J, Angel P, Lafyatis R, Hattori K, Kim KY, Sporn MB, Karin M, Roberts AB (1990): Autoinduction of transforming growth factor β-1 is mediated by the AP-1 complex. Mol. Cell. Biol. 10:1492–1497.

Kindy MS, Verma I (1988): Inhibition of c-*fos* gene expression does not alter the differentiation pattern of PC12 cells. In Melton DA (ed): Antisense RNA and DNA. Cold Spring Harbor, NY: Cold Spring Harbor Laboratory, pp 129–133.

König H, Ponta H, Rahmsdorf U, Büscher M, Schönthal A, Rahmsdorf JH, Herrlich P (1989): Autoregulation of *fos:* The dyad symmetry element as the major target of repression. EMBO J. 8:2559–2566.

Kouzarides T, Ziff E (1988): The role of the leucine zipper in the fos-jun interaction. Nature (London) 336:646–656.

Kozak M (1986): Point mutation define a sequence flanking the AUG initiator codon that modulates translation by eukaryotic ribosomes. Cell 44:283–292.

Kruijer WJ, Cooper JA, Hunter T, Verma IM (1984): Platelet-derived growth factor induces rapid but transient expression of c-*fos* gene and protein. Nature (London) 312:711–716.

Leal F, Robbins K, Aaronson S (1985): Evidence that the v-*sis* gene product transforms by interaction with the receptor for platelet-derived growth factor. Science 230:327–330.

Ledwith B, Manam S, Kraynak AR, Nichols WW, Bradley MO (1990): Antisense-*fos* RNA causes partial reversion of the transformed phenotypes induced by the c-Ha-*ras* gene. Mol. Cell. Biol. 10:1545–1555.

Lee WMF, Lin C, Curran T (1988): Activation of the transforming potential of the human *fos* proto-oncogene requires message stabilization and results in increased amounts of partially modified *fos* protein. Mol. Cell. Biol. 8:5521–5527.

Levi B-Z, Ozato K (1988): Constitutive expression of c-*fos* antisense RNA blocks c-*fos* gene induction by interferon and by phorbol ester and reduces c-myc expression in embryonal carcinoma cells. Gene Dev. 2:554–566.

Lockett TJ, Sleigh MJ (1987): Oncogene expression in differentiating F9 mouse embryonal carcinoma cells. Exp. Cell Res. 173:370–378.

Lucibello FC, Lowag C, Neuberg M, Müller R (1989): Trans-repression of the mouse c-*fos* promoter: A novel mechanism of fos-mediated trans-repression. Cell 59:999–1007.

Marx JL (1989): The *fos* gene as "Master Switch." Science 237:854–856.

Mason L, Murphy D, Hogan BLM (1985): Expression of c-*fos* in parietal endoderm, amnion and differentiating F9 teratocarcinoma cell. Differentiation 30:76–81.

Matrisian LM, Hogan B (1990): Role of growth factors in development. Curr. Topics Dev. Biol. 24:219–259.

Matsui T, Heidmaran M, Miki T, Popescu N, La Rochelle N, Kraus M, Pierce J, Aaronson S (1989): Isolation of a novel receptor cDNA establishes the existence of two PDGF receptor genes. Science 243:800–804.

Maxwell M, Nabey S, Wolf H, Galanopoulas T, Block P, Antoniades H (1989): Expression of PDGF and PDGF-receptor genes by metastatic tumors. J. Cell Biol. 107:40a.

McDonnell SE, Lawrence DK, Matrisian LM (1990): Epidermal growth factor stimulation of stromelysin mRNA in rat fibroblasts requires induction of proto-oncogenes c-*fos* and c-*jun* and activation of protein kinase C. Mol. Cell. Biol. 10:4284–4293.

McGarry T, Lindquist S (1986): Inhibition of heat shock protein synthesis by heat-inducible antisense RNA. Proc. Natl. Acad. Sci. U.S.A. 83:399–403.

Meijlink F, Curran T, Miller DA, Verma I (1985): Removal of a 67-base-pair sequence in the noncoding region of protooncogene *fos* converts it to a transforming gene. Proc. Natl. Acad. Sci. U.S.A. 82:4987–4991.

Mercola D (1986): The proto-oncogene c-*fos* encodes a potential regulatory site that is disrupted by viral transduction. J. Theor. Biol. 126:243–246.

Mercola D, Rundell A, Westwick J, Edwards S (1987): Antisense RNA to the c-*fos* gene: Restoration of density-dependent growth arrest in a transformed cell line. Biochem. Biophys. Res. Commun. 147:854–856.

Mercola D, Rundell A, Westwick J, Adamson E, Edwards S (1988): Analysis of a transformed cell line using antisense c-*fos* RNA. Gene 72:253–265.

Miller AD, Curran T, Verma I (1984): c-*fos* protein can induce cellular transformation: A novel mechanism of activation of a cellular oncogene. Cell 36:51–60.

Mitchell RL, Henning-Chubb C, Huberman E, Verma IM (1986): c-*fos* expression is neither sufficient nor obligatory for differentiation of monomyelocytes to macrophages. Cell 45:497–504.

Mölders H, Jenuwein T, Adamkiewicz J, Müller R (1987) Isolation and structural analysis of a biologically active chicken c-*fos* cDNA: Identification of evolutionarily conserved domains in *fos* protein. Oncogene 1:377–385.

Morgan J, Curran T (1989): Stimulus-transcription coupling in neurons: Role of cellular immediate-early genes. Trends Neurosci. 12:459–462.

Morris DR, Allen ML, Radler-Pohl A, Karin M (1991): Functional specialization within families for transcription factors: Dominant roles of *jun*-B and *fos*-B in regulation of interleukin 2 expression. J. Cell. Biochem. Suppl. 15E:198.

Moses HL, Yang EY, Pietenjpol JA (1990): TGF-β stimulation and inhibition of cell proliferation: New mechanistic insights. Cell 63:245–247.

Müller R (1983): Differential expression of cellular oncogenes during murine development and in teratocarcinoma cell lines. Cold Spring Harbor Conf. Cell Prolif. 10:451–468.

Müller R, Wagner EF (1984): Induction of differentiation after transfer of c-fos genes into F9 teratocarcinoma stem cell. Nature (London) 311:438–442.

Müller R, Curran T, Müller D, Guilbert L (1985): Induction of c-fos during myelomonocytic differentiation and macrophage proliferation. Nature (London) 314:546–548.

Nakabeppu Y, Ryder K, Nathans D (1988): DNA binding activities of three murine Jun proteins: Stimulation by Fos. Cell 55:907–915.

Nishikura K, Murray JM (1987): Antisense RNA of proto-oncogene c-fos blocks renewed growth of quiescent 3T3 cells. Mol. Cell. Biol. 7:639–649.

Nishina Y, Nakagoshi H, Imamoto F, Gonda T, Ishii S (1989): isolation and characterization of fra-2, an additional member of the fos gene family. Nucl. Acid Res. 17:107–117.

Nister M, Wedell B, Betsholtz C, Bywater M, Petersson M, Westermark B, Mark J (1987): Evidence for progressional changes in the human malignant glioma line U-343 MGa: Analysis of karyotype and expression of genes encoding the subunit chains of platelet-derived growth factor. Cancer Res. 47:4953–4960.

Nister A, Libermann T, Betsholtz C, Petersson M, Claesson-Welsh L, Heldin C-H, Schlessinger J, Westermark B (1988a): Expression of messenger RNAs for platelet-derived growth factor and transforming growth factor-α and their receptors in human malignant glioma cell lines. Cancer Res. 48:3910–3918.

Nister A, Mannacher A, Messström K, Siegbahn A, Rönnstrand L, Westermark B, Heldin C-H (1988b): A glioma-derived PDGF A chain homodimer has different functional activities from a PDGF AB heterodimer purified from human platelets. Cell 52:791–799.

Ovitt CE, Rüther U (1989): The proto-oncogene c-fos: Structure, expression, and functional aspects. Oxford Surv. Eukaryotic Genes 6:33–51.

Pantazis P, Pelicci P, Dalla-Favera R, Antoniades H (1985): Synthesis and secretion of proteins resembling platelet-derived growth factor by human glioblastoma and fibrosarcoma cells in culture. Proc. Natl. Acad. Sci. U.S.A. 82:2404–2408.

Paulsson Y, Hammacher A, Heldin C-H, Westermark B (1987): Possible positive autocrine feedback in the prereplicative phase of human fibroblasts. Nature (London) 328:715–717.

Pine M (1990): Developments in antisense technology for commercial application in human therapy and agriculture. Pine M (organizer), second annual conference, Philadelphia, PA, November 5–6.

Rahmnsdorf HJ, Schönthal A, Angel P, Litfin M, Rüther U, Herrlich P (1987): Posttranscriptional regulation of c-fos mRNA expression. Nucl. Acid Res. 15:1643–1659.

Rauscher III FJ, Cohen DR, Curran T, Bos TJ, Bogt PK, Bohmann D, Tjian R, Franza BR Jr (1988): Fos-associated protein (p39) is the product of the jun protooncogene. Science 240:1010–1016.

Raymond V, Atwater JA, Verma I (1989): Removal of an mRNA Destabilizing element correlates with the increased oncogenicity of proto-oncogene fos. Oncogene Res. 5:1–12.

Riabowol KT, Vosatka RJ, Ziff EB, Lamb NJ, Feramisco JR (1988): Microinjection of fos-specific antibodies blocks DNA synthesis in fibroblastic cells. Mol. Cell. Biol. 8:161–166.

Rivera VM, Sheng M, Greenberg M (1990): The inner core of the serum response element mediates both the rapid induction and subsequent repression of c-fos transcription following serum stimulation. Genes Dev. 4:255–268.

Ro J, Bresser J, Ro JY, Brasfield F, Hortobagyi G, Blick M (1989): Sis/PDGF-B expression in benign and malignant human breast lesions. Oncogene 4:351–354.

Robbins PD, Horowitz HM, Mulligan RC (1990): Negative regulation of human c-fos expression by the retinoblastoma gene product. Nature (London) 346:668–671.

Rorsman F, Bywater M, Knott TJ, Scott J, Betsholtz C (1988): Structural characterization of the human platelet-derived growth factor A-chain cDNA and gene: Alternative exon usage predicts two different precursor proteins. Mol. Cell. Biol. 8:571–577.

Rozengurt E, Sinnett-Smith J, Taylor-Papadimitriou J (1985): Production of PDGF-like growth factor by breast cancer cell lines. Int. J. Cancer 36:247–252.

Rüther U, Wagner EF, Müller R (1985): Analysis of the differentiation-promoting potential of inducible c-fos genes introduced into embryonal carcinoma cell. EMBO J. 4:1775–1781.

Rüther U, Gaber C, Komitowski D, Müller R, Wagner EF (1987): Deregulated c-fos expression interferes with normal bone development in transgenic mice. Nature (London) 325:412–416.

Rüther U, Müller R, Sumida T, Tokuhisa T, Rajewsky K, Wagner EF (1988): c-fos expression interferes with thymus development in transgenic mice. Cell 53:847–856.

Rüther U, Komitowski D, Schuber FR, Wagner EF (1989): c-fos expression induces bone tumors in transgenic mice. Oncogene 4:861–865.

Ryan WA Jr, Franza BR Jr, Gilman M (1989): Two distinct cellular phospholipids bind to the c-*fos* serum response element. EMBO J. 8:1785–1792.

Ryder K, Lau LF, Nathans D (1988): A gene activated by growth factors is related to the oncogene v-*jun*. Proc. Natl. Acad. Sci. U.S.A. 85:1487–1497.

Ryder K, Lanahan A, Perex-Albuerne E, Nathans D (1989):*Jun*-D: A third member of the *Jun* family. Proc. Natl. Acad. Sci. U.S.A. 86:1500–1503.

Sariban E, Nikolaos MS, Antoniades HN, Dufe DW, Pantazis P (1988): Expression of platelet-derived growth factor (PDGF)-related transcripts and synthesis of biologically active PDGF-like proteins by human malignant epithelial cell lines. J. Clin. Invest. 82:1157–1164.

Sassone-Corsi P, Verma, I (1987): Modulation of c-*fos* gene transcription by negative and positive cellular factors. Nature (London) 326:507–510.

Sassone-Corsi P, Sisson JC, Verma I (1988): Transcription autoregulation of the proto-oncogene c-*fos*. Nature (London) 334:314–319.

Scanlon KJ, Kashani-Sabet M, Miyachi H, Sowers L, Rossi J (1989): Molecular basis of cisplatin resistance in human carcinomas: Model systems and patients. Anticancer Res. 9:1301–1312.

Schönthal A, Herrlich P, Rahmsdorf HJ, Ponta H (1988a): Requirement for *fos* gene expression in the transcriptional activation of collagenase by other oncogenes and phorbol esters. Cell 54:325–334.

Schönthal A, Gebel S, Stein B, Ponta H, Rahmsdorf HJ, Herrlich P (1988b): Nuclear oncoproteins determine the genetic program in response to external stimuli. Cold Spring Harbor Symp. Quant. Biol. 53:779–787.

Schönthal A, Büscher M, Angel P, Rahmsdorf HJ, Ponta H, Hattori K, Chiu R, Karin M, Herrlich P (1989): The Fos and Jun/AP-1 proteins are involved in the downregulation of Fos transcription. Oncogene 4:629–636.

Schütte J, Viallet J, Nau M, Segal S, Fedorko J, Minns J (1989):*Jun*-B inhibits and c-*fos* stimulates the transforming and trans-activating activities of c-*jun*. Cell 59:987–997.

Sellers JW, Struhl K (1989): Changing Fos oncoproteins to a *jun* independent DNA-binding protein with GCN4 dimerization specificity by swapping "leucine zippers." Nature (London) 341:74–76.

Shaw PE, Grasch S, Nordheim A (1989): Repression of c-*fos* transcription is mediated through p67[SRF] bound to the SRE. EMBO J. 8:2567–2574.

Shyu A-B, Greenberg ME, Belasco JC (1989): Growth factors and membrane depolarization activate distinct programs of early response gene expression: dissociation of *fos* and *jun* induction. Genes Dev. 3:60–72.

Silver LM, Martin GA, Strickland S (1983): Teratocarcinoma Stem Cells, Vol. 10. New York: Cold Spring Harbor Laboratory Publications, Conferences on cell proliferation.

Smeal T, Angel P, Meek J, Karin M (1989): Different requirements for formation of Jun:Jun and Jun:Fos complexes. Genes Dev. 3:2091–2100.

Smith CC, Aurelian L, Reddy MP, Miller PS, Ts'o POP (1986): Antiviral effect of an oligo (nucleoside methylphosphonate) complementary to the splice junction of herpes simplex virus type I immediate early pre-mRNAs 4 and 5. Proc. Natl. Acad. Sci. U.S.A. 83:2787–2792.

Stacey DW, Watson T, Kung H-F, Curran T (1987): Microinjection of transforming *ras* protein induces c-*fos* expression. Mol. Cell. Biol. 7:523–527.

Stevens C, Brondyk W, Burgess H, Manoharan T, Häne B, Fahl W (1988): Partial transformed anchorage-independent human diploid fibroblasts result from over-expression of c-*sis* oncogene: Mitogenic activation of an apparent monomeric PDGF-2 species. Mol. Cell. Biol. 8:2089–2096.

Struhl K (1988): The Jun oncoprotein, a vertebrate transcription factor, activates transactivation in yeast. Nature (London) 322:649–650.

Tong B, Auer DE, Jaye M, Kaplow JM, Ricca G, McConthany E, Drohan W, Deuel TF (1987): cDNA clones reveal differences between human glial and endothelial cell platelet-derived growth factor A-chains. Nature (London) 328:619–621.

Treisman R (1985): Transient accumulation of c-*fos* sequences. Cell 42:889–902.

Treisman R (1986): Identification of a protein-binding site that mediates transcriptional response of the c-*fos* gene to serum factors. Cell 46:567–574.

Treisman R (1987): Identification and purification of a polypeptide that binds to the c-*fos* response elements. EMBO J. 6:2711–2717.

Van Beveran C, Van Straaten F, Curran T, Verma I (1983): Analysis of FBJ-MuSV provirus and c-*fos* (mouse) gene reveals that the proviral and cellular *fos* gene products have different carboxytermini. Cell 32:1241–1255.

Versnel M, Hagemeijer A, Bouts M, der Kwast van-T, Hoogsteden H (1988): Expression of c-*sis* (PDGF B-chain) and PDGF A-chain gene in ten human malignant mesothelioma cell lines derived from primary and metastatic tumors. Oncogene 4:601–605.

Vogt PK, Bos TJ (1989): The oncogene *jun* and nuclear signalling. TIBS 14:172–177.

Wagner RW (1989): A double stranded RNA unwinding activity introduces structural alteration by means of adenine to inosine conversions in mammalian cells and *Xenopus* eggs. Proc. Natl. Acad. Sci. U.S.A. 86:9139–9143.

Wagner RW, Nishikura K (1988): Cell cycle expression of RNA duplex unwindase activity in mammalian cells. Mol. Cell. Biol. 8:770–777.

Weima SM, van Rooijen MA, Mummery CL, Feijm A, de Laat S, van Zoelen MA (1989): Biosynthesis and processing of the receptor for PDGF in a human teratocarcinoma cell line producing a PDGF-like growth factor. J. Cell Biochem. Suppl. 13b:148.

Westermark B, Johnsson A, Paulsson Y, Betsholtz C, Heldin C-H, Herlyn M, Rodeck U, Koprowski H (1986): Human melanoma cell lines of primary and metastatic origin express the genes encoding the chains of platelet-derived growth factor (PDGF) and produce a PDGF-like growth factor. Proc. Natl. Acad. Sci. U.S.A. 83:7197–7200.

Westin E, Reitz M, Wong-Staal F (1984): Transforming potential of the c-sis nucleotide sequences encoding platelet-derived growth factor. Science 225:636–638.

Wilson T, Treisman R (1988): Removal of poly(A) and consequent degradation of c-fos mRNA facilitated by 3' AU-rich sequences. Nature (London) 336:396–399.

Womer R, Frick K, Mitchell CD, Ross AH, Bishayee S, Scher CD (1987): PDGF induces c-myc mRNA expression in MG-63 human osteosarcoma cells but does not stimulate cell replication. J. Cell Physiol. 132:66–72.

Wu J-X, Carpenter P, Gresens C, Keh R, Niman H, Morris JWS, Mercola D (1990): The proto-oncogene c-fos is over-expressed in a majority of human osteosarcomas. Oncogene 5:989–1000.

Yang-Fen H-F, Chiu R, Karin M (1990): Elevation of AP1 activity during F9 cell differentiation is due to increased c-jun transcription. The New Biol. 2:351–361.

Zarbl H, Latreille J, Jolicoeur P (1987): Revertants of v-fos transformed fibroblasts have mutations in cellular genes essential for transformation by other oncogenes. Cell 51:357–369.

Zerial M, Toschi L, Ryseck RP, Schuermann M, Müller R, Bravo, R (1989): The product of a novel growth factor activated gene interacts with Jun proteins enhancing their DNA binding activity. EMBO J. 8:805–813.

Ras, an Inner Membrane Transducer of Growth Stimuli

Esther H. Chang and Paul S. Miller

Departments of Pathology and Surgery, Uniformed Services University of the Health Sciences, Bethesda, Maryland 20814-4799 (E.H.C.) and Department of Biochemistry, School of Hygiene and Public Health, The Johns Hopkins University, Baltimore, Maryland 21205 (P.S.M.)

RAS ONCOGENES

Properties of the *ras* p21

The *ras* protein functions in the transduction of proliferative signals generated upstream by growth factor receptors (platelet-derived growth factor receptor, epidermal growth factor receptor), or membrane bound tyrosine kinases (*fes* and *src*), to the downstream cytosolic serine/threonine kinases (*raf, mos,* protein kinase C), and then the nuclear target proteins (*myc, fos,* c-*jun*) that regulate transcription and cell division. It seems that functional *ras* proteins are essential for the effective transduction of growth signals (Milburn et al., 1990; Bourne et al., 1990a; Satoh et al., 1990). Ha-, Ki-, and N-*ras* can be considered the prototypes of a *ras* gene superfamily whose members may share a common ancestral gene (Chardin, 1988). The *ras* gene product, a 21,000-Da phosphoprotein, has been localized to the inner side of the plasma membrane (Willingham et al., 1980), binds the guanine nucleotides GDP and GTP (Scolnick et al., 1979; Papageorge et al., 1982), and exhibits GTPase activity (McGrath et al., 1984; Sweet et al., 1984).

Ras p21 proteins resemble G proteins in sequence homology and function, and have been considered to be potential intermediaries in the signal transduction pathway (Marshall, 1987; McCormick, 1989; Barbacid, 1990; Bourne et al., 1990b). The GTPase activity of *ras* p21 is elevated by its interaction with a GTPase-activating protein, GAP, whose molecular weight is 110,000 (Trahey and McCormick, 1987; Vogel et al., 1988). The ability of GAP activity to deactivate GTP–p21 complexes is apparently affected by phosphorylation of its tyrosine residues (McCormick, 1989).

GAP has recently been found to inhibit the function of normal *ras* p21, but not that of the mutated *ras* p21 (Zhang et al., 1990). Furthermore, in activated T lymphocytes, GAP activity decreased upon stimulation of protein kinase C activity, whereas the GTP binding activity of *ras* p21 was significantly increased. These studies clearly demonstrate that *ras* p21 is a component in a signal transduction pathway (Downward et al., 1990). More recently, a protein encoded by the neurofibromatosis-1 (NF-1) gene has also been shown to accelerate GTPase activity (Herman and Hedgecock, 1990; Xu et al., 1990; Martin et al., 1990). Another interesting interdisciplinary finding revealed that the requirement of farnesyl modification for *ras* protein to be biologically functional (Schafer et al., 1989; Casey et al., 1989).

Implications in Human Neoplasia

Ras oncogenes have been previously implicated in the initiation, maintenance, and progression of many neoplasms. Mutated *ras* genes have been detected in 90% of human pancreatic cancers, over 50% of colorectal tumors, 30% of lung cancers, and 30% of

Prospects for Antisense Nucleic Acid Therapy of Cancer and AIDS, pages 115–124
©1991 Wiley-Liss, Inc.

keratoacanthomas, and occur in high incidences in various other cancers (Bishop, 1987; Weinberg, 1988; Barbacid, 1990). *Ras* viral oncogenes (v-*ras*) were originally identified through studies of several murine sarcoma retroviruses, such as Harvey murine sarcoma virus (Defeo et al., 1981), Kirsten murine sarcoma virus (Ellis et al., 1981), rat sarcoma virus (Rasheed et al., 1978), and Balb murine sarcoma virus (Peters et al., 1974). These acute transforming retroviruses, regardless of their phylogenetic origin of isolation, all share a similar transforming sequence, *ras*. Via a very different approach, *ras* tumor oncogenes have also been identified as the active transforming sequences in DNA from many tumors (Tabin et al., 1982; Reddy et al., 1982; Taparowsky et al., 1982) in transfection assays (Graham and van der Eb, 1973). Both viral and tumor *ras* oncogenes are derived originally from host cell sequences.

c-*ras* protooncogenes can be activated to become active transforming oncogenes through quantitative and qualitative changes. A single point mutation in the 12th or 61st amino acid codon region has been shown to confer transforming properties to normal c-*ras* protooncogenes and reduces the intrinsic GTPase activity 10- to 100-fold. Elevated expression of normal c-*ras* has been detected in a variety of tumors (Filmus and Buick, 1985; Heighway and Haselton, 1986) and shown to induce cellular transformation in vitro (Chang et al., 1982; Pulciani et al., 1985). The GTP bound p21 is the active form of the *ras* protein. Thus, the accumulated active form of *ras* protein produces a constant transduction of growth signal that may lead to cellular transformation. Intervention at various steps of this signal transduction pathway may vary the transforming phenotype and affect the outcome of cellular manifestations.

ANTISENSE OLIGONUCLEOSIDE METHYLPHOSPHONATES

To be used successfully as antisense reagents, oligonucleotides should have the following properties: (1) they should be able to bind specifically to their cellular nucleic acid target; (2) they should be resistant to hydrolysis by exo- and endonucleases; and (3) they should be taken up by cells in culture. Normal β-anomeric oligodeoxyribonucleotides meet requirements (1) and (2); however, they are readily degraded by nucleases (Goodchild, 1989; Neckers, 1989, Uhlmann and Peyman, 1990). These oligomers can be used in cell culture if precautions are taken to reduce or eliminate the nuclease activity usually found in cell culture medium. Generally this involves using heat-inactivated or serum-free medium for the experiments.

Nucleic acid chemists have created a variety of oligonucleotide analogs in an effort to increase the nuclease resistance of the oligomer without perturbing its ability to interact with complementary nucleic acids. For example, α-anomeric oligomers have been synthesized, which are both resistant to nuclease hydrolysis and form very stable complexes with RNA (Rayner et al., 1989). Oligomer analogs containing phosphorothioate linkages in place of the naturally occurring phosphodiester linkage show increased resistance to nucleases and biological activity in cultured cells (Stein and Cohen, 1989).

Nonionic oligonucleotide analogs, such as oligonucleoside methylphosphonates and oligonucleoside phosphoramidates, are totally resistant to nuclease hydrolysis and are thus very attractive candidates for use as antisense reagents (Miller, 1989). Rather extensive studies have been carried out with the oligodeoxyribonucleoside methylphosphonates (Miller and Ts'o, 1988; Miller, 1989, 1990).

Oligonucleoside methylphosphonates are quite hydrophobic and are taken up intact by mammalian cells in culture after addition of the oligomer to the cell culture medium. These oligomers form stable duplexes with complementary single-stranded DNA and RNA. The oligomers have also been conjugated with the photoreactive crosslinking group, psoralen. These psoralen-conjugated oligomers are able to form covalent adducts with complementary RNA and DNA when duplexes are irradiated with long wavelength ultraviolet light (Lee et al., 1988a,b; Kean et al., 1988; Bhan and Miller, 1990).

The methylphosphonate oligomers have been shown to have inhibitory activity against a number of viruses in cell culture including vesicular stomatitis virus (VSV) (Agris et al., 1986), herpes simplex virus (HSV) (Smith et al., 1986; Kulka et al., 1989), and human immunodeficiency virus (HIV) (Sarin et al., 1988; Zaia et al., 1988). In these studies the oligomers have been mainly targeted against the initiation codon regions of mRNA or the donor or acceptor splice sites of precursor mRNA. It is not certain which if either region is superior as a target for the oligomers. The oligomers inhibit in a dose-dependent and sequence-dependent manner. Rather high concentrations of the oligomers, between 50 and 150 μM, are required to give significant levels of inhibition. This concentration can be reduced approximately 10-fold by use of the psoralen-conjugated methylphosphonate oligomers. The mechanism by which the oligomers inhibit is not clearly understood. It is known that duplexes formed between the methylphosphonate oligomer and its RNA target are not substrates for ribonuclease H activity. Thus it appears likely that the oligomers inhibit as a result of physically blocking the interaction of important cellular components such as ribosomes with the mRNA or by causing conformational changes in the RNA that result in loss of activity to the RNA.

CONTROL OF *RAS* GENE EXPRESSION BY ANTISENSE OLIGONUCLEOTIDES

In Vitro Experiments

A comparison of a series of antisense oligodeoxyribonuclotides (11-mers) having different internucleoside linkages and various degrees of complementarity with the initiation codon region (first 11 coding nucleotides) of Balb-*ras* p21 mRNA has been performed in a rabbit reticulocyte lysate in vitro translation assay (Chang et al., 1989). The oligomers had either unmodified phosphodiester linkages, alternating methylphosphonate and phosphorothioate linkages, con-

tiguous methylphosphonate linkages, or contiguous phosphorothioate linkages. There was a $10^3 - 10^4$ molar excess of the oligomers over *ras* mRNAs in the assay. The inhibition of *ras* protein synthesis was, in general, dependent upon the dose and the sequence of the oligomer. At low concentrations (< 25 μM) the phosphorothioate oligomers were most effective in inhibiting *ras* expression. At higher concentration ($50-100$ μM) methylphosphonate oligomers were most sequence-specific (Chang et al., 1989).

Anti-*ras* ONMPs complementary to the initiation codon (IC) region of human c-Ha-*ras* gene have been tested for their efficacy and specificity on *ras* p21 translation using this cell-free assay.

The *in vitro* run-off transcripts of human c-Ha-*ras* were used as the source of mRNA. An ONMP (IC-0) precisely complementing the first 11 nucleotides of the c-Ha-*ras* initiation codon region acted in a dose-dependent manner to inhibit p21 translation in a rabbit reticulocyte lysate. At 100 μM, IC-0 nearly completely inhibited the cell-free translation of p21. The two control oligomers containing one (IC-1) or two (IC-2) nucleotide mismatches were significantly less effective than IC-0 at equivalent concentrations (Fig. 1). When an ONMP, anti-VSV, targeted against a sequence unrelated to *ras* was used to replace the anti-*ras* IC ONMPs, there was no inhibition of *ras* p21 synthesis observed at 100 μM (Fig. 2). These data suggest strongly that the inhibition of *ras* p21 synthesis is highly sequence-specific.

The inhibitory effects of a series of 8-mer ONMPs complementary to the twelfth amino acid codon in normal human c-Ha-*ras* or activated c-Ha-*ras*-T24 in a cell-free translation system have also been studied (Yu et al., 1989). Even though normal and mutated c-Ha-*ras* differ by only a single nucleotide, the corresponding antisense ONMPs displayed specific inhibitory activities. The ONMP targeted against the normal c-Ha-*ras* inhibited synthesis of the normal *ras* gene product more efficiently than did the ONMP complementary to the mutated Ha-*ras*. The opposite was also true (data not shown).

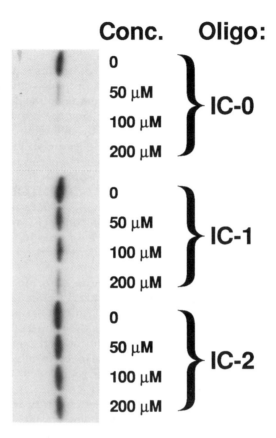

Conc.   Oligo:

0
50 μM   } IC-0
100 μM
200 μM

0
50 μM   } IC-1
100 μM
200 μM

0
50 μM   } IC-2
100 μM
200 μM

RNA: Normal c-Ha-ras

Fig. 1. The specificity of the inhibitory effect of anti-*ras* ONMPs on *ras* p21 translation. The ONMPs, IC-1, and IC-2, represent one and two mismatches at the center of the 11-mer, respectively, whereas IC-0 complements the Ha-*ras* initiation codon region precisely. The ONMP were stored as 1 μM solutions in autoclaved water at 20°C. The entire coding region for human c-Ha-*ras* p21 was cloned into the vector, pGEM-4 (Promega Biotec), in the 5′ and 3′ orientation relative to the SP6 promoter. Linearized plasmid DNAs were used as templates for in vitro run-off transcription reactions using SP6 polymerase (Promega Biotec). Transcription reactions contained 40 mM Tris-Cl, pH 7.5, 6 mM MgCl$_2$, 2 mM spermidine, 10 mM NaCl, 0.5 mM each of ATP, CTP, GTP, and UTP, 10 mM DTT, and 1 unit/ml RNasin. The reaction mixtures were incubated at 37°C for 1–2 hr. RNAs were purified by phenol:chloroform extraction and ethanol precipitation and analyzed on 1% formaldehyde agarose gels. Each 20 μl translation reaction contained 0.12 μg of template. The nuclease-treated rabbit reticulocyte lysate (Promega Biotec) contained 1.0 mCi/ml [^{35}S]methionine, 20 μM amino acid mixture (minus methionine), and RNasin (1 unit/μl). Prior to translation, the RNAs and anti-*ras* ONMP were preannealed by heating together at 90°C then allowed to cool slowly at room temperature prior to translation. Samples were incubated at 30°C for 1 hr and then treated with RNase A (100 μg/ml) at 37°C for 10 min. Monoclonal antibody to p21 plus protein A Sepharose (Pharmacia) was added. Immune complexes were washed, eluted, and applied to a 13% acrylamide gel crosslinked with Acrylaide on gel bond film (FMC Bioproducts) using the buffer system of Laemmli (1970). After fixing, the gels were soaked in Enlightning for 20 min (New England Nuclear), dried, and fluorographed at −70°C.

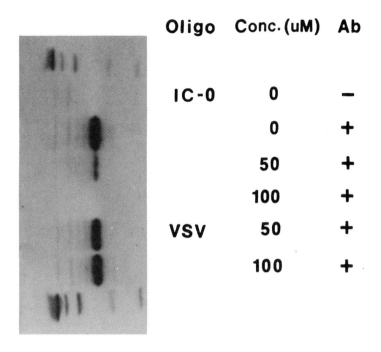

Oligo   Conc.(uM)   Ab

Oligo	Conc.(uM)	Ab
IC-0	0	–
	0	+
	50	+
	100	+
VSV	50	+
	100	+

RNA: C-Ha-ras, EJ

Fig. 2. Lack of inhibitory effect of anti-VSV ONMPs on *ras* p21 translation. An anti-VSV ONMP was tested in the cell-free assay described in the legend of Figure 1. The in vitro-translated mixtures were immunoprecipitated, then analyzed by SDS-polyacrylamide gel electrophoresis, followed by autoradiography.

Experiments in Cells

In a recent report (Daaka and Wickstrom, 1990), normal antisense pentadecadeoxynucleotides complementary to the 5'-cap region, a sequence further downstream in the untranslated region, and the initiation codon region of human c-Ha-*ras* were compared for their inhibitory efficacy on the *ras* p21 synthesis and colony formation. The oligomer complementary to the 5'-cap region was found to be most effective among the three. At 50 μM, an approximately 75% inhibition of *ras* p21 expression was observed. The colony size of transformed cells in semisolid agar culture also decreased significantly following anti-*ras* oligomer treatment.

The efficacies of various backbone modified oligonucleotides on the inhibition of *ras* expression have also been investigated. Anti-*ras* ONMPs have been shown to be effective and specific inhibitors of *ras* expression at the mRNA levels both in cell-free and cell culture systems (Brown et al., 1989; Yu et al., 1989; Chang et al., 1989). For example, an anti-*ras* ONMP complementary to the first 11 nucleotides of the initiation codon region of Balb-*ras* has been shown to inhibit p21 translation in a cell-free system (Brown et al., 1989). At a moderate ONMP concentration of 100 μM, p21 synthesis was almost completely inhibited. The sequence dependence of the inhibition was evaluated using two additional ONMPs, whose sequences were identical to the first ONMP with the exception of a single or double base mismatch. The inhibitory efficacies of these ONMPs were reduced in proportion to the number of base mismatches. In the same report the ONMP directed against the first 11 nucleotides of the initiation codon region of Ha-*ras* was tested in cell culture on RS485, a transformed NIH

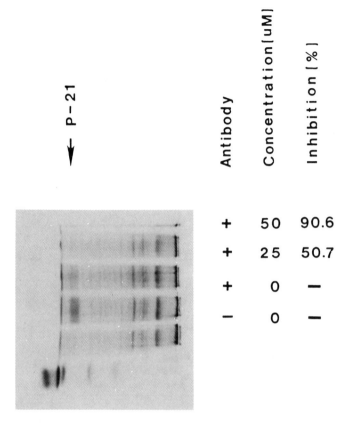

Cell line: RS 485

Oligomer: IC − 0

Fig. 3. Inhibition of p21 expression by a sequence-specific anti-*ras* oligonucleoside methylphosphonate (IC-0) in the *ras*-transformed NIH 3T3 cell line, RS485. Cells were treated with the indicated concentrations of IC-0 for 48 hr, then incubated in the presence of 150 μCi/ml [³⁵S]methionine in methionine-free media containing fresh IC-0 for an additional 12 hr.

3T3 cell line that overexpresses the normal Ha-*ras* p21 (Brown et al., 1989). After 60 hr of treatment, *ras* p21 was reduced by over 90% using only 50 μM ONMP (Fig. 3).

The use of anti-*ras* ONMPs complementary to the first acceptor splice site of *ras* p21 precursor mRNA has been studied. An octameric ONMP, SJ-O (5'-GAATCCTG-3') was tested in *ras*-transformed NIH 3T3 RS485 cells, inhibiting p21 synthesis approximately 90%

at an oligomer concentration of 150 μM (Fig. 4). This oligomer had no effect on the level of p21 in another transformed NIH 3T3 line, RS13B, which contains a single copy of the entire Harvey murine sarcoma virus genome (data not shown). Although p21 is overexpressed in RS13B, v-*ras* contains no introns and therefore expression was not affected by an anti-c-*ras* oligomer, which was directed against a *ras* mRNA that requires splicing.

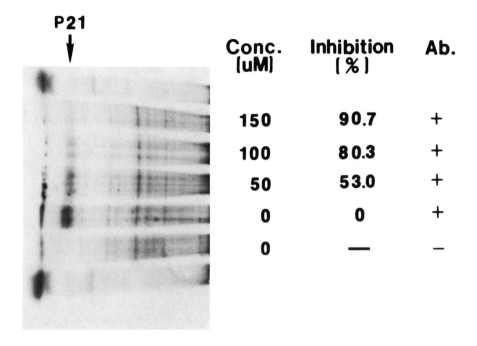

Conc. (uM)	Inhibition (%)	Ab.
150	90.7	+
100	80.3	+
50	53.0	+
0	0	+
0	—	−

Cell line: RS485

Oligomer: SJ-0, GAATCCTG

Fig. 4. Inhibition of p21 expression by a sequence-specific anti-*ras* oligonucleoside methylphosphonate (SJ-0) in the *ras*-transformed NIH 3T3 cell line, RS485. Cells were treated with the indicated concentrations for SJ-0 for 48 hr, then incubated in the presence of 150 μCi/ml [35S]methionine in low methionine media containing fresh SJ-0 for an additional 16 hr. TCA precipitable counts (2×10^7 cpm) were immunoprecipitated, then analyzed by SDS-polyacrylamide gel electrophoreses, followed by autoradiography.

POTENTIAL THERAPEUTIC USE OF ANTISENSE OLIGOMERS

It would appear that in order for antisense oligomers to be used in animals or in humans, the oligomer must be resistant to degradation to nucleases found in plasma and the blood stream. This requirement is met by oligonucleoside methylphosphonates. Oligonucleoside methylphosphonates appear to be distributed to most of the organs and tissues of mice when injected into the tail vein (Chen et al., 1990). The methylphosphonate linkages of the oligomer remain intact.

The general toxicity and immunogenicity, if any, of oligonucleoside methylphosphonates remain to be investigated. Extensive studies of this type will require large quantities of these high-molecular-weight compounds. Although this requirement will further challenge the synthetic capabilities of organic chemists, there is no theoretical reason that would prevent large-scale syntheses of these materials.

Assuming that antisense oligonucleoside methylphosphonates do not produce undue toxic or immunogenic responses, how might they be used as therapeutic agents in the treatment of neoplasia? Phenotypic reversion of transformed cells and control of tumor growth are the ultimate aims of reducing expression of an overexpressed protooncogene or eliminating expression of a mutated oncogene. Specifically, we have chosen to investigate the basic behavior of the sequence-specific anti-*ras* oligomer in *ras*-transformed cells. We hope to revert phenotypically the mouse transformants by selectively inhibiting *ras* oncogene expression. It is well recognized that activation of the *ras* gene represents only a single step in the multistep process of carcinogenesis. Inhibition of a single event may not be sufficient to generate a therapeutic effect. However, based upon a recent study on *ras* expression and natural killer (NK) cell sensitivity (Anderson et al., 1989), the anti-*ras* ONMPs, by inhibiting the p21 translation, may indirectly activate host NK-mediated cytolysis. Further studies in animal models should be able to better evaluate the potential of these sequence-specific ONMPs as anticancer therapeutic agents.

SUMMARY

As our understanding of the involvement of oncogenes in the cancer process increases, so too must our ability to design antitumor therapies that can be aimed at the specific genetic and metabolic basis of the particular neoplasms that oncogenes produce. Toward this end, we have tested the feasibility of using antisense-methylphosphonate-modified deoxyribonucleoside oligomers (ONMP) that are complementary to RNA transcripts of specific *ras* oncogene as antitumor agents. Nuclease resistance, ready uptake by living cells by passive diffusion, and nontoxicity at moderate concentrations are features of the nonionic ONMP, which offer potential in cell culture and in animals. We have shown that the *ras*-specific ONMPs bind to these target sequences and arrest the synthesis of the encoded p21 in both cell-free and cell culture systems.

REFERENCES

Agris CH, Blake BR, Miller PS, Reddy MP, Ts'o POP (1986): Inhibition of vesicular stomatitis virus protein synthesis and infection by sequence-specific oligodeoxyribonucleoside methylphosphonates. Biochemistry 25:6268–6275.

Anderson SK, Stankova J, Roder JC (1989): Decreased p21 levels in anti-sense *ras* transfectants augments NK sensitivity. Mol. Immunol. 26:985–991.

Ballester R, Marchuk D, Boguski M, Saulino A, Letcher R, Wigler M, Collins F. (1990): The *NFI* locus encodes a protein functionally related to mammalian gap and yeast *IRA* proteins. Cell 63:851–859.

Barbacid M (1990): *Ras* oncogenes: Their role in neoplasia. Eur. J. Clin. Invest. 20:225–235.

Bhan P, Miller PS (1990): Photo-cross-linking of psoralen-derivatized oligonucleoside methylphosphonates to single-stranded DNA. Bioconjugate Chem. 1:82–88.

Bishop JM (1987): The molecular genetics of cancer. Science 235:305–311.

Bourne HR, Wrischnile L, Kenyon C (1990a): Some signal developments. Nature (London) 348:678–679.

Bourne HR, Sanders DA, McCormick F (1990b): The GTPase superfamily: A conserved switch for diverse cell functions. Nature (London): 348:125–132.

Brown D, Yu ZP, Miller P, Blake KR, Wei C, Kung HF, Black R, Ts'o POP, Chang EH (1989): Modulation of *ras* expression by antisense, nonionic deoxyoligonucleotide analogs. Oncogene Res. 4:243–252.

Capon DJ, Chen EY, Levinson AD, Seeburg PH, Goeddel DV (1983): Complete nucleotide sequences of the T24 human bladder carcinoma oncogene and its normal homologue. Nature (London) 302:33–37.

Casey PJ, Solski PA, Der CJ, Buss JE (1989): p21*ras* is modified by a farnesyl isoprenoid. Proc. Natl. Acad. Sci. U.S.A. 86:8323–8328.

Chang EH, Furth ME, Scolnick EM, Lowy DR (1982): Tumorigenic transformation of mammalian cells induced by a normal human gene homologous to the oncogene of harvey murine sarcoma virus. Nature (London) 297:479–483.

Chang EH, Yu ZP, Shinozuka K, Wilson WD, Strekowska A, Zon G (1989): Comparison of efficacy of modified anti-*ras* oligodeoxynucleotides. Anti-Cancer Drug Design 4:221–232.

Chardin P (1988): The *ras* superfamily proteins. Biochimie 70:865–868.

Chen T-L, Miller PS, Ts'o POP, Colvin OM (1990): Disposition and metabolism of oligodeoxynucleoside methylphosphonates following a single IV injection in mice. Drug Metab. Dispos. 18:815–818.

Daaka Y, Wickstrom E (1990): Target dependence of antisense oligodeoxynucleotide inhibition of c-Ha-*ras* p21 expression and focus formation in T24-transformed NIH3T3 cells. Oncogene 5(4):267–275.

DeFeo D, Gonda MA, Young HA, Chang EH, Lowy DR, Scolnick EM, Ellis RW (1981): Analysis of two divergent rat genomic clones homologous to the transforming gene of Harvey murine sarcoma virus. Proc. Natl. Acad. Sci. U.S.A. 78:3328–3332.

Downward J, Graves JD, Warne PH, Rayter S, Cantrell DA (1990): Stimulation of p21 *ras* upon T-cell activation. Nature (London) 346:719–723.

Ellis RW, Defeo D, Shih TY, Gonda MA, Young HA, Tsuchida N, Lowy DR, Scolnick EM, Ellis RW (1981): The p21 *src* genes of Harvey and Kirsten sarcoma viruses originate from divergent members of a family of normal vertebrate genes. Nature (London) 292:506–511.

Filmus JE, Buick RN (1985): Stability of c-K-*ras* amplification during progression in a patient with adenocarcinoma of the ovary. Cancer Res. 45:4468.

Furth ME, Davis LJ, Fleurdelys B, Scolnick EM (1982): Monoclonal antibodies to the p21 products of the transforming gene of Harvey murine sarcoma virus and of the cellular *ras* gene family. J. Virol. 43:294–304.

Goodchild J (1989): Inhibition of gene expression by oligonucleotides. In Cohen J (ed): Oligodeoxynucleotides. Antisense Inhibitors of Gene Expression, Topics in Molecular and Structural Biology Vol. 12. Boca Raton, FL: CRC Press, pp 53–77.

Graham GL, van der Eb AJ (1973): A new technique for the assay of infectivity of human adenovirus 5 DNA. Virology 52:456–467.

Harford J (1984): An artifact explains the apparent association of the transferrin receptor with a *ras* gene product. Nature (London) 311:673–675.

Heighway J, Haselton PS (1986): c-Ki-*ras* amplification in human lung cancer. Br. J. Cancer 53:285.

Herman RK, Hedgecock EM (1990): Limitation of the size of the vulval primordium of *Caenorhabditis elegans* by *Lin*-15 expression in surrounding hypodermis. Nature (London) 348:169–171.

Kean JM, Murakami A, Blake KR, Cushman CD, Miller PS (1988): Photochemical cross-linking of psoralen-derivatized oligonucleoside methylphosphonates to rabbit globin messenger RNA. Biochemistry 27:9113–9121.

Kulka M, Smith C, Aurelian L, Fishelevich R, Meade K, Miller P, and Ts'o POP (1989): Site specificity of the inhibitory effects of oligo(nucleoside methylphosphonates) complementary to the acceptor splice junction of herpes simplex virus type 1 immediate early mRNA 4. Proc. Natl. Acad. Sci. U.S.A. 86:6868–6872.

Laemmli UK (1970): Cleavage of structural proteins during the assembly of the head of bacteriophage T4. Nature (London) 227:680–685.

Lee BL, Blake KR, Miller PS (1988a): Interaction of psoralen-derivatized oligodeoxyribonucleoside methylphosphonates with synthetic DNA containing a promoter for T7 RNA polymerase. Nucl. Acids Res. 16:10681–10697.

Lee BL, Murakami A, Blake KR, Lin S-B, Miller PS (1988b): Interaction of psoralen-derivatized oligodeoxyribonucleoside methylphosphonates with single-stranded DNA. Biochemistry 27:3197–3203.

Marshall CJ (1987): Meeting report: Oncogenes and growth control 1987. Cell 49:723–725.

Martin GA, Viskochil D, Bollag G, McCabe PC, Crosier WJ, Haubruck H, Conroy L, Clark R, O'Connell P, Cawthon RM, Innis MA, McCormick F (1990): The GAP-related domain of the neurofibromatosis type I gene product interacts with *ras* p21. Cell 63:843–849.

McCormick F (1989): *Ras* GTPase activating protein: Signal transmitter and signal terminator. Cell 56:5–8.

McGrath JP, Capon DJ, Goeddel DV, Levinson AD (1984): Comparative biochemical properties of normal and activated human *ras* p21 protein. Nature (London) 310:655.

Milburn MV, Tong L, deVos AM, Brunger A, Yamaizumi Z, Nishimura S (1990): Molecular switch for signal transduction: Structural difference between active and inactive forms of protooncogenic *ras* proteins. Science 247:939.

Miller PS, Ts'o POP (1988): Oligonucleotide inhibitors of gene expression in living cells: New opportunities in drug design. Annu. Rep. Med. Chem. 23:295–304.

Miller PS (1989): Non-ionic antisense oligonucleotides. In Cohen J (ed): Oligodeoxynucleotides. Antisense Inhibitors of Gene Expression. Topics in Molecular and Structural Biology, Vol. 12. Boca Raton, FL: CRC press, pp 79–95.

Miller PS (1990). Antisense nucleic acid analogures as potential antiviral agents. In Aurelian L (ed): Herpesviruses, the Immune System and AIDS. Boston: Kluwer Academic Publishers, pp 343–360.

Neckers LM (1989): Antisense oligodeoxynucleotides as a tool for studying cell regulation: Mechanism of uptake and application to the study of oncogene function. In Cohen J (ed): Oligodeoxynucleotides. Antisense Inhibitors of Gene Expression, Topics in Molecular and Structural Biology, Vol. 12. Boca Raton, FL: CRC Press, pp 211–231.

Papageorge A, Lowy D, Scolnick EM (1982): Comparative biochemical properties of p21 *ras* molecules coded for by viral and cellular *ras* genes. J. Virol. 44:509.

Peters RL, Rabstein LS, Van Vleck R, Kelloff GJ, Huebner RJ (1974): Naturally occurring sarcoma virus of the Balb/c Cr mouse. J. Natl. Cancer Inst. 53:1725–1729.

Pulciani S, Santos E, Long LK, Sorrentino V, Barbacid M (1985): *Ras* gene amplification and malignant transformation. Mol. Cell. Biol. 5:2836.

Rasheed SM, Gardner B, Heubner RJ (1978): *In vitro* isolation of stable rat sarcoma viruses. Proc. Natl. Acad. Sci. U.S.A. 75:2972–2976.

Rayner B, Malvy C, Paoletti J, Lebleu B, Paoletti C, Imbach JL (1989): α-Oligodeoxynucleotide analogues. In Cohen J (ed): Oligodeoxynucleotides. Antisense Inhibitors of Gene Expression, Topics in Molecular and Structural Biology, Vol. 12. Boca Raton, FL: CRC Press, pp 119–136.

Reddy EP, Reynolds RK, Santos D, Barbacid M (1982): A point mutation is responsible for the acquisition of transforming properties by the T24 human bladder carcinoma oncogene. Nature (London) 300:149–152.

Sarin PS, Agrawal S, Civeira MP, Goodchild J, Ikeuchi T, Zamecnik PC (1988): Inhibition of acquired immunodeficiency syndrome virus by oligodeoxynucleoside methylphosphonates. Proc. Natl. Acad. Sci. U.S.A. 85:7448–7451.

Satoh T, Endo M, Nakafuku M, Nakamura S, Kaziro Y (1990): Platelet-derived growth factor stimulates formation of active p21 *ras*. GTP complex in swiss mouse 3T3 cells. Proc. Natl. Acad. Sci. U.S.A. 87:5993–5997.

Schafer WR, Kim R, Sterne R, Thorner J, Kim SH, Rine J (1989): Genetic and pharmacological suppression of oncogenic mutations in *ras* genes of yeast and humans. Science 245:379.

Scolnick EM, Papageorge AG, Shih TY (1979): Guanine nucleotide binding activity as an assay for *src* protein of rat-derived murine sarcoma viruses. Proc. Natl. Acad. Sci. U.S.A. 76:5355–5359.

Smith CC, Aurelian L, Reddy MP, Miller PS, Ts'o POP (1986): Antiviral effect of an oligo(nucleoside methylphosphonate) complementary to the splice junction of herpes simplex virus type 1 immediate early pre-mRNAs 4 and 5. Proc. Natl. Acad. Sci. U.S.A. 83:2787–2791.

Stein CA, Cohen JS (1989): Phosphorothioate oligodeoxynucleotide analogues. In Cohen J (ed): Oligodeoxynucleotides. Antisense Inhibitors of Gene Expression. Topics in Molecular and Structural Biology, Vol. 12. Boca Raton, FL: CRC Press, pp 97–117.

Sweet RW, Yokoyama S, Kamata T, Feramisco JR, Rosenberg M, Gross M (1984): The product of *ras* is a GTPase and the T24 oncogenic mutant is deficient in this activity. Nature (London) 311:273.

Tabin CJ, Bradley SM, Bargmann CI, Weinberg RA, Papageorge AG, Scolnick ED, Dhar R, Lowy DR, Chang EH (1982): Mechanism of activation of a human oncogene. Nature (London) 300:143–149.

Taparowsky E, Suard Y, Fasano O, Shinizu K, Golfarb M, Wigler M (1982): Activation of the T24 bladder carcinoma transforming gene is linked to a single amino acid change. Nature (London) 300:762–765.

Trahey M, McCormick F (1987): Acytoplasmic protein stimulates normal N-*ras* p21 GTPase but does not affect oncogenic mutants. Science 238:542–547.

Uhlmann E, Peyman A (1990): Antisense oligonucleotides: A new therapeutic principle. Chem. Rev. 90:544–579.

Vogel US, Dixon R, Schaber MD, Diehl RE, Marshall MS, Scolnick EM, Sigal LS, Gibbs JB (1988): Cloning of bovine GAP and its interaction with Oncogenic *ras* p21. Nature (London) 335:90–93.

Weinberg RA (1988): The genetic origins of human cancer. Cancer 61:1963–1973.

Willingham MC, Pastan I, Shih TY, Scolnick EM (1980): Localization of the *src* gene product of the Harvey strain of MSV to plasma membrane of transformed cells by electron microscopic immunocytochemistry. Cell 19:1005–1014.

Xu G, Lin B, Tanaka K, Dunn D, Wood D, Gesteland R, White R, Weiss R, Tamanoi F (1990): The catalytic domain of the neurofibromatosis type I gene product stimulates *ras* GTPase and complements *IRA* mutants of *S. cerevisiae.* Cell 63:835–841.

Yu ZP, Chen DF, Black R, Blake K, Miller P, Ts'o POP, Chang EH (1989): Sequence-specific inhibition of in vitro translation of mutated or normal *ras*-p21. J. Exp. Pathol. 4:97–108.

Zaia JA, Rossi JJ, Murakawa GJ, Spallone PA, Stephens DA, Kaplan RE, Eritja R, Wallace RB, Cantin EM (1988): Inhibition of human immunodeficiency virus by using an oligonucleoside methylphosphonate targeted to the tat-3 gene. J. Virol. 62:3914–3917.

Zhang K, Dellue JE, Vass WC, Papageorge AG, McCormick F, Lowy DR (1990): Suppression of c-*ras* transformation by GTPase-activating protein. Nature (London) 346:746, 754–756.

Retinoids and Retinoid Receptors in Malignant Disease: Clinical Significance of Their Expression and the Alteration of Disease Course With Antisense DNA

Frederick O. Cope, John J. Wille, and L. David Tomei

Ross Laboratories, Columbus, Ohio 43215 (F.O.C.), Convatec-Squibb, Skillman, New Jersey 08558 (J.J.W.), and Comprehensive Cancer Center, Ohio State University, Columbus, Ohio 43210 (L.D.T.)

BACKGROUND

General

Vitamin A comprises a group of polyiso-prenoid compounds including vitamin A alcohol (retinol), vitamin A acid (retinoic acid), and numerous esters (retinyl palmitate, retinyl stearate, retinyl oleate, etc.). Vitamin A was shown to cause cell differentiation and was required for the maintenance of terminally differentiated cellular phenotypes in most epithelial cells; these initial observations were made by Wolbach and Howe (1925). The mechanism of action of retinoids was dealt with predominantly through two theories: (1) the transduction of the presence of a retinoid compound by cytosolic binding proteins (Chytil, 1985), and (2) the regulation of glycosylation of proteins through the conversion of retinoid metabolites to retinol phosphate (De Luca et al., 1985). Moreover, the consensus was that retinoids induced only terminal differentiation in malignant phenotypes (Sporn et al., 1985). None of these bits of evidence or any ideology dealt with the realities or the theories of cell biology. Indeed, the continued notions regarding the mechanism of action of vitamin A ignored the elucidation of three nuclear receptors, RAR-α, RAR-β, and RAR-γ, by Cope et al. (1984a).

The elucidation of these particular receptors suggested that not only did vitamin A have cytosolic receptors but that cells also contained several receptors that were unique to the cell nucleus. This model supported the action of retinoids as steroid hormone-like molecules. Inferred from this concept for the mechanism of action of retinoids was that these steroid hormone receptors themselves acted as transcriptional elements when activated in the presence of the retinoid ligands. Thus, the number of genes that can be controlled by any receptor either in terms of a single receptor or permutations with other receptors was extremely high. For example, it was shown that retinoic acid activates a homeobox gene cluster that regulates the morphogenic capacity of embryonic cells (Mavillo, 1988). Recently these receptors (RAR-α, RAR-β, and RAR-γ) have been cloned (Brand et al., 1988; de The, 1987; Gaub et al., 1989; Giguere et al., 1987). These receptors bear remarkable homology to the steroid hormone receptors, e.g., progesterone and estrogen, and hepatocellular carcinoma specific steroid hormone receptor. More recently another class of retinoid receptor transcriptional elements has been identified (Mengelsdorf et al., 1990). These proteins are not structurally related to the steroid hor-

Prospects for Antisense Nucleic Acid Therapy of Cancer and AIDS, pages 125–141

mone-like proteins of the RAR family and are designated RXR-α and RXR-β. Thus, the concept of a retinoid-regulated cascade of gene expression as it controls the cellular phenotype and terminal differentiation is one which is very attractive in terms of defining the mechanism of action of the retinoids.

In conjunction with the expression of specific genes that are under the regulation of the retinoid-activated transcriptional elements, one intuitively seeks a primary signal aspect of a cell that is similar to the cyclic AMP-dependent protein kinase pathway coupled to the steroid hormone receptor action (Cope, 1989). For a model system that would be linked to the retinoid pathway, one can look to the visual cycle where the vitamin A compound, retinaldehyde, is linked to the transduction of light in the eye and the generation of a neuroelectrical impulse. This particular pathway is remarkably similar to the steroid hormone receptor, and suggests that such a system, albeit not related to the covalent binding of retinaldehyde to rhodopsin, is conserved as a signal transduction mechanism coupled to the transcriptional response elements of vitamin A (Fig. 1).

It is these particular pathways, both the epigenetic or cell membrane signaling system, and the transcriptional elements, that present themselves for the greatest potential for exploitation in the diagnosis and treatment in human malignant disease. Thus, by either probing or regulating (up or down) particular receptors at key temporal phases in tumor growth or development, one can expect to discover or modulate the phenotype of the tumor. Certainly, antisense DNA presents itself as one of the major methods for perturbing this particular system. In the following sections we would like to present a summation of the signal aspect of cellular response to retinoids, a summary of genes that are important or may be important in the regulation of tumor phenotype, and in particular a new concept that relates to the function of vitamin A in its ability to maintain fidelity of terminal differentiation and/or proliferation; this is called *apoptosis* and this process appears to play a key role in the explanation of the mechanism of vitamin A. More importantly, it provides an attractive avenue for the

clinician to evaluate findings that may be important in determining whether to use retinoid therapy (retinoic acid) or antiretinoid therapy (retinoid receptor antisense DNA).

Primary Signal Aspect of Cellular Response to Retinoids

The extent of popular knowledge about the function of vitamin A and its metabolites generally relates to the function of vitamin A in the visual process. The transduction event, converting light into electrical signals to the brain, is conducted by the vitamin A metabolite, retinaldehyde (RAL) in the visual cycle and is dependent upon its covalent binding to opsin to form rhodopsin (RO). RO captures a photon and initiates a cascade of signal-generating enzymes: (1) a phosphodiesterase (PDE) that hydrolyzes cGMP to $5'$-GMP; (2) transducin (TS), a multipeptide GTPase activated by GTP binding, that couples the photon activation of RO to PDE; (3) rhodopsin kinase (ROK) that modulates RO cycling by phosphorylation; (4) a 45-kDa protein that controls TS coupling to RO; and (5) phosphatase 2A (P2A) that dephosphorylates opsin and functions coordinately with ROK (Mueller and Pugh, 1983). This cascade effectively deactivates a cGMP-dependent protein kinase that controls Na^+ gating, and generates the nerve signal. Additionally, Ca^{2+} translocation from the disc to the plasma membrane is also induced (Hollenberg and Cuatrecasas, 1975), possibly resulting from a change in inositol phosphate (IP) metabolite perturbations (Basu and Schnetkamp, 1989).

This signal system is remarkably similar to the hormone-dependent signal systems in other cells, in which Ca^{2+}, cyclic nucleotides, and G-proteins play major roles. Such primary signal systems, though not genetic, *are* coupled to nuclear receptor proteins and, coordinately, these systems define expression of gene sets as a matter of milieu adaptation and/or differentiation, both in normal as well as malignant cells.

Genes Under The Regulation of Retinoids and Their Receptors

A number of genes have been shown to be under the regulatory control of retinoic acid,

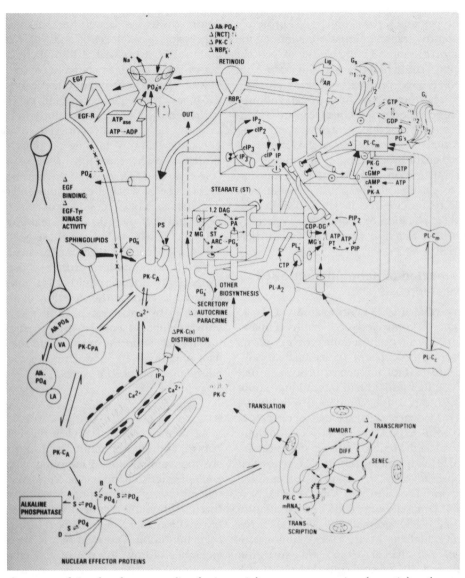

Fig. 1. Summary of signal pathway coupling (epigenetic) to gene expression (genetic) pathways.

retinol, and other metabolites of vitamin A. These include (1) alkaline phosphatase (AP; Reese et al., 1981, 1983), (2) transglutaminase (Yuspa et al., 1982), (3) epidermal growth factor receptor, (4) growth hormone (Bedo et al., 1989), (5) c-*erb*-B, (6) c-*ras*-K, (7) c-*myc*, (8) c-*src*, (9) c-*neu* (*erB*-2), and (10) c-*fos* (Saurat, 1985). Additionally, retinoids have been shown to induce the transcription of the progesterone and estrogen receptors; this latter obsevation is ex-

tremely important in considering the course of action in the treatment of breast malignancy and especially in the potential use of antisense DNA to retinoid receptors and responsive transcriptional elements which might be found in malignant breast tissue.

When viewing the above list, one can see that the list of genes currently controlled by retinoids, as well as the signal cascade outlined above, fall into three main categories: (1) control signal elements (G-proteins, both *i*-

and s-types, and K-ras, a mutated G-protein), (2) control transcriptional elements (c-fos and c-myc), and (3) regulation of protein phosphorylation [c-neu, c-src, and c-erb-B, all cell membrane protein (tyrosine) kinases]. Growth hormone, and the estrogen and progesterone receptors themselves, are essentially mixed function proteins, controlling these same elements. Thus, such elements of the malignant process and of the normal cell differentiation process may be used successfully to evaluate the responsiveness of neoplastic tissue cultured in vitro to either retinoid (application of retinoic acid) therapy or retinoid deletion (antisense DNA) therapy.

The Function of Retinoid-Controlled Genes: Apoptosis

Kerr et al. (1973) formulated the basic concept of controlled cell death in mammalian cells and proposed the term apoptosis to describe this concept. In mammalian cells, there are several important processes that involve either initiation or inhibition of apoptosis (Cowan, 1984; Saunders, 1966; Wyllie, 1980). Evidence suggests that apoptosis is involved in expression of drug cytotoxicity (Duncan and Heddle, 1984), radiation lethality (Hendy and Potten, 1982), target cell responses involved in cell-mediated immune processes (Clouston and Kerr, 1985), heat shock (Lee and Dewey, 1986), differentiation (Martin, 1988), normal cell renewal in tissues (Columbano et al., 1985), tumor promotion (Bursch et al., 1984), and expression of EBV- and HIV-infected cells (Valerie, 1988). Most important is the notion that the inhibition of apoptosis, by suppression of apoptotic genes, may be required for maintenance of a tumor phenotype. This concept lends itself to exploitation through the induction of apoptosis in apoptotic-resistant cells (tumor cells) by selection of appropriate antisense DNAs that may be needed to repress or derepress systems, thus allowing the expression of normal cell phenotypes.

Several apoptosis genes have been described and include the ced-3 and ced-4 genes. These are required for a total program of cell death while in "deathless" mutants the expression of ced-1, ced-2, and Nuc-1 are predominantly expressed. Additionally, a number of other cdc genes have been the center of attention. These include the cdc-13, and CK-II, and wee-1 genes (Pelech, 1990). This entire family of proteins is tightly integrated in controlling the phosphorylation and dephosphorylation of histone proteins as well as that of each other, thus determining the "on-ness" and "off-ness" of the mitotic cycle in single cells. What must be imposed on this system is a set of factors that coordinate "on-ness" and "off-ness" in functional groups of cells. As we will show later, these genes, too, fall into one of the three main events controlled by retinoid, i.e., protein phosphorylation. Hence, the use of either retinoids such as retinoic acid or the deletion of retinoid receptors by antisense DNA in the face of either malignant conversion or established tumors may suppress or eradicate the malignant phenotype by modulation of the phosphorylation of the cdc genes.

METHODS

As discussed above, it is clear that cell cycle control and differentiation (or lack thereof) are the culmination of coupling of genetic and epigenetic characters and events. No system represents a more appropriate model than that of retinoid-dependent induction of differentiation and/or apoptosis or, conversely, their inhibition. Outlined below are several examples of how the use of antisense DNA to retinoid receptors might be used to achieve clinical results. It should be kept in mind that, like the evaluation of breast tissue for hormone receptors, the evaluation of tumor tissue for specific retinoid receptors prior to clinical application of either antisense DNA or the direct use of retinoids, such as retinoic acid, is strongly advised. Such evaluations can be accomplished by northern blot analysis or by in situ hybridization using antisense DNA to the retinoid receptors. Antisense DNAs may be synthesized using any sequence against the translated regions of the receptor mRNAs. However, it has been our experience with the retinoid receptors that sequences over-lapping the 5'-untranslated region, the initi-ation codon, and a segment of the 5'-trans-lated region are extremely effective in blocking translation and accumulation of any com-pleted protein. I will not go into all the per-mutations in this review; readers are en-couraged to consult the references (Brand et al., 1988; de The, 1987; Gaub et al., 1989; Giguere et al., 1987) for making decisions on what oligonucleotides may be best for diag-nostic and treatment protocols. For in situ hybridization and northern blot analysis in tumor tissue there are many existing proto-cols using oligonucleotides (Schmechel et al., 1988); most are a matter of practice and maintaining appropriate quality control dur-ing sample processing. The readers are en-couraged to consult these references.

Other Considerations—Chemicals

[^{3}H]Retinoic acid (48 Ci/mmol) and [^{3}H]retinol (36 Ci/mmol) were the gift of E. I. DuPont, Inc. Unlabeled RA and retinol were purchased from Sigma Chemical Co., and pur-ified by reverse-phase HPLC as previously de-scribed (Cope et al., 1983). Antisense DNAs complementary to or of the same sequence as cRBP and RAR-α (hereinafter referred to as RAR) were prepared with the following sequences as previously described (Cope and Wille, 1989): sense cRBP = 5'-GTCACTCCCGAAATG-3', antisense cRBP = 5'-CATTTCGGGAGTGAC- 3', antisense RAR = 5'-CATGCCAAGCGGCCA-3'. Addition-ally, we used, as controls, antisense DNA that had been annealed to sense DNA, antisense DNAs that were degraded by endonuclease, and an antisense sequence to a *Drosophila* sp.-specific protein.

Retinoid Receptor Assay

The cold-acetone binding procedure of Ashendel and Boutwell (1981) was modified to permit independent assay of retinoid bind-ing to cRBP, cRABP, and RAR, based on their ligand binding and selective extraction from cytosolic and nuclear fractions, respectively (Cope et al., 1984b; Cope and Wille, 1989).

Cell Clonal Growth Assay and Viability

After treatment with cells were removed from the initial treatment dishes and re-plated. Conditions for this experiment are described in Table 3.

Other Methods

AP (EC 3.1.3.1) was assayed according to the method of Reese et al. (1981) and protein was determined by the method of Bradford (1976). Autoradiography and immunoflu-orescence were accomplished according to the method of Hames and Rickwood (1981).

THE SYSTEMS AND ANTISENSE DNA TEST RESULTS

The Systems

As discussed above, the action by which retinoids function can be somewhat com-plex. Nevertheless, a starting point(s) must be chosen if clinical progress is to be achieved. Picking several points is not always advan-tageous. However, in this case, we chose sev-eral points because of their logical interac-tion. System 1 was the human squamous cell carcinoma culture, SCC-25 (Wille et al., 1985). This system expresses high levels of the α-type receptor and treatment with antisense DNAs to specific receptors demonstrates what may be expected to happen clinically to this tumor type. Additionally, this system reveals the significance of the mechanism of apop-tosis in relieving the patient of such a tumor. We have contrasted these results with a look at the primary signal systems (G-proteins and protein phosphorylation) in malignant hu-man squamous cells. Lastly we have viewed the coupling of the receptors and signal sys-tems to the expression of oncogenes, and how this may be used to provide a better diagnosis in conjunction with the use of an-tisense DNAs in tissue culture, and in situ hybridization of biopsy material.

Time-Course Effect of DNAs in SCC-25

The direct addition of antisense DNAs to cell culture media and their uptake and reg-ulation of target proteins has been described

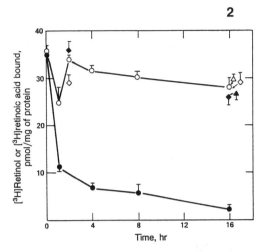

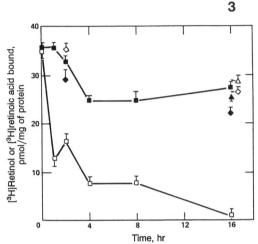

Fig. 2. Time-dependent effect of antisense cRBP DNA on cRBP and RAR levels in human malignant keratinocytes. SCC-25 cells were grown as described (Table 1). The cells received the following 30 μM DNA treatments: (●), antisense cRBP DNA→assay of cytoplasm for cRBP; (○), antisense cRBP DNA→assay of nucleus for RAR; (♦), sense cRBP DNA→assay of cytoplasm for cRBP; (◇), sense→assay of nucleus for RAR; (▲), no treatment with 16-hr incubation→assay of cytoplasm for cRBP; and (△), no treatment with 16-hr incubation→assay of nucleus for RAR. At the end of the treatment periods, the cells were washed three times with 5 ml of ice-cold 25 mM Tris-HCl buffer (pH 7.8), drained free of solution, and were quick frozen on solid CO$_2$. Both a cytoplasmic and a nuclear- pellet extract fraction were prepared and assayed for cRBP and/or RAR by the cold acetone binding assay. Values are the means of three replicate determinations ± SD.

Fig. 3. Time-dependent effect of antisense RAR DNA on cRBP and RAR levels in human malignant keratinocytes. The cells received the following 30 μM DNA treatments: (■), antisense RAR DNA→assay of cytoplasm for cRBP; (□), antisense RAR DNA→assay of nucleus for RAR; (♦), sense cRBP DNA→assay of cytoplasm for cRBP; (◇), sense cRBP DNA treatment→assay of nucleus for RAR; (▲), no treatment with 16-hr incubation→assay of cytoplasm for cRBP; and (△), no treatment with 16-hr incubation→assay of nucleus for RAR. At the end of the treatment periods, the cells were assayed as described in the legend to Figure 2. Values are the means of three replicate determinations ± SD.

previously (Wickstrom et al., 1988). We show in Figure 2 that 30 μM antisense DNA to cRBP resulted in a 90% loss of [³H]retinol binding the cytoplasm over a period of 16 hr with a 78% loss even at 4 hr. Antisense-cRBP DNA had no effect on the binding of [³H]RA to RAR. Additionally, exposure of the cells to the sense DNA of cRBP mRNA did not alter either the cRBP or the RAR level. Treatment of cells with antisense DNA to RAR (Fig. 3), for 16 hr resulted in a >90% loss of RAR in nuclear extracts as determined by [³H]RA binding compared to untreated controls. Binding of [³H]retinol in the cytoplasmic frac-

tion was not significantly changed in either sense-cRBP DNA-treated cells or in the untreated controls over the time course of exposure to antisense-RAR DNA. An approximate half-life value, $t_{1/2}$, for cRBP and RAR of 1–1.5 hr was calculated from the above data.

Effect of DNA Concentration on SCC-25 Retinoid Receptor Levels

In order to determine the optimal concentration and specificity of the DNAs that resulted in a loss of either cRBP or RAR, we tested both antisense DNAs in the concentration range of 0.5 to 40 μM, and a sense/antisense double-strand hybrid as well as DNase-treated DNA controls. Figure 4 indicates that antisense-cRBP DNA effectively reduced cytoplasmic cRBP by 70–92% in a concentration range of 10–40 μM over a 4.5 hr treatment period; the levels of RAR were not changed by this treatment. Sense-cRBP DNA did not significantly affect the cRBP level in the cytoplasmic fraction. Moreover, the level of cRBP in the cytoplasmic fraction was not reduced in cells treated with antisense-cRBP DNA hybridized to the complementary sense DNA, nor was this antisense DNA effective when treated with DNase, while the efficacy of 30 μM antisense-cRBP DNA pretreated with inactivated DNase was not diminished.

Similarly, antisense-RAR DNA reduced (>90%) the level of nuclear RAR (Fig. 5), while antisense-cRBP DNA had no effect on the level of this receptor. The hybridized sense/antisense-cRBP DNAs and DNase-treated antisense-RAR DNA were not effective in reducing the level of [³H]RA binding in the nuclear extract fraction, while pretreatment of 30 μM antisense-RAR DNA with inactive DNase did not impair its efficacy (91% loss of RAR). Additionally, the level of cRABP, the cytoplasmic RA counterpart of cRBP, was not reduced by treatment with either antisense DNA (data not shown).

Retinoid Target Responses

We identified three targets that are involved in or a product of the retinoid signal process in human malignant keratinocytes: (1) retinol-dependent induction of AP (Reese et al., 1983), (2) retinoid-dependent morphogenesis, and (3) the clonogenic potential and rescue of cells from antisense DNA effects. Table 1 indicates that, when cells were treated with an antisense DNA to either cRBP or RAR, retinol-dependent induction of AP was not observed. Induction of AP in cells treated only with retinol was >245% of the antisense DNA-treated cells. Neither antisense-cRBP nor antisense-RAR treatment resulted in a reduction of basal enzyme levels.

In order to demonstrate the coupling of antisense inhibition of AP induction by retinoic acid to a primary signal response, in this case a G-protein response, we cultured cells as described above. We then treated these cells as outlined in Table 2 with both cholera

toxin [a G-stimulatory (G$_s$)-protein pathway inhibitor] or pertussis toxin [a G-inhibitory (G$_i$)-protein pathway inhibitor]. We found that blocking either protein reduced the induction of AP by retinoic acid. However, when correcting for the base level of AP, pertussis toxin was far more effective, suggesting that the primary coupling pathway for the signal to the modulation of gene expression by retinoic acid is through the G$_i$ system. Moreover, when the phosphorylation of proteins, both cytoplasmic and nuclear were evaluated (Fig. 6) we found that cells treated with a tumor promoter (okadaic acid) exhibited phosphorylation of both the CK-II protein (band at 62 kDa, Lane 2) and histone H1 (band at 15 kDa, Lane 2). However, when cells were pretreated with retinoic acid (Lanes 3 and 4) such phosphorylation was inhibited (other lanes are control systems). These data provide a strong link between the control of the retinoid receptors by antisense DNA and the *cdc* genes described earlier; this is strong evidence for retinoid regulation in cell division control and apoptosis.

In order to complete the paradigm of retinoid gene regulation and signal coupling, we looked at two oncogene responses in the system. We found that the induction of malignancy of normal human keratinocyes (Fig. 7A) to squamous cell carcinoma resulted in the polarization and up-regulation of c-*neu* (Fig. 7B), a tyrosine kinase. This response was paralleled by the modulation of protein phosphotyrosine (Fig. 7C) and polarization of cytoskeletal protein phosphotyrosine (Fig. 7D).

The treatment of SCC-25 with DNAs produced distinct morphological changes in malignant cells treated for 3.5 hr with either 30 μM antisense-RAR (Fig. 8A) or antisense-cRBP (Fig. 8B) relative to untreated controls. Cells treated with antisense-cRBP rounded up, and >90% lost cell–cell attachments with decreased adhesion to the substratum; cells treated with antisense-RAR went through similar changes but numerous large cells were refractory to the effects of this antisense DNA. Sense-cRBP (Fig. 8C) had no effect on normal cell morphology (Fig. 8D), supporting the specific effect of antisense DNAs. These

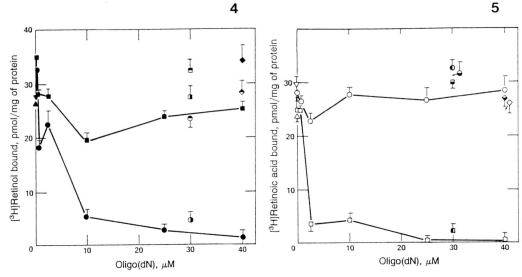

Fig. 4. Concentration-dependent effect of antisense cRBP and antisense RAR DNAs on cytoplasmic levels of cRBP in human malignant keratinocytes. Cells received the following treatments for 4.5 hr: (■), antisense RAR DNA; (●), antisense cRBP DNA; (♦), sense cRBP DNA; (◆), sense cRBP–antisense cRBP double-stranded DNA hybridized at 17% molar excess of sense cRBP (1.17/1.00 ratio of sense cRBP ODN to antisense cRBP DNA was heated to 95°C in phosphate-buffered saline and allowed to cool slowly to 25°C for a period of 3 hr, and the resulting hybrid was used to treat cells at a final concentration of 40 μM); (◐), inactive DNase-treated antisense cRBP DNA (DNase (EC 3.1.21.1) was boiled in buffer for 2 hr; 10 units of this enzyme was used to treat antisense cRBP at 37°C for 2 hr; this reaction mixture was used to treat cells at a final concentration of 30 μM); (◓), DNase-treated antisense cRBP DNA (10 units of DNase in buffer was used to treat antisense cRBP at 37°C for 2 hr; the reaction mixture was boiled for 2 hr and used to treat cells at a final concentration of 30 μM); (◩), DNase-treated antisense RAR DNA (treatment protocol was as for ◐); (◪), DNase-treated antisense hnRAR ODN (treatment protocol as for ◓); (▼), boiled DNase (final concentration of 10 units/ml of buffer-boiled DNase); and (▲), no treatment. At the end of the treatment period, cells were washed and frozen, and a cytoplasmic fraction was prepared and cRBP was assayed as described in the legend to Figure 2. Values are the means of three replicate determinations ± SD.

Fig. 5. Concentration-dependent effect of antisense cRBP and antisense RAR DNAs on nuclear levels of RAR in human malignant keratinocytes. Cells were grown as described; the nuclear pellet fraction was assayed for RAR as described in the legend to Figure 2. The cells received the following treatments for 4.5 hr: (□), antisense RAR DNA; (○), antisense cRBP DNA: (◇), sense cRBP DNA; (◆), sense cRBP–antisense-cRBP double-stranded DNA hybridized at 17% molar excess of sense cRBP DNA (see legend to Fig 4); (◐), inactive DNase-treated antisense cRBP DNA (see legend to Fig. 4); (◓), DNase-treated antisense cRBP DNA (see legend to Fig. 4); (■), inactive DNase-treated antisense RAR (treatment protocol was as for ◐): (◩), DNase- treated antisense RAR DNA (treatment protocol was as for ◓); (▽), boiled DNase (as described in legend to Fig. 4); and (△), no treatment. At the end of the treatment period, the cells were assayed as described in the legend to Figure 2. Values are the means of three replicate determinations ± SD.

changes were characteristic of the induction of apoptosis, as these cells excluded vital dyes

for periods exceeding 72 hr. Nevertheless, at some point, typically between 72 and 96 hr, these cells died.

We questioned whether the observed changes in cell morphology induced by antisense DNA treatments resulted in a *cell viability*-dependent receptor loss. The viability of cultures after exposure to the DNAs was not significantly reduced at the end of the 3.75 hr treatment period (Table 3). In contrast, clonal growth of cells treated with either antisense-cRBP or antisense-RAR (Table

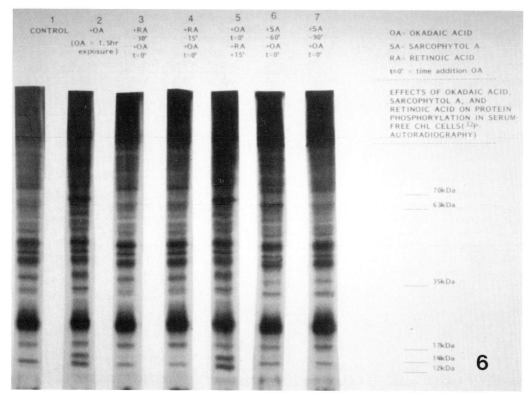

Fig. 6. Autoradiography of complete proteins treated with the tumor promoter okadaic acid (OA) and/or retinoic acid (RA). Lane 1, control; lane 2, OA; lane 3, RA prior to OA for 30 min; lane 4 RA prior to OA for 15 min; lane 5 OA prior to RA for 15 min; lane 6 OA prior to RA for 30 min; lane 7, sarcophytol A (SA) prior to OA as positive control.

3) and replated in either serum-containing or serum-free medium was reduced by ≥83% when compared to sense-cRBP treated cells or untreated controls. The choice of media for replating was a determinant, however, in the final colony count, where the presence of serum in the medium increased clonogenic potential by approximately 100%. In addition, Table 3 shows that by stringently washing antisense DNA-treated cells with serum-free medium at time points up to 1.75 hr after antisense DNA exposure, there was no loss of clonogenic potential. However, when the antisense DNA treatment period was >2.75 hr, washing was ineffective in preventing the loss of clonogenic potential. Moreover, when cell rounding was induced

using cytochalasin B or when cell stasis was induced in these cells with 0.05% NaN_3, no change in cRBP, cRABP, or RAR levels was observed (data not shown), further suggesting that the effect of the DNAs was specific and not the result of simple cytokinetic or cytostatic phenomena.

DISCUSSION

General

Our results strongly support the conclusion that antisense DNAs against portions of the 5′-untranslated regions of human cRBP and RAR mRNAs specifically down-regulate their respective retinoid receptor levels in a human malignant keratinocyte cell line.

TABLE 1. The Effect of Antisense-cRBP, Antisense-RAR, and Sense-cRBP DNAs on Retinol-Dependent Alkaline Phosphatase (AP; EC 3.1.3.1) Induction in SCC-25 Cells

DNA treatment[a]	Level of AP induction in SCC-25 cells, nmol PO₄ hydrolyzed/30 min/mg protein[b]	
	+ Retinol	− Retinol
Antisense-cRBP	28.8 ± 0.8	48.9 ± 0.1
Antisense-RAR	48.3 ± 4.2	47.9 ± 2.5
Sense-cRBP	105.4 ± 5.8	48.9 ± 2.6
Antisense-cRBP/sense-cRBP double-strand hybrid	97.1 ± 4.1	44.2 ± 3.2
Vehicle	115.0 ± 5.2	46.9 ± 7.3

[a]Standard AP induction was as follows: SCC-25 cells were cultured in 1.0 ml of serum-free medium and treated with retinol dissolved in 10 µl of ethanol. The cells were incubated for 3.5 hr at 37°C. At the end of the induction period, the treatment medium was removed, the cells were washed with 3 × 5 ml of ice-cold 0.85% NaCl solution, pH 7.2, and resuspended in 250 µl of AP assay buffer (50 mM Tris-HCl/2 mM Mg²⁺, pH 9.7). The cell suspension was sonicated at a setting of 40 with a *Sonic Sonifier* for 20 sec on ice followed by centrifugation for 8 × 10³ g-min. AP was assayed in 50–100 µl of the supernatant fraction using *p*-nitrophenyl phosphate substrate, 6 mM, in a 2 ml reaction volume (Reese et al., 1981). The DNA listed (30 µM) or the vehicle was also added to the cell culture 90 min prior to retinol treatment. The cells were incubated in the dark at 37°C for 3.5 hr. The retinol exposure period was terminated by aspiration of the treatment medium and washing of the cells as described above.
[b]Values were obtained by treatment of cells with the optimum level of retinol (5 × 10⁻⁶ M) and are the means ± SD of three replicate determinations.

Moreover, each receptor class is down-regulated independently in the absence of exogenous ligands to those receptors. Corresponding to the loss of either receptor class is the loss of retinol-inducible AP, a gene product shown to be under de novo regulation by retinoids (Ng et al., 1988). We also conclude that the expression of transglutaminase, a gene known to be under the control of the retinoic acid receptor holo-complex and assumed to be a marker for apoptosis (Yuspa et al., 1982), is in fact a gene that is expressed only when coupling exists between differentiation and apoptosis. This conclusion is based on the fact that in the SCC-25 cells used here, we have *eliminated* a functional receptor complex on which the transglutaminase gene depends and, yet, concomitantly induced apoptosis. Additionally, we suggest that the evidence (AP activity, G-protein requirements, and phosphorylation of CK-II and histone H1) for the coupling of the primary signal response in cells and gene expression (antisense DNA response and inhibition of AP gene induction) provides numerous key points for combining treatment modalities using chemotherapy and antisense DNA in combating malignant disease.

The selection of retinol for the induction of AP addresses the possible irreversible oxidation of retinol to RA in these cells. In-

TABLE 2. The Effect of G-Protein Modulating Agents on Retinoic Acid Inducibility of Alkaline Phosphatase (AP; EC 3.1.3.1) in Cultured Cells

Agent/protocol[a]	nmol *p*-NPP hydrolyzed/min/ mg protein[b]	
	2.5 hr	4.5 hr
No treatment	11.6	11.9
Retinoid acid (5 × 10⁻⁶ M)	—	46.6
PT(3.0 hr) > RA (4.5 hr)[c]	—	14.2
CT(1.0 hr) > RA (4.5 hr)	—	20.5
GDP-β-S(1.0 hr) > RA (4.5 hr)	—	15.9

[a]Abbreviations: *p*-NPP, *para*-nitrophenylphosphate; RA, retinoic acid; PT, pertussis toxin; CT, cholera toxin.
[b]All dishes contained 2 × 10⁶ cells ± 8%.
[c]Control values are as follows: PT (10 µg/plate), 3.0 hr = 14.2; CT (10 µg/plate), 1.0 hr = 20.5; GDP-β-S (20 µM), 1 hr = 11.3.

volvement of this metabolic pathway may be required for the ultimate transduction of the retinoid signal cascade and may involve interaction between cRBP with RAR or either receptor acting alone. In fact, the apparent differential morphological effect of antisense-cRBP DNA versus -RAR may be evidence of such pathway divergences. Conversely, one may predict the operation of nuclear receptors for retinol whose function is consolidated with RAR, also accounting for the differential morphogenic response or that RAR requires both retinol and RA or modifying its interaction with chromatin. Whether AP expression is an integral part of the retinoid signal cascade and, thus, necessary for maintenance of the differentiated state of cells remains to be determined.

Our results also indicate that clonal growth of human malignant keratinocytes requires sustained levels of both the cytoplasmic and nuclear retinoid receptors. We estimate from the above data that the critical level of receptors is about 20–25% that of untreated cells after a >2.75 hr exposure period beyond which a cell commitment pathway is induced (apoptosis), which ultimately results in irreversible terminal differentiation and/or cell death (Tomei et al., 1988). Thus, assay for such receptors, like steroid hormone receptors, may be critical in defining the selection of antisense DNA for the treatment of malignancy. Like the human SCC-25 cells here (which, contrary to dogma, require the presence of RAR-α type receptor for the malignant phenotype) we suggest this may well be the case for certain types of breast cancer, as there is an established link between the expression of the progesterone receptor and vitamin A response elements.

Previously, we reported that the response of Chinese hamster lung cells to retinoids was composed of at least two distinct phases, an immediate signal-transduction or *primary response* phase, coupled with a later *cell-commitment* phase. The primary response phase appeared to occur within 5 min when these cells were exposed to physiological levels of either retinol ($1 \times 10^{-6} M$) or retinoic acid ($5 \times 10^{-9} M$). This phase was marked by

changes in the accumulation (400%) of IP_4, phosphorylation of nucleolin, and a sensitivity to G-protein inhibitors (Cope et al., 1988; Fujiki et al., 1988; Issinger et al., 1988). We have shown in this study that cRBP (and probably cRABP) compartmentalized in the cytoplasm is not just a shuttle system for retinol (or RA), but is intimately involved with or is a component feature of a retinoid-dependent cell maintenance system, as evidenced by cRBP down-regulation and the resulting refractoriness of malignant human keratinocytes cells to retinol-dependent induction of AP. Additionally, these data suggest a tight coupling of the primary response with the cell-commitment phase in these cells. Thus, these observations add clinical significance to the assay of these proteins in biopsy material. Moreover, these proteins, too, become a target for combination antisense DNA therapies.

There are also strong implications from our data about the role of retinoids in the process of carcinogenesis. Retinoids are particularly potent antipromoting agents. They are antagonists to both the phorbol ester-type tumor promoters as well as the non-phorbol ester-type tumor promoters, e.g., okadaic acid (Cope et al., 1988). Recent data also demonstrate that retinoids are highly effective against initiating agents such as dimethylbenz[a]anthracene (Zile et al., 1986). We propose that retinoid binding proteins such as cRBP and cRABP are factors in the suppression of tumor promotion where the primary response is central to retinoid antipromoter activity. We further suggest that this phase is most likely coupled to the cell-commitment phase involving RAR proteins, providing the efficacy of retinoids against initiators or complete carcinogens such as dimethylbenz[a]anthracene or methylnitrosourea. Thus, for example, if the expression of distinct homeobox gene sets (or some subsets) (1) is repressed by carcinogen treatment, or (2) not completely integrated with the activity of cRBP, cRABP, or RAR, cells will be refractory to the morphogenic or differentiating potential of retinoids. Conversely, we suggest that treatment of cells with retinoids may

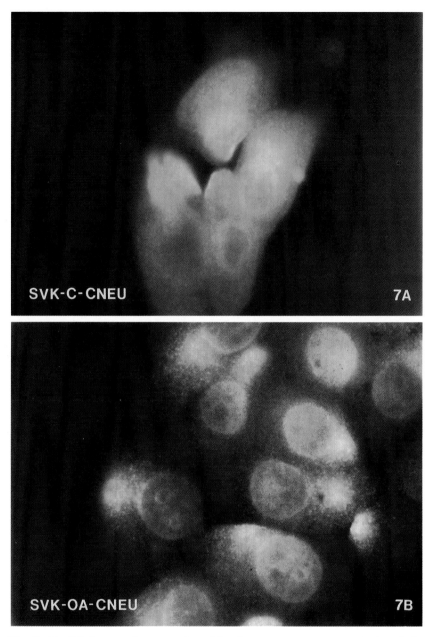

Fig. 7. (A–D) Immunofluorescent probing for c-*neu* oncogene and phosphoty-rosine protein residues (PTYR) in transformed human keratinocyte (SVK) control cultures (C) and okadaic acid-treated cells (OA).

induce coupling of the primary response and the cell-commitment phase, which may then result in the constitutive expression of "anticarcinogenic" homeobox genes and their persistent messages (a feature of retinoid bi-ological $t_{1/2}$). The presence and/or amount of such messages, their products, or their control of suppressor elements may be the defining feature(s) of retinoid-responsive malignancy. Additionally, the deletion of retinoid

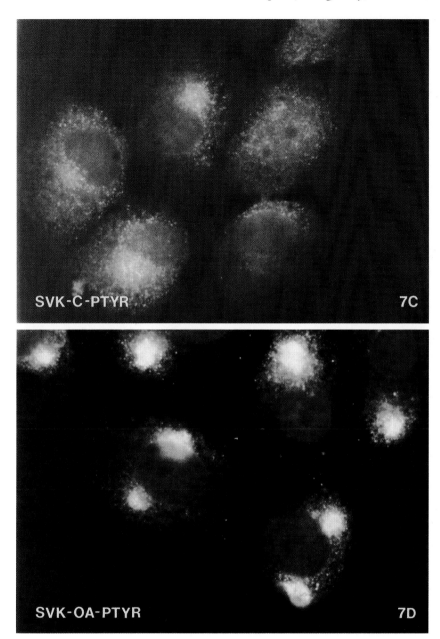

receptor genes from immortalized cells results in an apoptotic effect in human squamous carcinoma cells, suggesting that at least one *"apoptosis gene"* is temporally under the regulation of the holo-RA receptor in these cells and that de novo gene expression for maintenance of the *"immortality"* of these cells is required. Additionally, RA function has been shown to be highly integrated with epidermal growth factor receptor (EGF) maintenance and the effect of phorbol esters, known protein kinase-C (PK-C) activators, as well as non-PK-C protein phosphorylation-inducing agents, such as okadaic acid (OA).

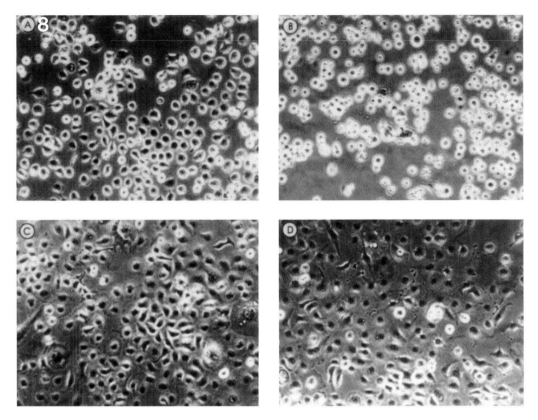

Fig. 8. (A–D) Morphological changes induced in human squamous cell carcinoma (SCC-25) by antisense DNAs to RAR and CRBP. (**A**) 30 μM antisense RAR; (**B**) 30 μM antisense CRBP; (**C**) 30 μM sense CRBP; (**D**) untreated control.

The action of all these modulators is coupled to changes in gene cassette expression through primary signal systems involving G-proteins and inositol phosphate (IP) metabolism, and thus, probably through calcium translocation signaling or abrogation of the calcium signal, which appears to play a significant role in the apoptotic process.

What is most important about such observations is that the integration points of the retinoid receptors with signal systems increases the potential sites for multidrug and antisense DNA therapies. What remains to be established is the extent of expression of each of the retinoid receptors in the normal vs. malignant tissue, and how the selection for either a retinoid or antisense DNA will proceed. Nevertheless, current technology is

more than adequate to determine both quantitative and qualitative values for retinoid receptors, protein phosphorylation, and integral proteins such as c-*ras* and c-*neu* (and other oncogenes and apoptotic proteins). Stable antisense derivatives and clinical evaluation are primary goals of our efforts and we currently await our first in vivo results.

SUMMARY

Retinoid control of cytodifferentiation is dependent upon the coupling of a retinoid-dependent signal cascade system to the temporal expression of one or more retinoid receptor proteins. These include the cytoplasmic binding proteins retinol(I)-binding protein and retinoic acid-binding protein

TABLE 3. Clonogenic Potential and Viability of SCC-25 Cells Following Exposure to Antisense DNAs

	Colonies formed relative to untreated control[a]		
	Complete media with serum	Complete media without serum	
DNA treatment	3.75 hr exposure[b]	3.75 hr exposure[c]	1.75 hr exposure washed out and additional 1.0 hr incubation[d]
Antisense-cRBP	0.14 ± 0.06 (97)	0.03 ± 0.02 (89)	0.96 ± 0.05 (90)
Antisense-RAR	0.08 ± 0.04 (94)	0.17 ± 0.08 (89)	0.96 ± 0.07 (94)
Sense-cRBP	1.08 ± 0.04 (97)	0.93 ± 0.04 (94)	0.94 ± 0.05 (95)
No treatment	1.00 ± 0.05 (97)	1.00 ± 0.03 (96)	1.00 ± 0.08 (89)

[a]Values are expressed as the mean decimal fraction of the control for the experimental protocol in each column for six replicate determinations ± SD. Values in parentheses are percentage viable cells at plating where viability was determined by trypan blue exclusion.

[b]Subconfluent (70–80% confluent) cultures of cells were treated as described with 40 μM DNA or received no treatment. At the end of the 3.75-hr treatment period, cells were washed with 3 × 5 ml of isotonic saline, pH 7.4; cells were removed from the dish using 20 mM EDTA/0.5% trypsin in Eagle's balanced salt solution. The cell suspension was centrifuged for 3 × 10^3 g-min and the supernatant was aspirated from the vial. The cells were washed two times with complete MCDB 153 medium containing 10% fetal bovine serum, counted, and plated with this medium in a 60-mm Falcon dish at a density of 3000 cells/dish. The cells were cultured for 7 days, rinsed with isotonic saline, pH 7.4, fixed in the dish with 5% formaldehyde, and stained with 0.2% crystal violet (Wille et al., 1985). The colonies were counted using an automated Artek model 870 colony counter set for colony inclusion at greater than seven cells per colony. The "no treatment" control value (1.00) = 610 ± 3 colonies per dish.

[c]The protocol for these cells was identical to footnote b above. However, at a time point of 3.75 hr of treatment, the DNA was washed out of the dish using 3 × 5 ml of basal MCDB 153 medium without serum. The cells were incubated an additional 2.0 hr in the DNA-free medium, removed from the plate, replated, terminated at 7 days, and colonies were counted as described above. The "no treatment" control value (1.00) = 298 ± 9 colonies/dish.

[d]The protocol for these cells was identical to footnote b above. However, at a time point of 1.75 hr of treatment, the DNA was washed out of the dish using 3 × 5 ml of basal MCDB 153 medium without serum. The cells were incubated an additional 2.0 hr in the DNA-free medium, removed from the plate, replated, terminated at 7 days, and colonies were counted as described above. The "no treatment" control value (1.00) = 250 ± 20 colonies/dish.

(cRBP-I and cRABP), a membrane-associated RABP (mRABP), and a family of nuclear retinoic acid (RA) steroid hormone-like receptors (RAR-α, -β, and -γ). Additionally, an altenative pathway involves two retinoic acid-sensitive transcription factors (RXR-α and -β) that are unrelated to the aforementioned family of proteins. The action of the cytoplasmic proteins appears to be associated with two aspects of cell function and differentiation programs: (1) the primary signal cascade, and (2) the transport of retinoids to the cell nucleus with resultant "activation" of the nuclear retinoid receptors. Conversely, the nuclear receptors appear to transduce and coordinate a retinoid-dependent fraction of the genetic program required for either terminal differentiation *or* proliferation. The path of any cell, as a function of exposure to a retinoid, would thus depend on which gene sets were appropriately expressed and coordinated with other regulated gene sets in the cell at any given time point. Clinically, this concept can be exploited for evaluating tumors and malignant disease with regard to the possibility of response by a tumor to defined therapeutic protocols. These therapies may vary depending upon the temporal aspects within a given tumor type or with the variance in gene expression between tumor types, e.g., whether or not markers for malignancy and tumor type are activated or repressed by an activated retinoid receptor. We have presented one practical case in which antisense-DNA deletion of the α-type retinoic aid receptor from a human squamous cell carcinoma resulted in the induction of programmed cell death (*apoptosis*) in this tumor. These data suggest that at least one gene regulating the terminal differentiation and/or ultimate gene-directed death phase of this tumor type is under the regulation of or is tightly coupled to the α-type receptor. Additionally these results were discussed in light of the suggested role of retinoids in modulating malignant disease, both positively (malignant melanoma) and negatively (malignancy of the breast). It is clear that the use of stable antisense DNA in vivo, focused on the expression of temporally and spatially selected retinoid receptors, can be of clinical value for both the treatment and diagnosis of malignant disease.

REFERENCES

Ashendel A, Boutwell R (1981): Direct measurement of specific binding of highly lipophilic phorbol diester to mouse epidermal membranes using cold acetone. Biochem. Biophys. Res. Commun. 99:543–551.

Basu DK, Schnetkamp P (1989): Cation fluxes via Na-Ca exchange in isolated bovine rod outer segments. In Hatuse O, Wang J (eds): Bioinformatics. New York: Elsevier, pp 137–140.

Bedo G, Santisteban P, Aranda A (1989): Retinoic acid regulates growth hormone gene expression. Nature (London) 339:231–234.

Bradford M (1976): A rapid and sensitive method for the quantitation of microgram quantities of protein utilizing the principle of protein-dye binding. Anal. Biochem. 72:248–257.

Brand N, Petkovich M, Gaub M, Chambon P (1988): Identification of a second human retinoic acid receptor. Nature (London) 332:850–852.

Bursch W, Lauer B, Timmermann-Trosiener I, Barthel G, Schuppler J, Schulte-Hermann R (1984): Controlled death (apoptosis) of normal and putative preneoplastic cells in rat liver following withdrawal of tumor promoters. Carcinogenesis 5:453–459.

Chytil F (1985): Retinoid intracellular binding proteins and mechanism of action. In Saurat JH (ed): Retinoids—New Trends in Research and Therapy. Basel, Switzerland: Karger, pp 20–27.

Clouston W, Kerr J (1985): Apoptosis, lymphotoxicity and the containment of viral infections. J. Med. Hypotheses 18:399–410.

Columbano A, Ledda-Columbano G, Coni P, Faa G, Liguori C, Santa Cruz G, Pani P (1985): Occurrence of cell death (apoptosis) during the involution of liver hyperplasia. Lab. Invest. 52:670–676.

Cope F (1989): The cytodifferentiating activity of retinoic acid is regulated by IP₄, protein phosphorylation, and G-protein-dependent transduction systems: Keys to the universal function of retinoids. In Hatuse O, Wang J (eds): Bioinformatics: Signal Processing. New York: Elsevier, pp 141–144.

Cope F, Wille J (1989): Retinoid receptor antisense DNA's inhibit alkaline phosphatase induction and clonogenicity in malignant keratinocytes. Proc. Natl. Acad. Sci. U.S.A. 86:5590–5594.

Cope F, Knox K, Hall R (1983): The *in vitro* metabolism of retinoids in the presence of their binding proteins by rat testes interstitial cell fractions. Nutr. Rep. Int. 29:1209–1217.

Cope F, Knox K, Hall R Jr. (1984a): Rat testes interstitial cell nuclei exhibit three distinct receptors for retinoic acid. Experientia 40:276–278.

Cope F, Knox K, Hall R (1984b): Retinoid binding to nuclei and microsomes of rat testes interstitial cells: I-Mediation of retinoid binding by cellular retinoid-binding proteins. Nutr. Res. 4:289.

Cope F, Fujiki H, Issinger O (1988): The antitumorigenic action of retinoids is medicated through a G-protein-dependent signal system in CHL cells. Proc. Am. Assoc. Cancer Res. 29:156.

Cowan W (1984): Regressive events in neurogenesis. Science 225:1258–1264.

de The H (1987): A novel steroid thyroid hormone receptor-related gene inappropriately expressed in human hepatocellular carcinoma. Nature (London) 330:667–669.

DeLuca LM, Creek K, Clifford AJ, Rimoldi D (1985): Similarities and differences between retinylphosphate mannose and dolichyl phosphate mannose biosynthetic processes. In Saurat J (ed): Retinoids-New Trends in Research and Therapy. Basel, Switzerland: Karger, pp 28–34.

Duncan A, Heddle J (1984): The frequency and distribution of apoptosis induced by three non-carcinogenic agents in mouse colonic crypts. Cancer Lett. 23:307–314.

Fujiki H, Cope F, Issinger O (1988): Sarcophytol A, a novel tumor promoter inhibitor, prevents phosphorylation of cell cycle protein-23 induced by the non-TPA-type tumor promoter, okadaic acid. Proc. Am. Assoc. Cancer Res. 29:155.

Gaub M, Lutz Y, Ruberte E, Petkovich M, Brand N, Chambon P (1989): Antibodies specific to the retinoic acid human nuclear receptors alpha and beta. Proc. Natl. Acad. Sci. U.S.A. 86:5618–5622.

Giguere V, Ong E, Sugui P, Evans R (1987): Identification of a receptor for the morphogen retinoic acid. Nature (London) 330:624–626.

Hames BD, Rickwood D (1981): Gel Electrophoresis of Proteins. London: IRL Press, pp 23–73.

Hendy J, Potten C (1982): Intestinal cell radiosensitivity: A comparison of cell death assayed by apoptosis or by loss of clonogenicity. J. Radiat. Biol. 42:621–629.

Hollenberg MD, Cuatrecasas P (1975): Inositol phosphate changes in outer rod segments. J. Biol. Chem. 250:3845–3535.

Issinger O, Cope F, Fujiki H (1988): Retinoic acid blocks the phosphorylation of nuclear cell cycle protein-23 induced by the non-TPA-type tumor promoter okadaic acid in Chinese hamster lung cells cultured in serum-free medium. Proc. Am. Assoc. Cancer Res. 29:156.

Kerr J, Wyllie A, Currie A (1973): Apoptosis: A basic biological phenomenon with wide-ranging implications in tissue kinetics. Br. J. Cancer 25:239–247.

Lee Y, Dewey W (1986): Protection of CHO cells from hyperthermic killing by cycloheximide or puromycin. Radiat. Res. 106:98–104.

Martin D (1988): Inhibitors of protein synthesis and RNA synthesis prevent neuronal death caused by nerve growth factor deprivation. J. Cell. Biol. 106:829–836.

Mavillo F (1988): Activation of four homeobox gene clusters in human embryonal carcinoma cells induced to differentiate by retinoic acid. Differentiation 37:73–79.

Mengelsdorf DJ, Ong E, Dyck J, Evans RM (1990): Nuclear receptor that identifies a novel retinoic acid response pathway. Nature (London) 345:224–228.

Mueller P, Pugh EN (1983): Phosphorylation-dephosphorylation in the visual cycle. Proc. Natl. Acad. Sci. U.S.A. 80:1892–1896.

Ng KW, Gummer P, Michelangeli V, Bateman J, et al (1988): Regulation of alkaline phosphatase expression in a neonatal rat clonal calvarial cell strain by retinoic acid. J. Bone Mineral Res. 3:53–61.

Pelech S (1990): CDC genes-universal pacemakers in cells. The Sciences 3:23–27.

Reese D, Fiorentino G, Claflin A, Malinin T (1981): Rapid induction of alkaline phosphatase activity by retinoic acid. Biochem. Biophys. Res. Commun. 102:315–319.

Reese D, Gordon B, Gratzne RH, Claflin (1983): Effect of retinoic acid on the growth and morphology of a prostatic adenocarcinoma cell line cloned for the retinoid inducibility of alkaline phosphatase. Cancer Res. 43:5443–5448.

Saunders J (1966): Death in embryonic systems. Science 154:604–606.

Saurat J (1985): Retinoids: New Trends in Research and Therapy. Basel, Switzerland: Karger, pp 59–62, 68–75, 314–320.

Schmechel D, Goldgaber D, Burkhart D, Gilbert J (1988): Alzheimer's Disease. Los Angeles: West Geriatric Res Grp, pp 96–111.

Sporn M, Roberts AV, Heine UI, Roche NS (1985): Retinoids and differentiation of cells of mesenchymal origin. In Saurat J (ed): Retinoids—New Trends in Research and Therapy. Basel, Switzerland: Karger, pp 35–39.

Tomei LD, Kanter P, Wenner C (1988): Inhibition of radiation-induced apoptosis in vitro by tumor promoters. Biochem. Biophys. Res. Commun. 155:324–330.

Valerie K (1988): Activation of human immunodeficiency virus type 1 by DNA damage in human cells. Nature (London) 333:78–85.

Wickstrom EL, Bacon TA, Gonzalez A, Freeman DL, Lyman GH, Wickstrom E (1988): Human promyelocytic leukemia HL-60 cell proliferation and c-myc protein expression are inhibited by an antisense pentadecadeoxynucleotide targeted against c-myc mRNA. Proc. Natl. Acad. Sci. U.S.A. 85:1028–1032.

Wille J, Pittelkow M, Scott R (1985): Normal and transformed human prokeratinocytes express divergent effects of a tumor promoter on cell cycle-mediated control of proliferation and differentiation. Carcinogenesis 6:1181–1189.

Wolbach SB, Howe PR (1925): Tissue changes following deprivation of fat-soluble vitamin A. J. Exp. Med. 42:753–777.

Wyllie A (1980): Glucocorticoid-induced thymocyte apoptosis is associated with endogenous endonuclease activation. Nature (London) 284:555–557.

Yuspa S, Ben T, Steinhart P (1982): Retinoic acid induction of epidermal transglutaminase inhibits cornification. J. Biol. Chem. 257:9906–9908.

Zile, MH, Cullum M, Roltech I, deHoog J, Welsch C (1986): Effect of moderate vitamin supplementation and lack of dietary vitamin A on the development of mammary tumors in female rats treated with low dose DMBA. Cancer Res. 257:3495–3503.

Antisense Oligonucleotides: A Possible Approach for Chemotherapy of AIDS

Sudhir Agrawal

Worcester Foundation for Experimental Biology, Shrewsbury, Massachusetts 01545

INTRODUCTION

Gene expression in cells is normally regulated by DNA-binding proteins, repressors, and activators. In the early days of protein synthesis, a new finding indicated that complementary nucleic acids could also modulate gene expression. The observation was based on the fact that tRNA can hybridize with the mRNA by a trinucleotide at the anticodon loop (an antisense trinucleotide) and translate the language of the gene into that of protein (Hoagland et al., 1958, 1959). This was again confirmed by the finding of chain termination anticodon triplets.

Later, it was realized that this type of antisense genetic strand plays an important role in regulating gene expression in some natural systems. It was found that RNA complementary to primer RNA regulates plasmid replication in bacteria (Lactena and Cesarani, 1981; Tomizawa and Itoh, 1981). Then "antisense" RNA complementary to regions of mRNA was demonstrated to suppress translation and, hence, gene expression in prokaryotes (Mizuno et al., 1983).

The concept that synthetic oligonucleotides might regulate gene expression was first suggested in studies carried out using diribonucleotides carrying alkylating agents (Belikova et al., 1967) and short nonionic oligonucleotides (Barrett et al., 1974; Miller et al., 1974). Subsequently, it was reported that cell-free translation could be inhibited by annealing full length, complementary DNA to the mRNA (Hastie and Held, 1978).

However, Zamecnik and Stephenson were the first to propose the use of synthetic oligonucleotides for therapeutic purposes (Zamecnik and Stephenson, 1978; Stephenson and Zamecnik, 1978). They showed that a 13-mer oligonucleotide, complementary to Rous sarcoma virus RNA, inhibited the growth of this virus in cell culture.

From 1978 to 1984, not much progress was made in the use of synthetic oligonucleotides to regulate viral genes or cellular gene expression. The main reasons were (1) there was doubt that oligonucleotides get into the cell and, if so, what the mechanism was, and (2) the chemistry for synthesizing oligonucleotides was not established fully for everyday oligonucleotide use. Solid phase chemistry for synthesizing oligonucleotides was at a developmental stage.

However, since 1985, there have been steadily increasing reports of the use of synthetic oligonucleotides to regulate gene expression, for both viral as well as cellular genes. This review will focus on the use of synthetic oligonucleotides, both unmodified (normal) as well as phosphate backbone-modified analogues, to inhibit human immunodeficiency virus replication in tissue culture. Also acute animal toxicity of three types of oligonucleotides: an unmodified, and its two phosphate backbone-modified analogues, phosphorothioate and phosphomorpholidate will be discussed. In general, the antisense oligonucleotide field has been reviewed elsewhere (Van der Krol et al., 1988; Uhlmann and Peyman, 1990).

Prospects for Antisense Nucleic Acid Therapy of Cancer and AIDS, pages 143–158

ANTISENSE OLIGONUCLEOTIDE TARGETS FOR AIDS THERAPY

The targets for antisense oligonucleotides are in general single-stranded RNA molecules such as mRNA or viral genome RNA. Keeping this in mind one can look at the HIV viral cycle and the genetic map for targets. HIV, a member of the retrovirus family, is an enveloped single-stranded RNA virus, the genome of which exists as a dimer of identical molecules. At least nine viral genes have been identified thus far, including those encoding structural function (*gag, pol, env*); the *pol* gene encodes several enzymatic functions critical for HIV replication including proteinase, integrase, RNase H, and reverse transcriptase), those regulating transcription and RNA nuclear transport (*tat, rev, nef*), and those serving ancillary functions in modulating viral infectivity (*vif, vpr, vpu*).

A simplified version of the HIV replication cycle and the steps involved is shown in Figure 1. After HIV enters the cell, the virus releases its functional RNA genome into the cytoplasm as a ribonucleoprotein complex— a process known as uncoating. This single-stranded viral genomic RNA becomes the first target for an antisense oligonucleotide.

After viral entry and uncoating, viral RNA is used as a template for proviral DNA synthesis, catalyzed by a viral DNA polymerase called reverse transcriptase. RNase H specifically degrades the RNA template of an RNA–DNA hybrid to permit subsequent synthesis of double-stranded viral DNA. Reverse transcribed single-stranded DNA is a possible target for "sense oligonucleotide."

After synthesis of double-stranded DNA, the viral DNA can migrate into the nucleus by an as yet poorly characterized mechanism. Some double-stranded DNA (proviral) may be circularized, although circularization apparently plays little or no role in integration.

After integration of the HIV genome into the host cell genome and a variable latency period, the proviral DNA is transcribed to mRNA by host RNA polymerase. Viral RNA can be translated directly, or undergoes splicing before translation, to produce viral proteins. The mRNAs are the main targets for antisense oligonucleotides. Other possible target sites are RNA sequences that are involved in viral maturation, and viral packaging.

Based on the availability of single-stranded nucleic acids that are involved in the viral replication cycle, several targets have been selected to examine the inhibition of HIV replication by antisense oligonucleotides. Specifically the sites that have been studied so far are (1) 5'-cap site; (2) 5'-noncoding region near the cap structure, (3) AUG initiator codons for different genes, (4) splice acceptor sites, (5) splice donor sites, (6) poly(A) signal sites, (7) binding site for tRNA primer, (8) immediately upstream of tRNA binding site, (9) middle of the 5'-repeated regions, and also several areas in the TAR region. TAR is the position of a transactivator responsive region that is essential for transcription stimulation by the tat protein.

IN VITRO SYSTEMS FOR ASSAYING HIV-1 INHIBITION

The screening of oligonucleotides and their various analogues was facilitated by the availability of rapid and sensitive assay systems (Sarin et al., 1985, 1987). The level of inhibition of HIV-1 expression in the presence of oligonucleotides was monitored by four parameters: (1) syncytial formation, (2) p17 expression, (3) p24 expression, and (4) reverse transcriptase activity. The cytotoxic effect of oligonucleotides was studied by the trypan blue dye exclusion method.

In short, 5×10^5 cells ml^{-1} (H-9 or MOLT-30 cells) were infected with $2.5–5 \times 10^8$ virus particles of HIV-1 (HTLV-IIIb or HTLV-IIIc). Oligonucleotides were added at different concentrations only once, either at 0 hr (simultaneously with the virus) or 4 to 48 hr postinfection. Cells were washed with PBS to wash off free virus particles. The cells were cultured for a further 96 hr in a humidified atmosphere containing 5% CO_2 at 37°C. The cells and supernatant were then examined for the level of HIV-1 expression by measuring syncytial formation, viral p17 and p24

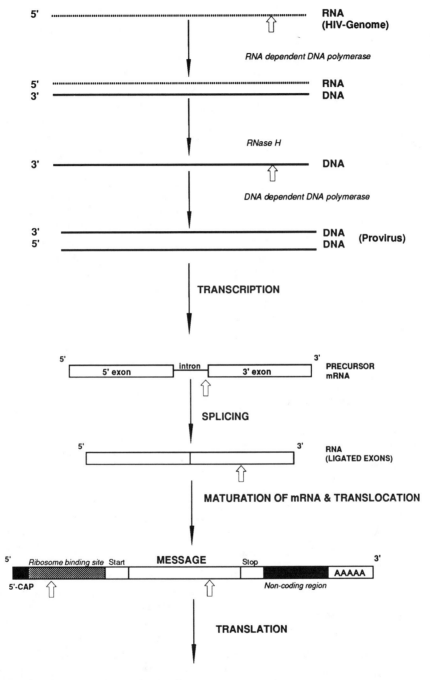

Fig. 1. Simplified version of important steps in the HIV life cycle and potential targets for antisense oligonucleotides. The arrow shows the steps at which single-stranded RNA or DNA is available for binding to antisense oligonucleotides.

antigens expression, and reverse transcriptase activity, as well as cell viability.

The above mentioned four parameters give similar levels of inhibition of HIV expression, which ensures reliability and reproducibility (Figs. 3 and 5). The inhibition of HIV-1 expression has been reported in percentage, obtained by dividing the value of the parameter in control-infected cells (no oligonucleotide added) to infected cells in the presence of oligonucleotide.

In independent studies, other in vitro systems have been used, which include plaque assay (Shibahara et al., 1989), radioimmunoprecipitation assay (RIPA) of viral antigens, and RNA analysis (Matsukura et al., 1989).

PROPERTIES OF ANTISENSE OLIGONUCLEOTIDES

Antisense oligonucleotides must have the following properties.

1. *Specificity*: The interaction between oligonucleotide and its target nucleic acid must be by specific Watson–Crick base pairing. Also the duplex formed between oligonucleotide and its target must be sufficiently stable under physiological conditions.

2. *Stability*: The oligonucleotide must have a sufficiently long half-life under in vivo conditions for it to be able to hydridize to its target.

3. *Cellular Uptake*: The oligonucleotide must be able to pass through the cell membrane to reach its site of action.

INHIBITION OF HIV BY OLIGONUCLEOTIDES

Essentially three types of oligonucleotides have been studied to inhibit HIV replication in tissue culture.

1. Oligodeoxy- or oligoribonucleotides
2. Oligonucleotides having modified sugar-phosphate backbone
3. Oligonucleotides carrying "conjugates"

Oligodeoxy- or Oligoribonucleotides

Approximately 20 different areas of the HIV genome were selected as targets of highly conserved primary structure for the initial screening of oligonucleotides to inhibit HIV replication in tissue culture (Zamecnik et al., 1986; Goodchild et al., 1988). Oligonucleotides were synthesized using phosphoramidite chemistry. All oligonucleotides were 20-mers, considering that (1) an oligomer over 16 units in length would be unlikely to occur at random in human genome (one chance in 4×10^9) and (2) melting temperature (T_m) under physiological conditions would be around 54–58°C, well above the 37°C incubation temperature. In general, the ID_{50}, as assayed by syncytial formation and p24 expression, was in the range of 20–100 μg ml^{-1} of unmodified oligonucleotides (3.5 to 16 μM) when oligonucleotide was added simultaneously with the virus. The splice acceptor site for tat (5349–5368) was the most active site (Fig. 3).

A mismatched oligonucleotide showed no activity. In general, at this concentration oligonucleotides showed no cytotoxicity.

When unmodified oligonucleotides were added to tissue culture at postinfection time, the inhibitory dose was much higher, compared to simultaneous infection (Fig. 4). At 4 and 8 hr postinfection, the ID_{50} of oligonucleotide was about five times and eight times higher, respectively. At 24 and 48 hr postinfection, the inhibitory dose was more than 200 μg ml^{-1} oligonucleotide. At this dose the oligonucleotides were found to be cytotoxic.

Oligoribonucleotides have also shown inhibitory properties similar to those of oligodeoxynucleotides, but they are less effective in inhibiting HIV replication (Agrawal and Tang, 1990; Agrawal et al., 1991). The high effective dose of oligonucleotides required at postinfection time may be because oligonucleotides are susceptible to digestion by extra- or intracellular nucleases. To overcome that problem, 2′-OMe oligoribonucleotides have been studied to inhibit HIV replication (Shibahara et al., 1989). 2′-OMe oligoribonucleotides are less susceptible to several nucleases (Dunlap et al., 1971). However, 2'-OMe oligoribonucleotides, complementary to the tRNA primer binding site and the splice acceptor site of HIV *tat* gene, failed to inhibit HIV replication.

Oligonucleotides Having Modified Sugar Phosphate Backbone

As discussed above, the susceptibility of the phosphodiester linkage in unmodified oligonucleotides to degradation by nucleases would be expected to reduce their potency and in vivo persistence, limiting their effectiveness as antiviral agents (Wickstrom, 1986). To overcome this problem, several sugar-phosphate backbone analogues of oligonucleotides have been studied. However, the modification often has an adverse effect on solubility, specificity, duplex stability, cellular uptake, and mechanism of their inhibitory effect.

Several phosphate backbone-modified analogues of oligonucleotide have been studied for their anti-HIV activity.

Methylphosphonates. In methylphosphonates, the negatively charged phosphate oxygen is replaced by a methyl group, which is neutral and achiral (Fig. 2). This analogue is resistant to nucleases, and forms a less stable duplex with complementary DNA/RNA than unmodified DNA at physiological pH and salt concentration (Froehler et al., 1988; Agrawal et al., 1989a; Quartin and Wetmur, 1989). The duplex of methylphosphonate oligonucleotides and RNA is not a substrate for RNase H (Furdon et al., 1989, Agrawal et al., 1990a). Methylphosphonates can be assembled using methylphosphonamidite chemistry (Agrawal and Goodchild, 1987).

Several oligonucleotides (20-mers), containing methylphosphonate linkages have been tested for their anti-HIV activity. Since methylphosphonates are not a substrate for RNase H, oligonucleotides methylphosphonates were synthesized complementary to the splice acceptor site of HIV *tat* gene, to inhibit splicing (Sarin et al., 1988). Oligonucleotides containing all methylphosphonate linkages were effective at a concentration of 3 μM, inhibiting syncytia formation and p24 synthesis by 98 and 100%, respectively. The activity of oligonucleoside methylphosphonate was chain length dependent, a 15-mer being less effective than the 20-mer. The same oligonucleotide (20-mer) containing only a few methylphosphonate linkages was less effective.

However, in another study, a 16-mer methylphosphonate oligonucleotide was found inactive against HIV at a concentration of 25 μM (Matsukura et al., 1987), but an 8-mer was effective at 100 μM concentration against HIV (Zaia et al., 1988).

As it is evident from the results, the effectiveness of oligonucleoside methylphosphonates is chain length dependent. However, an oligonucleotide 20-mer containing all internucleotide linkages as methylphosphonate is not very soluble in aqueous media.

Phosphoramidates. Phosphoramidates have an internucleotide linkage containing a nitrogen-phosphorus bond (Fig. 2). These analogues have properties similar to methylphosphonates, described above, except the duplex stability of phosphoramidates is weaker than methylphosphonates, which depend on the size of the amine molecule attached at the internucleotide linkage (Froehler et al., 1988; Agrawal et al., 1989a).

Three different phosphoramidates—phosphomorpholidate, phosphobutylamidate, and phosphopiperazidate—have been studied for their anti-HIV activity (Agrawal et al., 1988). The oligonucleotides studied were complementary to a splice donor site and a splice acceptor site. Oligonucleotide phosphoramidates were effective in inhibiting HIV at a concentration of 0.5 to 1 μM. At higher concentrations phosphoramidates were cytotoxic. The inhibition of HIV was found to be dependent on chain length of phosphoramidates, a 20-mer being more effective than a 15-mer.

The phosphoramidates also have a drawback, similar to methylphosphonates, of poor solubility in aqueous media.

Phosphorothioates. In phosphorothioates, one of the phosphate oxygen atoms not involved in the bridge is replaced by a sulfur atom, with a negative charge being dis-

2

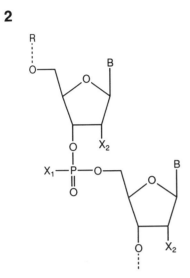

Fig. 2. General molecular structure of a normal oligonucleotide (R = H, X_2 = H, X_1 = H), oligonucleoside methylphosphonate (R = H, X_2 = H, X_1 = CH_3), oligonucleotide phosphorothioate (R = H, X_2 = H, X_1 = S), oligonucleoside phosphoramidates (R = H, X_2 = H, X_1 = NHR), 2′-OMe oligoribonucleotide (R = H, X_2 = OCH_3, X_1 = H), and 2′-OMe oligoribonucleotide phosphorothioate (R = H, X_2 = OCH_3, X_1 = S). In cases of oligonucleotide "conjugates," the R groups denotes the "conjugates."

tributed unsymmetrically and located mainly on sulfur (Fig. 2). Phosphorothioates are also somewhat nuclease resistant (Eckstein, 1985), and have lower duplex stability than their phosphodiester counterparts (Stein et al., 1988; Agrawal et al., 1989a). The duplex formed between phosphorothioates and RNA is a substrate for RNase H.

Phosphorothioates (20-mers) complementary to several conserved regions of the HIV-genome have been screened for their efficacy in tissue culture experiments. Here the result of a phosphorothioate complementary to a splice acceptor site of HIV *tat* gene will be discussed, but sites such as packaging, protease, primer binding site, and poly(A) signal site, have also shown promise (Agrawal and Zamecnik, unpublished data). Figure 5 shows the data on inhibition of HIV by the phosphorothioate, as assayed by syncytial forma-

3

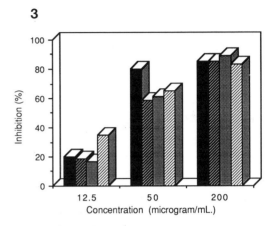

Fig. 3. Inhibition of HIV-1 expression by an unmodified oligonucleotide. The oligonucleotide is a 20-mer, complementary to the splice acceptor site (5849–5368) of HIV. Inhibition of HIV expression has been reported at three concentrations of oligonucleotides (μg ml^{-1}). The inhibition of HIV was assayed by four parameters: syncytial formation (solid bars), p17 expression (dark cross bars), p24 expression (dotted bars), and reverse transcriptase activity (light cross bars).

tion, p17 and p24 expression, and reverse transcriptase activity measuremnt. A dose-dependent inhibition was observed, with ID_{50} at about 1.5 μg ml^{-1} of phosphorothioates (approximately 0.25 μM). A self-complementary 20-mer phosphorothioate was ineffective at up to 15 μM concentration.

When phosphorothioates were added to tissue culture at postinfection times, they were still effective in inhibiting HIV replication (Fig. 4). The amount of phosphorothioate required to inhibit 50% syncytia formation when added at 4, 8, 24, and 48 hr postinfection was 2.4, 2.9, 3.4, and 4.7 μg ml^{-1}, compared to 1.4 μg ml^{-1} when added simultaneously with the virus. As discussed above, the dose of unmodified oligonucleotides required to inhibit 50% syncytia formation was more than 199 μg ml^{-1} at 24 hr postinfection.

The phosphorothioates inhibited HIV replication in tissue culture when added at the same time as the virus even though they were mismatched or homooligomers (Agrawal et al., 1988). At low concentrations, they were less effective than the complementary phos-

4

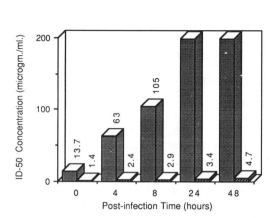

Fig. 4. Comparative inhibition of HIV replication as assayed by syncytial formation by an unmodified oligonucleotide (shaded bars) and a phosphorothioate oligonucleotide (crossed bars), when added to tissue cultures at 0 hr or at different times postinfection. The sequence is the same as in Figure 3. The figure shows the concentration of oligonucleotide (in $\mu g \; ml^{-1}$) required to inhibit 50% of syncytial formation at different times postinfection. At 24 and 48 hr postinfection, the amount of unmodified oligonucleotide required was more than 199 $\mu g \; ml^{-1}$.

5

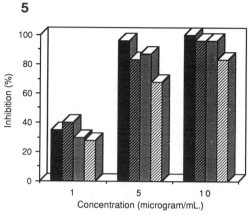

Fig. 5. Inhibition of HIV-1 replication by a phosphorothioate oligonucleotide. The oligonucleotide is a 20-mer, complementary to the splice acceptor site (5849–5368) of HIV. The inhibition of HIV was assayed by four parameters: syncytial formation (solid bars), p17 expression (dark cross bars), p24 expression (dotted bars), and reverse transcriptase activity (light cross bars).

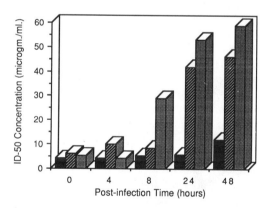

Fig. 6. Comparative inhibition of HIV replication by three oligonucleotide phosphorothioates: GCGTACTCACCAGTCGCCGC, complementary to splice donor site of HIV (280–299) (solid bars), CGAGATAATGTTCACACAAC, a mismatched oligomer (crossed bars), and d-A_{20} (shaded bars). The figure shows the concentration of oligonucleotide required to inhibit 50% syncytia formation. When the oligonucleotide was added at 0 hr, the concentration of oligonucleotide required to inhibit 50% of syncytial formation was the same for all three oligonucleotides; however, when added postinfection the ID_{50} of mismatched and homooligomers were much higher, compared to the complementary sequence. Data are from Agrawal et al. (1989b).

phorothioate, but at higher concentrations showed similar activity. Similar results were obtained by Matsukura et al. (1987). They observed that a homopolymer, a cytidine, phosphorothioate 28-mer, was as effective as a phosphorothioate complementary to the *rev* gene of HIV.

However, phosphorothioates, either mismatched or homooligomers, were not as effective when added to tissue culture at postinfection times. Figure 6 shows the data on inhibition of HIV by three different phosphorothioates complementary to splice donor sites of HIV, a mismatched and a homopolymer of adenine, a 20-mer. When phosphorothioates were added to tissue culture simultaneously with the virus, the concentration required to inhibit 50% of syncytial formation was almost the same for all three phosphorothioates. However, at postinfection times, the concentration required for mismatched or homooligomer was much higher than the complementary phosphorothioate. Similarly, when HIV chronically infected cells were cultured in the presence of various phosphorothioates, only phosphorothioates complementary to the *rev* gene of HIV inhibited p24 *gag* production at 10 μM concentration. Other phosphorothioates, either mismatched or homooligomers, showed no effect at this concentration (Matsukura et al., 1987). At higher concentrations of this series of oligonucleotides, a mechanism other than hybridization arrest may be operational. In contrast, sequence specific inhibition of influenza virus was obtained by phosphorothioate oligonucleotides (Leiter et al., 1990).

As has been discussed elsewhere (Agrawal and Sarin, 1991), when oligonucleotides are added simultaneously with the virus, the site of action is focused particularly on the steps leading from uncoating to integration (Fig. 1) in the HIV life cycle. It has been found that phosphorothioates are inhibitors of reverse transcriptase in particular, but other cellular enzymes, too (Mazumdar et al., 1989). Also, partial complementarity of the phosphorothioates with HIV RNA may activate RNase H. It is therefore difficult to sort out inhibition effects of phosphorothioates unrelated to hybridization arrest. However, complementary phosphorothioates are much superior to homooligomer or mismatched phosphorothioates in inhibiting HIV replication in previously infected and chronically infected cells. In previously infected cells, the site of action is focused on the steps leading from integrated viral DNA to viral packaging and release. The earlier described steps (uncoating to integration) are also present, but to an unknown extent.

Phosphorothioate analogues of 2'-OMe oligoribonucleotides have also been studied for inhibition effect against HIV in cultured cells. Their inhibitory activity was quite similar to that of phosphorothioate oligodeoxynucleotides (Shibahara et al., 1989). Phosphorothioate analogues of oligoribonucleotides were found to be less effective than their counterparts in the oligodeoxynucleotide series in inhibiting HIV expression (Agrawal et al., 1991). In both, the phosphorothioate analogue of oligodeoxynucleotide and 2'-OMe oligoribonucleotide, anti-HIV activity was chain length dependent of antisense oligonucleotide (Shibahara et al., 1989; Agrawal et al., 1990b; Zamecnik and Agrawal, 1991).

Oligonucleotide Carrying "Conjugates"

"Conjugates" have been attached to oligonucleotides either to increase the cell membrane permeability or to stabilize the duplex stability. Oligonucleotide-carrying conjugates that have been studied against HIV fall into the first category.

Two different conjugates have been attached to oligonucleotides; one is poly-L-lysine to methylphosphonates and the other is cholesterol to unmodified oligonucleotides as well as to phosphorothioates. It has been demonstrated that by attaching the poly-L-Lysine to a 17-mer methylphosphonate complementary to a splice donor site, the effectiveness of inhibition of HIV was improved 5-fold (Stevenson and Iverson, 1989). Similarly cholesteryl-conjugated phosphorothioate oligonucleotides showed complete inhibition of HIV at a concentration of 0.2 μM. However, by attaching cholesterol to oligonucleotides, there was an adverse effect on sequence specificity (Letsinger et al., 1989).

ACUTE TOXICITY STUDIES

Preliminary acute toxicity studies in mice and rats have been performed with an oligonucleotide and its phosphate backbone modified analogues, phosphorothioate and phosphomorpholidate.

Toxicity Study of Unmodified Oligonucleotide in Mice

First the study was carried out to establish the highest concentration range of the oligonucleotide (unmodified) at which the mouse host will not die or develop acute symptoms. The preferred route of administration was intraperitoneally (ip), for convenience. The test animals used in the study were Swiss–Webster male and female albino mice (Taconic Farms, Germantown, NY). The oligonucleotide selected for this study was a

20-mer, 5'-ACACCCAATTCTGAAAATGG. The oligonucleotide was synthesized using H-phosphonate chemistry and purified by the protocol described elsewhere, to obtain an oligonucleotide with triethylammonium as counterion.

The oligonucleotide dose was dissolved in 1 ml physiological saline and injected intra-peritoneally. The dose was from 640 mg/kg body weight to 2.5 mg/kg body weight (Table 1). The oligonucleotide was extremely toxic to mice when introduced ip at 640 mg/kg. One male mouse developed convulsions within 1 min and was dead in about 3 min. The second male mouse developed left hind leg paralysis (ip injection through left inguinal area), tilted and limped in 1–2 min, and by 2 min rolled over in convulsions and became unconscious. The hind legs experienced constant jerks and spasm until the mouse died after 13 min.

One female mouse receiving 640 mg/kg developed hind leg paralysis and became convulsant in less than 1 min. It rolled over and was dead in 4 min. The other female at that dose level showed the same symptoms except that convulsions commenced at 1.5 min and the mouse lay on its back with periodic gasps and spasm until it died at 16 min.

The male mice that received 160 mg/kg became lethargic, subdued, and drowsy. They developed an irregular gait and moved with difficulty. They retained their climbing, grasping and rightive reflexes and respiration appeared to be normal. The male mice began to show signs of recovery in about 30 min and appeared to have recovered completely by 24 hr.

The female mice on 160 mg/kg developed the same symptoms as the males and showed signs of recovery after about 45 min. By 24 hr, they were alert, eating, and defecating. At 48 hr, however, they were lethargic and appeared to be ill. One became comatose at around 54 hr and was dead by 58 hr. The other remained alive, barely moving, finally dying after about 82 hr. It would appear that females are more sensitive than males to the oligonucleotide, at least at a concentration of 160 mg/kg bodyweight.

Doses of 60, 10, and 2.5 mg/kg into males and females failed to evoke any toxic responses. Also average gain in body weights strongly indicated that there was no intoxication by oligonucleotide at these doses. At 160 mg/kg the two surviving males gained a considerable amount of weight (Table 2). This suggests that whatever toxic manifestation occured during the first 24 to 48 hr, the mice had long since recovered completely. The study was also performed by using 15-mers, homopolymers of $(dC)_{15}$ and $(T)_{15}$. At a concentration of 160 mg/kg, the mice appeared to have experienced transient, low level intoxication, from which all recovered completely. At 40 mg/kg and lower concentration, there was no observable intoxification.

Toxicity Study of Unmodified Oligonucleotide in Rats

From the studies carried out in mice, it was observed that 160 mg/kg body weight is a tolerable dose. This study in rats was performed using the 150 mg/kg body weight as the highest dose, and 25 mg/kg as the lowest (Table 3).

TABLE 1. Mouse Mortality Following Intraperitoneal Injection of Normal (Phosphodiester) Oligonucleotides

Oligonucleotide concentrations (mg/kg)	Deaths (time to death)		Total
	Male	Female	
640	2 (3 min; 13 min)	2 (4 min; 16 min)	4/4
160	0	2 (at 58 hr; at 82 hr)	2/4
40	0	0	0/4
10	0	0	0/4
2.5	0	0	0/4

TABLE 2. Average Body Weights of Mice Injected Intraperitoneally With Various Concentration of Normal Oligonucleotide

Oligonucleotide concentrations (mg/kg)	Average body weight—2 males, 2 females (g)		
	Day 0	Day 7	Day 14
640	21.18	—[a]	—[a]
160	20.95	29.45[b]	33.75[b]
40	21.30	27.18	29.90
10	21.38	26.80	29.01
2.5	21.25	28.45	30.38

[a]Two males, two females dead.
[b]Average of two males only; two females dead.

The sequence of oligonucleotides used in the study was the same as described above. The test animals were Sprague–Dawley male and female albino rats. Each dose of oligonucleotide was injected ip into two males and two females in 1 ml sterile, physiological saline. Control rats were injected ip with 1 ml of physiological saline.

The oligonucleotide was nontoxic to the young male and female rats at the four concentrations tested (Table 3). There was no evidence of lethargy or malaise at any time. The rats were active and alert and free of symptoms constantly; and food ingestion was noted within 15 min of treatment. None of the rats exhibited observable delayed toxicity and no hyperirritability was detected.

The slope of body weight gain appears to reflect a normal growth pattern (Table 4). After 14 days, one male and one female of each dose group was sacrificed and autopsied, as was a male and female of the control

TABLE 3. Rat Mortality Intraperitoneal Injection of Normal (Phosphodiester) Oligonucleotides

Oligonucleotide concentrations (mg/kg)	Deaths		
	Male	Female	Total
150	0	0	0/4
100	0	0	0/4
50	0	0	0/4
25	0	0	0/4
Controls	0	0	0/6

group. The heart, lung, liver, spleen, and kidneys were excised and examined for gross abnormalities (Table 5). The gross appearance of all organs suggested that they were normal.

Toxicity Study of the Oligonucleotide Phosphorothioate in Mice

A toxicity study was performed using a 20-mer oligonucleotide phosphorothioate containing the same sequence as the phosphodiester oligonucleotide used in the toxicity study describd above. The dose of the oligonucleotide tested was from 640 to 2.5 mg/kg body weight (Table 6). In this study also, each dose was dissolved in 1 ml of physiological saline and injected intraperitoneally to two male and two female mice (Swiss–Webster albino).

The oligonucleotide phosphorothioate was also extremely toxic to mice when introduced intraperitoneally at a dose of 640 mg/kg. One male mouse became lethargic and drowsy after about 10 min, moving by means of a slithering crawl. At 30 min posttreatment, the mouse went into convulsions and died within minutes. The second mouse at 640 mg/kg dose developed the same symptoms but experienced the convulsions after 50 min and died at 55 min.

The females receiving the 640 mg/kg dose developed the same symptoms as described above. One female exhibited constant body tremors after 20 min, followed by convulsions at 22 min and almost immediate death.

TABLE 4. Averge Body Weights of Rats Injected Intraperitoneally With Various Concentrations of Normal Oligonucleotide

Oligonucleotide concentrations (mg/kg)	Average body weight—2 males, 2 females (g)		
	Day 0	Day 7	Day 14
150	83.88	136.18	172.05
100	84.40	136.73	175.30
50	83.90	134.05	169.03
25	81.43	137.05	176.30
Controls[a]	88.10	143.58	184.87

[a]Average of three males and three females.

The other female mouse became convulsive after 26 min and died within 1 min.

The two males injected with 160 mg/kg became lethargic and drowsy after about 10–12 min. This was accompanied by writhing and extension of left hind legs (the oligonucleotide was introduced via the left inguinal area) and they appeared to experience discomfort. These mice recovered in about 2 hr and by 4 hr they were eating and alert. They appeared to behave normally thereafter.

The two females given 160 mg/kg doses became lethargic and drowsy after about 10 min posttreatment. They also developed the same symptoms described above for males. They recovered after about 2 hr and by 24 hr were eating and alert. However, on day 3, one mouse was subdued and inactive, and was found dead at about 80 hr posttreatment. The other female mouse continued to behave normally. However, the surviving female mouse gained very little weight during the first 7 days posttreatment and by day 14 gained

TABLE 5. Rat Organ Weights and Observations, 2 Weeks Posttreatment With Normal Oligonucleotide

	Test groups (mg/kg)				
	150	100	50	25	Control
Day 14					
body weight (g)					
M[a]	183.2	178.2	179.9	188.4	212.0
F[a]	140.5	160.5	129.5	154.3	149.2
Organ Weight (g)					
Heart					
M	1.03	0.78	0.73	0.89	0.94
F	0.66	0.75	0.66	0.74	0.67
Lung					
M	1.21	1.41	1.03	1.43	1.17
F	1.02	1.04	0.82	0.96	0.93
Liver					
M	10.68	8.67	10.67	9.96	9.32
F	6.85	7.56	5.09	7.77	6.40
Spleen					
M	0.72	0.74	0.62	0.85	0.81
F	0.41	0.64	0.37	0.66	0.56
Kidneys					
M	2.20	2.06	1.84	2.03	2.22
F	1.67	1.89	1.17	1.75	1.71

[a]M, male; F, female.

TABLE 6. Mouse Mortality Following Intraperitoneal Injection of Oligonucleotide Phosphorothioate

Oligonucleotide concentrations (mg/kg)	Deaths (time to death)		Total
	Male	Female	
640	2 (31 min; 55 min)	2 (22 min; 26 min)	4/4
160	0	1 (80 hr)	1/4
40	0	0	0/4
10	0	0	0/4
2.5	0	0	0/4

only about 3.5 g, considerably less than the other female mice treated with lower concentrations of the product. Thus, failure to gain weight normally was evidence of chronic toxicity.

There were no deaths and no evidence of toxicity among the male and female mice treated with a dose of 40 mg/kg or lower. The oligonucleotide phosphorothioate appeared to be nontoxic at these dose levels.

Table 7 lists the average body weight of mice treated with various doses of oligonucleotide phosphorothioate. Normal weight gain or failure to gain is an index of intoxication or its persistence.

Similarly, homooligonucleotides, $d(C)_{15}$ and $(T)_{15}$ phosphorothioate analogues, were also tested for their toxicity. At a dose of 150 mg/kg, there was no observable toxicity in mice.

The oligonucleotide phosphorothioate was not toxic if given daily to mice at a dose of

100 mg/kg for 15 days. This result is based on no mouse mortality at this dose.

Toxicity Study of the Oligonucleotide Phosphorothioate in Rats

The same oligonucleotide phosphorothioate 20-mer has also been studied for toxicity in rats. A dose of 150 mg/kg body weight or lower dose was injected intraperitoneally in Sprague-Dawley albino rats (Table 8). Each dose was injected in 1 ml physiological saline in two male and two female rats.

The oligonucleotide phosphorothioate showed no toxicity to young male and female rats at the four concentrations tested via the intraperitoneal route. The rats remained active and alert and free of symptoms constantly, and food ingestion resumed shortly after dosing. Also, none of the rats exhibited observable delayed toxicity.

The slope of the body weight gain appeared to reflect a normal growth pattern (Table 9). The control males gained more weight over the 2-week test period, but the data are too small to draw any conclusion.

TABLE 7. Average Body Weights of Mice Injected Intraperitoneally With Various Concentrations of Oligonucleotide Phosphorothioate

Oligonucleotide concentrations (mg/kg)	Average body weight—2 males, 2 females		
	Day 0	Day 7	Day 14
640	21.20	—[a]	—[a]
160	21.16	28.27[b]	31.87[b]
40	20.98	28.68	30.66
10	20.83	27.83	29.70
2.5	20.84	27.51	30.08

[a]Two males and two females dead.
[b]One female dead, survivor gained very little weight.

TABLE 8. Rat Mortality Following Intraperitoneal Injection of Oligonucleotide Phosphorothioate

Oligonucleotide concentrations (mg/kg)	Deaths		Total
	Male	Female	
150	0	0	0/4
100	0	0	0/4
50	0	0	0/4
25	0	0	0/4
Controls	0	0	0/6

TABLE 9. Average Body Weights of Rats Injected Intraperitoneally With Various Concentrations of Oligonucleotide Phosphorothioate

Oligonucleotide concentrations (mg/kg)	Average body weight—2 males, 2 females (g)		
	Day 0	Day 7	Day 14
150	89.80	135.43	175.03
100	84.03	136.55	176.65
50	86.40	136.00	172.13
25	88.58	142.36	181.23
Controls[a]	88.10	143.58	184.87

[a]Average of three males and three females

After 14 days posttreatment, one male and female of each dose group were sacrificed and autopsied. The gross appearance of all organs suggested that they were normal (Table 10). However, liver weights of three of the four male test rats were significantly greater than that of control males, even though the body weight of control rats was 14–25 g greater than test rats.

Toxicity Study of the Oligonucleoside Phosphomorpholidate in Mice

An oligonucleoside phosphomorpholidate tested for its toxicity in mice had the same sequence as for the unmodified oligonucleotide and its phosphorothioate analogue. The male mice at a dose of 150 mg/kg showed signs of some toxicity after dosing, but recovered completely after 24 hr. However, with a dose of 150 mg/kg, female mice showed no sign of recovery from the toxicity after dosing. By 24 hr, they appeared shrunken, sluggish, anorexic, and had ruffled coats. Both died at about 60 hr posttreatment. Lower doses of 40, 10, and 2.5 mg/kg into male and female mice failed to evoke any toxic responses. There was an average body weight gain.

Toxicity Study of Oligonucleoside Phosphomorpholidate in Rats

The same sequence, as in the case of mice, showed no sign of toxicity at doses of 150, 100, or 25 mg/kg as observed by no deaths and average body weight gain. However, in this case also, the liver weights of three out of four male test rats were significantly greater than that of the control male. The same was

true of the livers of female test rats at all concentrations.

CONCLUSION

In summary, oligonucleotide phosphorothioates are effective in inhibiting HIV replication in tissue culture (ID_{50}) at a concentration of about $1 \times 10^{-7} M$, which translates into a dose of 0.6 mg/kg body weight. There are several factors directly related to this effective dose. The main factor is the multiplicity of virus infection (moi). In the above experiments, the moi was 0.5 to 1, which was chosen to evoke a maximal effect by 96 hr. This high moi is far beyond what would be expected in a human patient with AIDS, so a lower dose might be effective. In a comparative study, an oligonucleotide phosphorothioate has been found to be, on a molar basis, at least five times more active than AZT in inhibiting HIV replication in postinfected cells (Agrawal, Sarin, and Zamecnik, unpublished data).

The toxicity study in mice and rats showed that a dose of 100 mg/kg of oligonucleotide phosphorothioate is not toxic. However, the gain in liver weight in rats receiving oligonucleotides (phosphodiester, phosphorothioate, or phosphomorpholidate) needs to be explained. At this time the data are insufficient to make any conclusive statement. The study of pharmacokinetics, bioavailability, and metabolic rate of oligonucleotide phosphorothioates is in progress.

Since the effective dose of antisense phosphorothioates against HIV in tissue culture is

TABLE 10. Rat Organ Weights and Observations, 2 Weeks Posttreatment With Oligonucleotide Phosphorothioate

	Test groups (mg/kg)				
	150	100	50	25	Control
Day 14					
bodyweight (g)					
M[a]	190.3	197.9	187.2	192.0	212.0
F[a]	151.3	141.2	137.3	162.1	149.2
Organ weight (g)					
Heart					
M	1.23	1.01	0.84	1.04	0.94
F	0.81	0.66	0.68	0.81	0.67
Lung					
M	1.39	1.29	1.74	1.11	1.17
F	1.06	0.92	0.90	1.00	0.93
Liver					
M	11.38	9.72	11.48	11.26	9.32
F	7.09	6.52	6.17	7.64	6.40
Spleen					
M	1.11	1.06	0.98	0.85	0.81
F	0.75	0.70	0.73	0.74	0.56
Kidney					
M	2.50	2.65	1.98	2.20	2.22
F	1.61	1.45	1.31	1.59	1.71

[a]M, male; F, female.

in the 10^{-7} to 10^{-8} M range (i.e., 0.6 mg/kg), and the dose tolerated in the mouse and rat has been found in the 100 mg/kg range, an apparent therapeutic window may be present, with the caveat that extrapolation from tissue culture to animal is perilous.

ACKNOWLEDGMENTS

I am very much indebted to Paul C. Zamecnik for his encouragement, and continuous guidance and support in this work. This work was supported by the National Cooperative Drug Discovery Group for the Treatment of AIDS Grant U01 A124846 from the National Cancer Institute and the National Institute of Allergy and Infectious Diseases, and by a grant by the G. Harold and Leila Y. Mathers Foundation. Animal toxicity tests were performed by Dr. Eugene Birnstein, University Laboratories, Inc., New Jersey. I am thankful to Ms. Rebecca Bell and Carol Tuttle for processing the manuscript and expert secretarial assistance.

NOTE ADDED IN PROOF

Preliminary pharmacokinetic study of oligonucleotide phosphorothioate in mice showed that, if oligonucleotide phosphorothioate is administered intravenously or intraperitoneally, it is rapidly cleared from the blood stream and enters the liver, kidney, muscle, heart, bladder, stomach, lung, testes, intestine and possibly brain, with significant retention of undegraded oligonucleotide therein for 48 hours. (Agrawal, S., Temsamani, J. and Tang, J.-Y. [1991]. Pharmacokinetics, biodistribution and half life of oligodeoxynucleotide phosphorothioate in mice. Proc. Natl. Acad. Sci. [USA], 88, in press.)

REFERENCES

Agrawal S, Goodchild J (1987): Oligodeoxynucleoside methylphosphonates: Synthesis and enzyme degradation. Tetrahedron Lett. 28(31):3539–3542.

Agrawal S, Sarin, PS (1991): Antisense oligonucleotides: Gene regulation and chemotherapy of AIDS. Advanced Drug Deliv. Revs. 6(3):251–270.

Agrawal S, Tang J-Y (1990): Efficient synthesis of oligoribonucleotide and its phosphorothioate analogue using H-phosphonate approach. Tetrahedron Lett. 31:7541–7544.

Agrawal S, Goodchild J, Civeira MP, Thornton AH, Sarin PS, Zamecnik PC (1988): Oligodeoxynucleoside phosphoramidates and phosphorothioates as inhibitors of human immunodeficiency virus. Proc. Natl. Acad. Sci. U.S.A. 85:7079–7083.

Agrawal S, Goodchild J, Civeira MP, Thornton AH, Sarin, PS, Zamecnik, PC (1989a): Phosphoramidates, phosphorothioates and methylphosphonate analogs of oligodeoxynucleotides: Inhibitors of replication of human immunodeficiency virus. Nucleosides Nucleotides 8:819–823.

Agrawal S, Ikeuchi T, Sun D, Sarin PS, Konopka A, Maizel T, Zamecnik PC (1989b): Inhibition of human immunodeficiency virus in early infected and chronically infected cells by antisense oligodeoxynucleotides and its phosphorothioate analogue. Proc. Natl. Acad. Sci. U.S.A. 86:7790–7794.

Agrawal S, Mayrand SM, Zamecnik PC, Pederson T (1990a): Site specific excision from RNA by RNase H and mixed phosphate backbone oligodeoxynucleotides. Proc. Natl. Acad. Sci. U.S.A. 87:1401–1405.

Agrawal S, Sun D, Sarin PS, Zamecnik PC (1990b): Inhibition of HIV-1 in early infected and chronically infected cells by antisense oligodeoxynucleotides and their phosphorothioate analogue. J. Cell. Biochem. 14D:145.

Agrawal S, Tang J-Y, Sun DK, Sarin PS, Zamecnik PC (1991): Inhibition of HIV-1 replication by antisense oligoribonucleotide and its phosphorothioate analogue. In preparation.

Barrett JC, Miller PS, Ts'o POP (1974): Inhibition effects of complex formation with oligodeoxynucleotide aminoacylation. Biochemistry 13:4897–4906.

Belikova AM, Zarytova VF, Grineva NI (1967): Synthesis of ribonucleosides and diribonucleoside phosphates containing 2-chloroethylamine and nitrogen mustard residues. Tetrahedron Lett. (37):3557–3562.

Bester AJ, Kennedy DS, Heywood SM (1975): Two classes of translational control RNA: Their role is regulation of protein synthesis. Proc. Natl. Acad. Acad. Sci. U.S.A. 72:1523–1527.

Dunlap BE, Friderici KH, Rottman F (1971): 2′-O-Methyl polynucleotides as templates for cell amino acid incorporation. Biochemistry 10:2581–2587.

Eckstein F (1985): Investigation of enzyme mechanisms with nucleoside phosphorothioates. Annu. Rev. Biochem. 54:367–402.

Froehler B, Ng P, Matteucci M (1988): Phosphoramidate analogues of DNA: Synthesis and thermal stability of heteroduplexes. Nucl. Acids Res. 16:4831–4839.

Furdon PJ, Dominski Z, Kole R (1989): RNase H cleavage of RNA hybridized to oligonucleotides containing methylphosphonate, phosphorothioate and phosphodiester bonds. Nucl. Acids Res. 17:9193–9204.

Goodchild J, Agrawal S, Civeira MP, Sarin PS, Sun D, Zamecnik PC (1988): Inhibition of human immunodeficiency virus replication by antisense oligodeoxynucleotides. Proc. Natl. Acad. Sci. U.S.A. 85:5507–5511.

Hastie ND, Held WA (1978): Analysis of mRNA populations by cDNA:mRNA hybrid-mediated inhibition of cell-free protein synthesis. Proc. Natl. Acad. Sci. U.S.A. 75:1217–1221.

Hoagland MB, Stephenson ML, Scott JF, Hecht LE, Zamecnik PC (1958): A soluble ribonucleic acid intermediate in protein synthesis. J. Biol. Chem. 231:241–256.

Hoagland MB, Zamecnik PC, Stephenson ML (1959): A hypothesis concerning the roles of particulate and soluble ribonucleic acids in protein synthesis. In A Symposium on Molecular Biology. Chicago: Univ. of Chicago Press, pp 105–114.

Lactena RM, Cesareni G (1981): Base pairing of RNA I with its complementary sequence in the primer precursor inhibits Col E 1 replication. Nature (London) 294:623–626.

Leiter JME, Agrawal S, Palese P, Zamecnik PC (1990): Inhibition of influenza virus replication by phosphorothioate oligodeoxynucleotides. Proc. Natl. Acad. Sci. U.S.A. 87:3430–3434.

Letsinger RL, Zhang G, Sun D, Ikeuchi T, Sarin PS (1989): Cholesteryl-conjugated oligonucleotides: Synthesis, properties and activity as inhibitors of replication of human immunodeficiency virus in cell culture. Proc. Natl. Acad. Sci. U.S.A. 86:6553–6556.

Matsukura M, Shinozuka K, Zon G, Mitsuya H, Reitz M, Cohen JS, Broder S (1987): Phosphorothioate analogs of oligodeoxynucleotides: Inhibitors of replication and cytopathic effects of human immunodeficiency virus. Proc. Natl. Acad. Sci. U.S.A. 84:7706–7710.

Matsukura M, Zon G, Shinozuka K, Robert-Guroff M, Shimada T, Stein CA, Mitsuya H, Wong-Staal F, Cohen J, Broder S (1989): Regulation of viral expression of human immunodeficiency virus in vitro by an antisense phosphorothioate oligodeoxynucleotide against rev (art/trs) in chronically infected cells. Proc. Natl. Acad. Sci. U.S.A. 86:4244–4248.

Mazumdar C, Stein CA, Cohen JS, Broder S, Wilson SH (1989): Stepwise mechanism of HIV reverse transcriptase: Primer function of phosphorothioate oligodeoxynucleotide. Biochemistry 28:1340–1346.

Miller PS, Barrett JC, Ts'o POP (1974): Synthesis of oligodeoxyribonucleotide ethyl phosphotriesters and their specific complex formation with transfer ribonucleic acid. Biochemistry 13:4887–4896.

Mizuno T, Chou M-Y, Inouye M (1983): A unique mechanism of regulating gene expression: Translational inhibition by a complementary RNA transcript (mic RNA). Proc. Natl. Acad. Sci. U.S.A. 81:1966–1970.

Quartin RS, Wetmur JG (1989): Effect of ionic strength on the hybridization of oligonucleotides with reduced charge due to methylphosphonate linkages to unmodified oligodeoxynucleotides containing the complementary sequence. Biochemistry 28:1040–1047.

Sarin PS, Taguchi Y, Sun D, Thornton A, Gallo RC, Oberg B (1985): Inhibition of HTLV-III/LAV replication by foscarnet. Biochem. Pharmacol. 34:4075–4079.

Sarin PS, Sun D, Thornton A, Muller WEG (1987): Inhibition of replication of the etiologic agent of acquired immune deficiency syndrome (human T-lymphotropic retrovirus/lymphoadenopathy-associated virus) by avarol and quarone. J. Natl. Cancer Inst. 78:663–666.

Sarin PS, Agrawal S, Civeira MP, Goodchild J, Ikeuchi T, Zamecnik PC (1988): Inhibition of acquired immunodeficiency syndrome virus by oligonucleoside methylphosphonates. Proc. Natl. Acad. Sci. U.S.A. 85:7448–7451.

Shibahara S, Mukai S, Morisawa H, Nakashima H, Kobayashi S, Yamamoto N (1989): Inhibition of human immunodeficiency virus (HIV-1) replication by synthetic oligo-RA derivatives. Nucl. Acids Res 17:239–252.

Stein CA, Subasinghe C, Shinozuka K, Cohen JS (1988): Physiochemical properties of phosphorothioate oligodeoxynucleotides. Nucl. Acids Res. 16:3209–3221.

Stephenson ML, Zamecnik PC (1978): Inhibition of Rous sarcoma viral RNA translation by a specific oligodeoxyribonucleotide. Proc. Natl. Acad. Sci. U.S.A. 75:285–288.

Stevenson M, Iverson PL (1989): Inhibition of human immunodeficiency virus type-I mediated cytopathic effects by poly(L-lysine)-conjugated synthetic antisense oligodeoxyribonucleotides. J. Gen. Virol. 70:2673–2682.

Tomizawa J-I, Itoh T (1981): Plasmid Col E 1 incompatibility determined by interaction of RNA I with primer transcript. Proc. Natl. Acad. Sci. U.S.A. 78:6096–6100.

Uhlmann E, Peyman A (1990): Antisense oligonucleotides: A new therapeutic principle. Chem. Rev. 90:544–584.

Van der Krol AR, Mol JNM, Stuitje AR (1988): Modulation of gene expression by complementary RNA or DNA sequences. Biotechniques 6:958–976.

Wickstrom E (1986): Oligodeoxynucleotide stability in subcellular extracts and culture media. J. Biochem. Biophys. Methods 13:97–102.

Zaia JA, Rossi JJ, Murakawa GJ, Spallone PA, Stephens DA, Kaplan BE, Eritza R, Wallace RB, Cantin EM (1988): Inhibition of human immunodeficiency virus by using an olignucleoside methylphosphonate targeted to the tat-3 gene. J. Virol. 62:3914–3917.

Zamecnik PC, Agrawal S (1991): The hybridization inhibition, or antisense approach to the chemotherapy of AIDS. In Koff WC, Wong-Staal F, Kennedy RC (eds): AIDS Research Reviews. Vol. 1. New York: Marcel Dekker, pp 301–313.

Zamecnik PC, Stephenson ML (1978): Inhibition of Rous sarcoma virus replication and transformation by a specific oligodeoxynucleotides. Proc. Natl. Acad. Sci. U.S.A. 75:280–284.

Zamecnik PC, Goodchild J, Taguchi Y, Sarin PS (1986): Inhibition of replication and expression of human T-cell lymphotrophic virus type III in cultured cells by exogenous synthetic oligonucleotides complementary to viral RNA. Proc. Natl. Acad. Sci. U.S.A. 83:4143–4146.

A New Concept in AIDS Treatment: An Antisense Approach and Its Current Status Towards Clinical Application

Makoto Matsukura, Hiroaki Mitsuya, and Samuel Broder

Department of Pediatrics, Kumamoto University Medical School, Kumamoto 860, and Department of Pediatrics, Ashikita Institute for Handicapped Children, Ashikita 869-54, Kumamoto Prefecture, Japan (M.M.) and Clinical Oncology Program, National Cancer Institute, National Institutes of Health, Bethesda, Maryland 20892 (H.M., S.B.)

INTRODUCTION

Oligoribonucleotides or oligodeoxynucleotides complementary to mRNA, which are called "antisense," could theoretically inhibit translation of mRNA and thereby block synthesis of an encoded protein (Izant and Weintraub, 1985; Stephenson and Zamecnik, 1978; Zamecnik and Stephenson, 1978). This process has been referred to as "translation arrest." However, experiments in whole-cell systems and cell-free systems to prove such phenomenon have not been uniformly successful (Hasan et al., 1988). The apparent inconsistency of the data has resulted in some questions as to application of this concept more widely. One of the reasons for the inconsistency is that, in some assays, methods were not developed for detection of truly specific inhibition of translation of the target mRNA. For instance, in experiments using viruses, almost all of the work was conducted in a *de novo* infection format (Agris et al., 1986; Gupta, 1987; Lemaitre et al., 1987; Sarin et al., 1988; Shibahara et al., 1989; Smith et al., 1986; Stephenson and Zamecnik, 1978; Wickstrom et al., 1986; Zaia et al., 1988; Zamecnik et al., 1986; Zamecnik and Stephenson, 1978; Zerial et al., 1987). In some cases, there was pretreatment of cells with virus for several hours prior to giving the inoculum (Agrawal et al., 1989), but chronically/persistently infected, virus-producing cells were not established. This complicates interpretations of the results. One of the other reasons is that, in antisense experiments with oncogenes, inhibitory effects on the growth of target cells were often used to assess inhibitory activities of antisense oligomers against oncogene expression (Heikkila et al., 1987; Wickstrom et al., 1988; McManaway et al., 1990). However, retardation of cell growth could also be due to toxicity of an oligomer and, in fact, we found that different sequences could have different profiles of toxicity, apparently independent of their intended antisense activity. We also thought that the sensitivity of "normal," unmodified oligomers to cellular nucleases complicated experiments using these oligomers (Matsukura et al., 1987, 1989; Shibahara et al., 1989).

We therefore decided to first focus on the clear demonstration of sequence specificity of antisense activity in ordinary culture conditions *without using special serum-free medium or special techniques such as microinjection of oligomers into cells.* We gave attention to (1) use of appropriate assay systems to show the selective inhibitory activity against gene expression (e.g., use of chronically infected cells instead of newly or *de novo* infected cells); (2) use of optimal culture conditions to measure antisense activity and distinguish possible effects of toxicity of

Prospects for Antisense Nucleic Acid Therapy of Cancer and AIDS, pages 159–178
©1991 Wiley-Liss, Inc.

oligomers on the target protein production versus real antisense activity; (3) use of various types of oligomer controls to establish the sequence specificity of the anti-HIV activity produced by an oligomer; and (4) use of both nuclease-sensitive and nuclease-resistant oligomers to get information about the requirement for nuclease resistance of oligomers for exhibiting antisense activity.

In this chapter, we first describe the methods for the synthesis and purification of two kinds of oligodeoxynucleotides, normal unmodified oligomers and phosphorothioate oligomers, and then describe antiviral activities of the oligomers, namely, sequence-nonspecific and sequence-specific activity against HIV replication and their mechanisms. We also discuss, briefly, some preliminary results for the uptake profile of phosphorothioate oligomers, and their pharmacokinetic behavior in animals, which relate to the prospects for clinical use of antisense oligomers against AIDS and related disorders.

SYNTHESIS AND PURIFICATION OF UNMODIFIED, NORMAL OLIGODEOXYNUCLEOTIDES AND PHOSPHOROTHIOATE ANALOGS

We employed the conventional phosphoramidite method for the synthesis of unmodified, normal oligomers and either the stepwise or single-step sulfurization methods for the phosphorothioate oligomers (Fig. 1a, b, and c, respectively), using in all cases an Applied Biosystems Model 380B DNA Synthesizer. Manufacturer-supplied reagents and solvents were used except for the sulfurization step and the pre(post)sulfurization wash steps, which used freshly prepared solution A and solution B, respectively: solution A = 2.5 g sulfur (Aldrich, >99.999%), 23.7 ml carbon disulfide (Aldrich, HPLC grade), 23.7 ml pyridine (Aldrich, highest purity, anhydrous), and 2.5 ml triethylamine (Aldrich, highest purity); solution B = 50 ml carbon disulfide and 50 ml pyridine.

Stepwise Sulfurization

The details for this method have been pub-

lished (Matsukara et al., 1988) as an improved version of the original procedure reported by Stec et al. (1984). As shown in Figure 1b, sulfur is incorporated in each chain-extension cycle by reaction of the phosphite intermediate with elemental sulfur (S_8) in a mixture of carbon disulfide, pyridine, and triethylamine. Stepwise sulfurization offers an advantage in synthetic targets where there is a mixture of unmodified linkages and/or other modifications in the phosphate linkages in the oligomer (P=O, P=S, P—Me, etc.). A new reagent to replace S_8 is now available ("TETD," Applied Biosystems).

Single-Step Sulfurization

This method of synthesis is called the "H-phosphonate method" (Fig. 1c). Hydrogen-phosphonate monomers (Applied Biosystems) were used with manufacturer-supplied cycles for "trityl-on" synthesis with either the 1-μmol or 10-μmol scale employing a new, stable activator (1-adamantanecarbonyl chloride) and isopropyl phosphite, an essential capping reagent (Andrus et al., 1988). The sulfurization step with S_8 is then performed only once, following automated synthesis on the DNA synthesizer (Marti et al., work in progress).

Convenient radiolabeling of the oligomer by ^{35}S is made possible by this method (Stein et al., 1990). Efficiency of sulfur incorporation into phosphorothioate oligomers is usually better in the single-step sulfurization method (100%) compared to stepwise sulfurization (ca. 95–99%). However, the latter is much more convenient, especially with the availability of TETD (*vide supra*).

Purification of Oligomers

The crude product synthesized on the automated synthesizer and sulfurized is purified by HPLC using a reversed-phase (C_8 or C_{18} or polystyrene) column and a linear gradient of increasing acetonitrile concentration (20–50% v/v) over 30 min in 0.1 M triethylammonium acetate, pH 7.0 (Stec et al., 1985). The purified product is obtained as the sodium salt by ethanol precipitation from aqueous NaCl.

SEQUENCES OF THE TARGET REGION IN THE HIV GENOME AND OLIGOMERS USED

Target Sequences

We initially employed the *rev* region (Fig. 2) as a target sequence for the following reasons. (1) In spite of the high mutational rate of HIVs, the *rev* sequence is well conserved among HIV isolates (Table 1). It is obviously beneficial if not necessary to use such a conserved sequence in effective treatment of patients infected with HIV. (2) *Rev* is known to be an essential region for efficient viral replication of HIV (Feinberg et al., 1986; Sodroski et al., 1986). It was expected that inhibition of rev protein synthesis by antisense oligomers against *rev* mRNA would profoundly affect HIV replication. (3) Since the production of HIV regulatory proteins appears to be very low (Feinberg et al., 1986), it was presumed that *rev* mRNA would not be abundant and/or the level of translation of *rev* mRNA into rev protein would not be high. In either case, the interference by the antisense oligomer against the translation of such mRNA might be more effective than against mRNA coding for abundant structural proteins (gag, pol, and env). On the other hand, a low concentration of mRNA target could lead to slower interception by the antisense oligomer.

In addition to the above noted *rev* sequence, we targeted the initiation site of *gag*, the frameshift regions between *gag* and *pol* (*G/P*) unspliced *tat*, and spliced *tat* sequences (Fig. 2).

Oligomers Used (Table 2)

For the reasons mentioned in the introduction, we synthesized antisense sequences and various types of control sequences relative to *rev*. Controls included the sense sequence, a random sequence (having the same net base composition as the antisense), a homooligomer, and an N^3-methylthymidine containing antisense sequence. The four N^3-methylthymidines [*N*-Me-TdR (Kyogoku et al., 1967)] in the *rev* antisense sequence are known to block hybridization to the target

mRNA (G. Zon, personal communication). The oligomers used were phosphorothioate oligomers (S-oligos) and unmodified "normal" oligomers (n-oligos).

SEQUENCE NONSPECIFIC ANTI-HIV ACTIVITY

Cytopathic Effect Inhibition Assay (Fig. 3)

Culture of target cells. ATH8 is a CD4$^+$ T cell line that was immortalized by infection of HTLV-I (human T-lymphotropic virus type 1), and is profoundly sensitive to the cytopathic effect of HIV (Mitsuya et al., 1984, 1987). ATH8 cells were kept in culture in complete medium (RPMI 1640 supplemented with 15% fetal calf serum, 4 mM L-glutamine, 50 nM 2-mercaptoethanol, and 50 units of penicillin and 50 μg of streptomycin per ml) in the presence of 20 units of recombinant interleukin-2 (IL-2) (Amgen Biologicals, Thousand Oaks, CA) per ml and 15% (v/v) of conventional IL-2 (Advanced Biotechnologies, Silver Spring, MD).

Assay procedure. The target ATH8 cells were pretreated with Polybrene for 45 min and centrifuged, pelleted, and incubated with HIV/III$_B$ for 1 hr at a virus inoculum of 500 virus particles per cell, which is 100–1000 times higher than the minimum cytopathic dose in this assay. Then, 2×10^5 ATH8 cells were cultured in each tube in either the presence or absence of oligomers in 2 ml of the complete medium mentioned above. On either day 7 or later, the number of viable cells was counted in a hemocytometer under a microscope by the trypan blue dye exclusion method. Virus nontreated samples were similarly cultured, except for exposure to the virus.

Sequences of oligomers tested. The sequences of oligomers tested in the cytopathic effect inhibition assay are shown in Table 2. They consisted of 14-mers that included antisense, sense, and random sequences, relative to *rev*, as well as homooligomers. To assess length effects, 5-, 14-, 18-, 21-, and 28-mers of deoxycytidine were tested. The effects of base composition were evaluated with

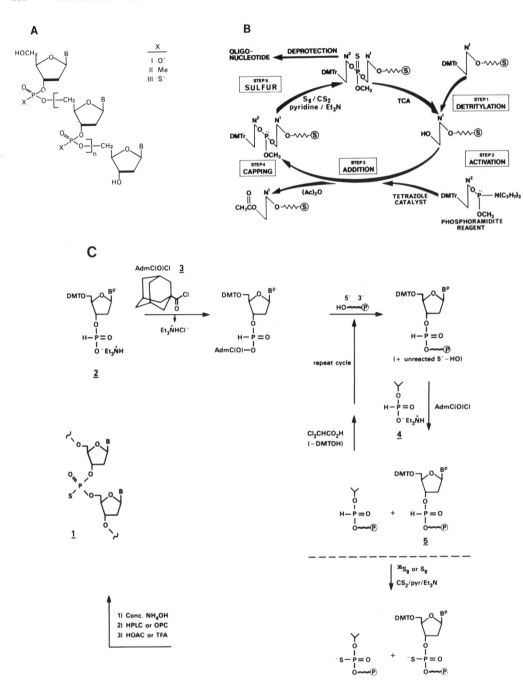

Fig. 1. (A) General molecular structure of normal (I), methylphosphonate (II), and phosphorothioate oligomers (III) (n-oligo, M-oligo, and S-oligo, respectively). B is adenine, guanine, cytosine, or thymine. **(B)** Schematic representation of step-wise sulfurization. Sulfur is incorporated in each chain-extension cycle by reaction of the phosphite intermediate with elemental sulfur (S_8) in a mixture of carbon disulfide, pyridine, and triethylamine. **(C)** Schematic representation of the preparation of polyphosphorothioates (1) by the use of hydrogen phosphonate monomers (2), adamantane carbonyl chloride (3) activator, and an isopropyl phosphite (4) capping reagent. Sulfur-

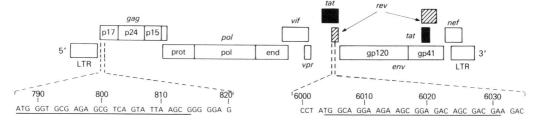

Fig. 2. Genomic structure of HIV-1. The structural genes are *gag*, *pol*, and *env* and the regulatory genes are *vif*, *vpr*, *tat*, *rev* and *nef*, respectively.

homooligomers of deoxycytidine, deoxyadenosine, and deoxythymidine in 5-, 14-, and 28-mers. We also evaluated base composition effects using oligomers that had different ratios of guanosine (G) and cytidine (C) deoxynucleotides (G + C% in the sequence).

Anti-HIV Activity of Oligomers

Sequence-nonspecific anti-HIV activity. Without a protective agent, most of the

ATH8 cells were killed by the viral cytopathic effect on day 7. As expected, a 14-mer unmodified antisense compound showed no detectable anti-HIV activity in the assay we used (Matsukura et al., 1987). In contrast, the corresponding 14-mer phosphorothioate showed potent anti-HIV activity against the cytopathic effect of HIV. However, it was unexpectedly found that phosphorothioate oligomers with sense and random sequences,

TABLE 1. Conserved Sequence of the HIV-1 *rev* Initiation Site Among Viral Isolates

HIV clones	Target sequences of *rev*	No. of Different Bases
BH10:	ATG GCA GGA AGA AGC GGA GAC AGC GAC GA	Prototype
HXB2:	ATG GCA GGA AGA AGC GGA GAC AGC GAC GA	0
HXB3:	ATG GCA GGA AGA AGC GGA GAC AGC GAC GA	0
BH102:	ATG GCA GGA AGA AGC GGA GAC AGC GAC GA	0
BH5:	ATG GCA GGA AGA AGC GGA GAC AGC GAC GA	0
PV22:	ATG GCA GGA AGA AGC GGA GAC AGC GAC GA	0
BRU:	ATG GCA GGA AGA AGC GGA GAC AGC GAC GA	0
SF2:	ATG GCA GGA AGA AGC GGA GAC AGC GAC GA	0
CDC42:	ATG GCA GGA AGA AGC GGA GAC AGC GAC GA	0
RF:	ATG GCA GGA AGA AGA GGA GAC AGC GAC GA	1(3.6%)
MN:	ATG GCA GGA AGA AGC GGA GAC AGC GAC GA	0
SC:	ATG GCA GGA AGA AGC GGA GAC AGC GAA GA	1(3.6%)
MAL:	ATG GCA GGA AGA AGC GGA GAC AGC GAC GA	0
ELI:	ATG GCA GGA AGA AGC GGA GAC AGC GAC GA	0
Z6:	ATG GCA GGA AGA AGC GGA GAC AGC GAC GA	0
NL43:	ATG GCA GGA AGA AGC GGA GAC AGA GAC GA	1(3.6%)

ization of support-bound, full-length, hydrogen phophonate polymer (**5**) with a solution of radioactive elemental sulfur ($^{35}S_8$) is performed once, manually, prior to ammoniolytic cleavage from the support and deblocking of protected base residues (B^p) to give the crude 5'-dimethoxytrityl (5'-DMT) material for either HPLC or, preferably, OPC pu-

rification, which utilizes in situ detritylation with dilute aqueous trifluoroacetic acid (TFA) to directly give the final product **1**. Detritylation after HPLC is generally performed with acetic acid (HOAc). (**C**) reproduced from Stein et al. (1990), with permission of the publisher.

**TABLE 2. Sequences of Oligomers Used in the Cytopathic Effect
Inhibition Assay and the Viral Expression Inhibition Assay[a]**

	For cytopathic effect inhibition assay using uninfected cells	
ODN-1	5′-TCG TCG CTG TCT CC-3′	Antisense
ODN-2	5′-GGA GAC AGC GAC GA-3′	Sense
ODN-3	5′-CAT AGG AGA TGC CT-3′	Antisense
ODN-4	5′-CTG GTT CGT CTC CC-3′	Random
dC_n	5′-$(dC)_n$-3′	Homooligomer
dA_n	5′-$(dA)_n$-3′	Homooligomer

	For viral expression inhibition assay using chronically infected cells
Anti-rev_{28}	5′-TCG TCG CTG TCT CCG CTT CTT CCT GCC A-3′
N-Me-anti-rev_{28}	5′-TCG T*CG CTG TCT* CCG CT*T CTT CCT* GCC A-3
	(* denotes N^3-methylthymidine)
Anti-rev_{27}	5′-TCG TCG CTG TCT CCG CTT CTT CCT GCC-3′
Sense-rev	5′-TGG CAG GAA GAA GCG GAG ACA GCG ACG A-3′
Random-rev	5′-TCG TCT TGT CCC GTC ATC GTT GCC CCT C-3′
Anti-gag	5′-CGC TTA ATA CTG ACG CTC TCG CAC CCA T-3′
Anti-tat(unspl)	5′-GTC GAC ACC CAA TTC TGA AAA TGG ATA A-3′
Anti-tat(spl)	5′-GTC GAC ACC CAA TTC AGT CGC CGC CCC T-3′
Anti-G/P-1	5′-TCT TCC CTA AAA AAT TAG CCT GTC TCT C-3′
Anti-G/P-2	5′-CCT GGC CTT CCC TTG TAG GAA GGC CAG A-3′
Anti-G/P-3	5′-TGG CTC TGG TCT GCT CTG AAG AAA ATT C-3′
dC_{28}	5′-CCC CCC CCC CCC CCC CCC CCC CCC CCC C-3′

[a]Data of experiments using some sequences (ODN-1, -2, -3, -4, and dA_n) in cytopathic effect inhibition assay were described previously (Matsukura et al., 1987).

and a 14-mer homooligomer of deoxycytidine (S-dC_{14}) exhibited similar anti-HIV activity in the cytopathic effect inhibition assay system (Matsukura et al., 1987). These data strongly suggested that protective effects of phosphorothioate oligomers against *de novo* infection of HIV, and the viral cytopathic effect, do not require a specific sequence, which we refer to as *sequence nonspecific anti-HIV activity*. Similar results using an indirect immunofluorescent assay were obtained in a *de novo* infection with H9 cells, which are resistant to the viral cytopathic effect (Matsukura, unpublished data).

Base composition effect on anti-HIV activity. There seemed to be some difference among the phosphorothioate oligomers initially tested, possibly due to different base compositions. Therefore, we tested other

homooligomers, namely deoxythymidine and deoxyadenosine phosphorothioates (S-dT_n and S-dA_n) in addition to S-dC_n. As reported (Matsukura et al., 1987), deoxycytidine phosphorothioate oligomers were significantly more active than deoxythymidine or deoxyadenosine phosphorothioate oligomers. In addition, phosphorothioate oligomers that had a wide variety of G + C content were tested in the cytopathic effect inhibition assay. Base composition seemed to be a more significant factor in shorter oligomers (Stein et al., 1989).

Length effect on anti-HIV activity. 14-, 18-, 21-, and 28-mers of deoxycytidine phosphorothioate oligomers (S-dC_{14}, -dC_{18}, -dC_{21}, -dC_{28}) were tested in parallel. The results in Figure 4 showed a clear length dependency in their anti-HIV activities. It is worthwhile to note that 0.5 μM S-dC_{28} showed more ac-

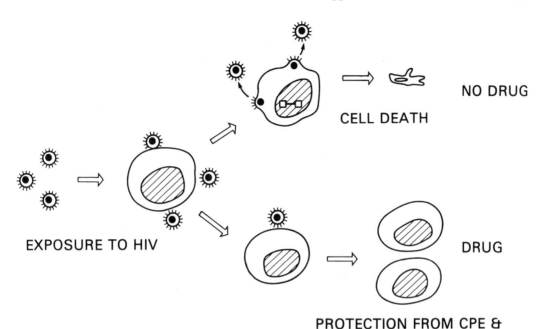

EXPOSURE TO HIV

CELL DEATH

NO DRUG

DRUG

PROTECTION FROM CPE &
CELL PROLIFERATION

Fig. 3. Schematic representation of a cytopathic effect inhibition assay which uses *de novo* infection of HIV. Without protective agents, cells sensitive to the cytopathic effect of HIV are dying. In the presence of potent antiviral drugs, cells are protected from the cytopathic effect and proliferate. Anti-HIV activity can be quantitated by counting the viable cells by the trypan blue dye exclusion method.

tivity than 1 μM S-dC$_{14}$, even though both have the same number of monomeric nucleoside residues, suggesting a real length effect. In another words, the oligomers probably functioned in their full length rather than substantially degraded, shorter lengths.

Combination of S-oligo and 2′,3′-dideoxynucleoside (ddN). A possible enhancement of anti-HIV activity of ddNs by S-oligos was tested with a combination of 2′,3′-dideoxyadenosine (ddA) and S-dC$_{14}$ in the cytopathic effect inhibition assay using ATH8 cells. As shown in Figure 5, 2 μM ddA and 5 μM S-dC$_{14}$ each exhibited a very marginal effect; however, the combination of both almost nullified the cytopathic effect of HIV, thus suggesting a synergistic enhancement of anti-HIV activity. The mechanism of such synergism is not established at this time but could likely result from different mechanisms of action for ddA and S-oligos.

Mechanism of Sequence Nonspecific Anti-HIV Activity

Since we had originally expected to find antisense activity (sequence-specific inhibition of virus at the level of viral expression), the results obtained in the cytopathic effect inhibition assay were surprising. Later, as described below, we found that phosphorothioate oligomers do have a sequence-specific inhibitory effect on HIV expression. Therefore we postulated that there are mechanisms other than translation arrest for this sequence-nonspecific antiviral activity. The effect on viral DNA synthesis by HIV infection was therefore studied by Southern blot analysis, and it was found that an S-oligo completely inhibited the synthesis of viral DNA (Fig. 6). This suggested but did not prove that intervention by S-oligo is prior to and/or at the reverse transcription step in the HIV

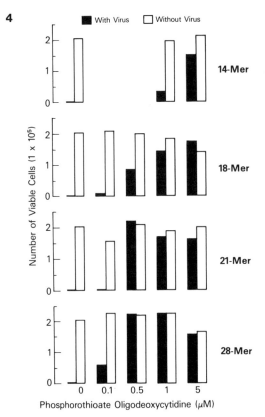

Fig. 4. Detailed comparison of anti-HIV activity between the 14-mer and 28-mer of S-dC$_n$. Filled columns represent virus-exposed cells and open columns represent non-virus-exposed cells. There is an obvious *length effect* in sequence-nonspecific anti-HIV activity even with an increase of nucleotide lengths as short as three nucleotides (see the data in 18-mer and 21-mer).

life cycle. Next we checked anti-RT (reverse transcriptase) activity *in vitro*, as RT is unique to retroviruses, and AZT and other ddNs are known to inhibit reverse transcription (as chain terminators). As shown in Figure 7, S-oligo is a sequence-nonspecific inhibitor of HIV-RT, which was also reported (Majumdar et al., 1989). In such an assay, the S-oligo is 100 to 1000 times better than the corresponding n-oligo. However, in the presence of albumin, the inhibitory effect of S-oligo against HIV-RT was almost nullified (Fig. 7). Moreover, we found a difference between the antiviral activity and anti-RT activity of

S-oligos in some systems: although S-oligo is inhibitory against viral replication of HIV and CAEV but not against AMuLV (J. Dahlberg, personal communication), S-oligo is inhibitory against RT of HIV and AMuLV but not against CAEV (Fig. 8). Since an S-oligo did

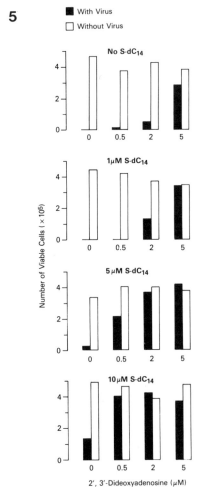

Fig. 5. Synergistic enhancement of sequence-nonspecific anti-HIV activity of ddAdo (2′,3′-dideoxyadenosine) by S-dC$_{14}$. Since this experiment involved a more potent viral inoculum for a longer duration than in the other experiments reported (Matsukura et al. 1987, Mitsuya et al. 1987),·5 μM S-dC$_{14}$ or 2 μM ddAdo alone exhibits very marginal protection compared to the data in Figure 4. In the combination with 2 μM ddAdo, 5 μM S-dC$_{14}$ exhibits 100% protective effect.

6

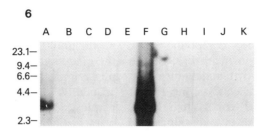

Fig. 6. S-dC$_{28}$ inhibits *de novo* HIV DNA synthesis, as a marker of viral replication in culture, in ATH8 cells exposed to the virus. On day 4 (lanes A–E) and day 7 (lanes F–J) following exposure to the virus, high-molecular-weight DNA was extracted, digested with Asp-718 (a *Kpn*I isoschizomer), and subjected to Southern blot analysis. The blot was hybridized with a labeled insert of a molecular clone of the *env* region of HIV/BH10 containing a 1.3-kb *Bgl*II fragment. Lanes A and F contain DNA from ATH8 cells that were exposed to the virus and not protected by S-dC$_{28}$. Lanes B and G, C and H, and D and I contain DNA from ATH8 cells cultured with 1, 5, and 7 μM S-dC$_{28}$, respectively. Lanes E and J contain DNA from ATH8 cells treated with 50 μM 2′,3′-dideoxyadenosine (ddAdo), and lane K contains DNA from ATH8 cells that was not exposed to the virus. Sizes are shown in kilobases. The 2.7-kb *env*-containing internal *Kpn*I fragment of the virus genome was detected only in lanes A and F.

8

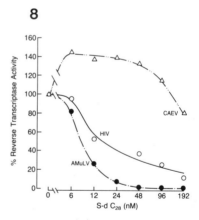

7

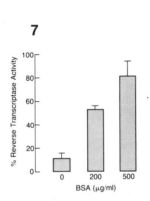

Fig. 7. Addition of bovine serum albumin (BSA) nullifies the inhibitory effect of phosphorothioate oligomer to HIV-1 reverse transcriptase in a cell-free assay system. The reverse transcriptase assay is performed as reported by Veronese and co-workers (Veronese et al., 1986) except for 0.6 μg/ml poly(rA)-oligo(dT) and 300 nM S-dC$_{28}$. In absence of proteins other than the enzyme (HIV-1 RT), 300 nM S-dC$_{28}$ inhibits HIV-1 RT activity by 89% but in the presence of 500 μg/ml BSA the inhibitory activity is reduced to 20%. The experiment was done in duplicate.

Fig. 8. Phosphorothioate oligomer (S-dC$_{28}$) inhibits RT activity of HIV and AMuLV (amphotropic murine leukemia virus) but not CAEV (caprine arthritis encephalitis virus) in vitro. Forty milliliters of culture supernatants of infected cells is pelleted by ultracentrifugation and pelleted viruses are dissolved in 400 μl of the buffer (25 mM Tris, pH 7.5, 5 mM DTT, 0.25 mM EDTA, 0.025% Triton X-100, 50% glycerol, 50 mM KCl). RT assays are performed similarly except for 0.1 mM MnCl$_2$ for AMuLV, 1 mM MgCl$_2$ for CAEV, and 0.6 mM MgCl$_2$ for HIV-1.

not significantly block the binding of HIV to CD4-positive cells at concentrations where it is effective as an anti-HIV compound (Matsukura et al., 1987), we postulate that S-oligos may inhibit viral entry by blocking fusion of the virus and cell membrane as one explanation for their anti-HIV effects. It is, however, likely that this class of compounds has complex and multifaceted effects on HIV replication, and moreover, may not work by the same mechanism in animal and human viruses. Experimental tests of this proposal require further studies.

SEQUENCE-SPECIFIC ANTI-HIV ACTIVITY

Viral Gene Expression Inhibition Assay

Culture of chronically infected cells. H9 cells were exposed to HIV-1 B (formerly known as HTLV-IIIB) at an inoculum of 500 virus particles per cell. Usually, the H9 cells were almost all infected on day 10 after the exposure to the virus. Since infected cells shortly after exposure to HIV tend to be fragile, the infected H9 cells (H9/III$_B$) were kept in culture for several months or more in complete medium (RPMI 1640 supplemented with 15% fetal calf serum, 4 mM L-glutamine, 50 nM 2-mercaptoethanol, and 50 units of penicillin, and 50 µg of streptomycin per ml) prior to the experiment.

Optimization of assay. To detect inhibitory activities of oligomers against the viral expression and toxicities of oligomers in a quantitative manner, we first assessed the effect of the number of inoculated cells on cell growth and viral production. We inoculated different numbers of H9/III$_B$ cells and then quantitated both cell growth (by [^{3}H]thymidine uptake) and the viral production in the supernatant (by p24 gag protein antigen assay, ELISA or RIA). Figures 9 and 10 show that an inoculum of H9/III$_B$ at 1250 cells per well in a 96-well culture plate gave a linear increase of p24 gag protein in the supernatant during a 5 day culture, and that [^{3}H]thymidine uptake was still increasing on day 5, suggesting that the culture was in good condition at the end of assay. Higher amounts

of the cells for the inoculum gave nonlinear curves in cellular growth and p24 gag protein production. Lower amounts of the cells for the inoculum also gave a nonlinear curve in p24 gag production (data not shown). Consequently, 1250 cells of H9/III$_B$ per well were employed for the assay. Each assay system using a different cell line needs to be optimized as the H9/III$_B$ assay system was used.

Inhibitory Effect Against Viral Production

First, we tested two representative sequences, namely S-anti-rev_{28} and S-dC$_{28}$. S-dC$_{28}$ had been found to be one of the most potent sequences tested in the cytopathic effect inhibition assay against *de novo* infection of HIV (*vide supra*). In chronically infected cells, however, S-dC$_{28}$ did not inhibit p24 gag protein production at the concentrations tested, whereas S-anti-rev_{28} inhibited viral protein production measured in the supernatant (~90% at 25 µM) in a dose-dependent manner without significant toxicity (Fig. 11). All of the control sequences tested (phosphorothioate sense, random, homocytidine, N-Me-TdR containing anti-rev_{28} S-oligomer, and nuclease-sensitive normal oligomer antisense) failed to inhibit the viral expression (Fig. 12) thus confirming both the sequence specificity of the antisense activity of S-anti-rev_{28} and the necessity of nuclease resistance for antisense activity in S-anti-rev_{28}. Another viral protein, envelope glycoprotein, was assayed using radioimmunoprecipitation in the presence of a monoclonal antibody against gp160 (Matsushita et al., 1988). The production of envelope was inhibited similar fashion as p24 gag protein and its precursor proteins, p55 and p38 (Fig. 13), which indicated that anti-rev could inhibit whole viral production rather than simply one viral protein. Since this assay system used cellular lysate labeled with [^{35}S]methionine and -cysteine, the results also indicated that S-anti-rev_{28} could inhibit not only viral production into supernatant but also the intracellular viral production *per se*.

It is worth noting that the systems used for sequence-specific anti-HIV activity can

9

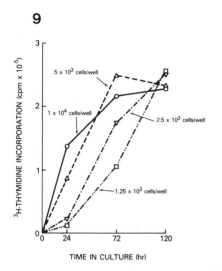

Fig. 9. Cells (1250) of H9/III$_B$ per well in 96-well flat culture plate as an inoculum are still growing on day 5 in the experiment. Cultures with inoculums more than 2500 cells per well reach plateau in growth on day 5.

10

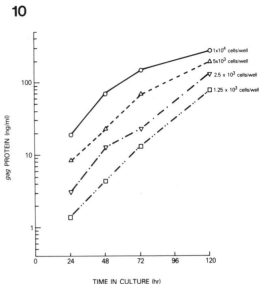

Fig. 10. The culture with 1250 cells per well as an inoculum seems to show an exponential increase of p24 gag production in supernatant during 5 day culture but other cultures with more cells do not.

produce variable data to a certain extent and, therefore, repeated experiments were necessary to produce the conclusive data such as we described here. The viral expression of different isolates and clones of HIV-1 other than HIV-1/III$_B$ should theoretically be inhibited by the same S-anti-rev_{28} for the reasons we discussed earlier in this chapter. However, recent data (H. Mitsuya, unpublished work) indicate that this may not be the case. For some strains of virus or cell types, the results using HIV-1/III$_B$ and H9 may not be seen. Because of the preliminary nature of such results, it is not possible to explain the apparent differences at this time. Possible factors include differences in the transport of synthetic oligonucleotides across cell or nuclear membrane, the amount of mRNA target, the level of expression, and the actual secondary structures of target mRNA at the *rev* initiation site among the isolates and the clones, all of which could affect the inhibitory activity of S-anti-rev_{28} against the viral expression of HIV-1. This problem will be studied further.

RNA Analyses

Functional viral gene, *rev*, in HIV is known to affect the profile of viral mRNA splicing, perhaps resulting from stabilization of the mRNA in its presence (Felber et al., 1989; Sadaie et al., 1988). In addition, we hypothesized that S-anti-rev_{28}, which inhibited viral expression in the assay we used, functioned via inhibition of rev protein production. Therefore, a Northern blot analysis and an RNA protection assay were performed using HIV-III$_B$ in H9 cells (the system with the most reproducible effects against chronic expression). As shown in Figure 14, none of the control sequences tested, except S-anti-rev_{28}, changed the mRNA pattern. Even after 28 days of continuous exposure to the compound, the mRNA pattern produced by S-anti-rev_{28} was observed, but not with S-dC$_{28}$, thus indicating no emergence of resistant virus against the antisense intervention during the time of culture (Matsukura et al., 1989).

11

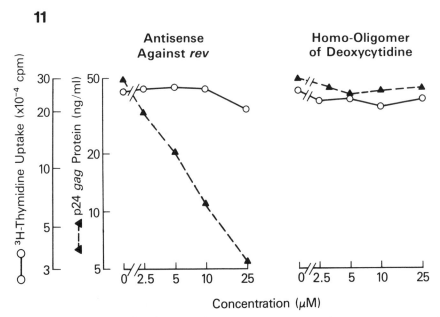

Fig. 11. Only anti-*rev* (28-mer) shows potent dose-dependent inhibition of the viral p24 gag protein production (left). S-dC$_{28}$ shows no inhibition even at 25 μ*M* highest concentration (right).

12

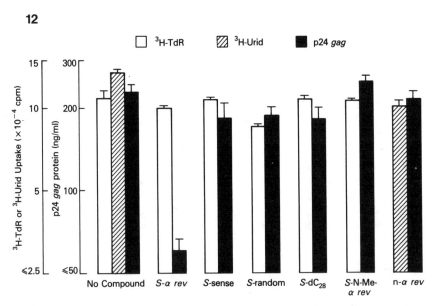

Fig. 12. Phosphorothioate oligomers (28-mer) including antisense sequence and control sequences show sequence-specific inhibition of p24 gag production. The concentrations of phosphorothioate oligomers used here were 10 μ*M*, which does not show any toxicity to cells. Error bars represent standard deviations of the data. □, [³H]thymidine; ▨, [³H]uridine; ■, p24 gag protein. S-anti-*rev* and n-anti-*rev* is denoted by S-α *rev* and n-α *rev* in this figure.

13

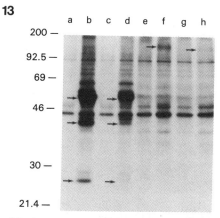

Fig. 13. S-anti-rev_{28} inhibits not only p24 gag protein but also gag protein precursors p55, p38, and envelope glycoprotein, gp120 in radioimmunoprecipitation assay (RIPA). Cell cultures of H9/III$_B$ are metabolically labeled with 2.5 mCi of [^{35}S]methionine and -cysteine for 4 hr. Equivalent trichloroacetic acid-precipitable radioactivity was treated with the following antibodies: 10 μl of control mouse ascites fluid generated by p3×63 cells (lanes a and c), 10 μl of mouse ascites fluid containing the monoclonal antibody to HIV-1 p24 (Veronese et al., 1985) (lanes b and d), 5 μg of control mouse IgG (lanes e and g), and 5 μg of mouse monoclonal IgG antibody to *env* (Matsushita et al., 1988) (lanes f and h). Lanes a, b, e, and f, no drug control; c, d, g, and h, samples treated with 25 μM S-anti-rev_{28}. The gag proteins p55, p38, and p24 indicated by arrows in lane d (S-anti-*rev*) are greatly reduced in comparison with those in lane b (control) as was gp120 env glycoprotein, indicated by the arrow in lane h (S-anti-*rev*), in comparison with that in lane f (control).

Time Course

To assess the dynamics of the inhibitory activity of S-anti-rev_{28}, a time course (24, 48, 72, 96, and 120 hr after the exposure to the oligo) was studied. Significant dose-dependent anti-HIV activity in p24 gag protein production was not observed until 96~120 hr. In the Northern blot assay, by as early as 72 hr there was a drastic change of HIV mRNA pattern (disappearance of genomic mRNA, 9.4 kb) (Fig. 15). Mechanistically, the time delay between these two time profiles could result from sequential effects, which are translation arrest of *rev* mRNA into rev pro-

tein by S-anti-rev_{28} → lower level of intracellular rev protein → change of mRNA splicing pattern → inhibition of production of viral structural proteins. This hypothesis, however, needs to be tested by further experiments using a particular system to quantitatively detect rev protein production in cell culture system.

Other Antisense S-Oligos

Other target sequences, including the initiation site of *gag*, the frameshift region between *gag* and *pol*(*G/P*), and the initiation site of *tat* exon 1 (Fig. 2) were tested. S-anti-*gag* shows marginal inhibitory effect in the assay. Even in combination with S-anti-rev_{28}, S-anti-*gag* exhibited comparatively little activity in antiviral activity against HIV (Fig. 16). There was significant inhibition of viral production by S-anti-*tat*(spl). By contrast, S-anti-*tat*(unspl) gave no inhibition of viral production, suggesting that the inhibition of the viral protein production by S-anti-*tat*(spl) may be from hybridization with the viral mRNA, but probably not with the genomic proviral DNA of HIV-1. S-anti-*G/P* -1, -2, and -3 showed inhibitory activity that was much less than that of S-anti-rev_{27} (Fig. 17).

UPTAKE OF OLIGOMERS BY CELLS

Since oligomers such as the S-anti-rev_{28} have a molecular weight of approximately 10,000, there has been concern whether these and other antisense oligomers, in general, could accumulate in cells at an effective concentration for antisense inhibition of expression. Some data concerning uptake of n-oligomers have been published (Goodchild et al., 1988; Wickstrom et al., 1988, Loke et al., 1989, Yakubov et al., 1989); however, n-oligomers are sensitive to nucleases and, theoretically, it would be very difficult to measure cellular uptake of the oligomer and the various fragments, and unambigously interpret the results. Indeed, as suggested by the data given in Figure 18, the n-oligomer was significantly degraded into monomers, and one of these, thymidine, competed with [^{3}H]thymidine thus resulting in lower counts at nontoxic oligo-

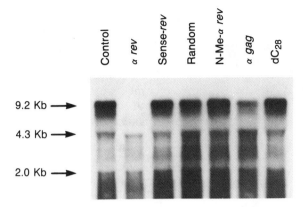

Fig. 14. Northern blot analysis shows reduction of genomic mRNA of HIV-1 in the presence of S-anti-rev_{28}. Cytoplasmic RNAs are extracted from H9/III$_B$ on day 5 in culture in the presence or absence of phosphorothioate oligomers. Note the change of mRNA profile in S-anti-rev_{28} (α rev in this figure) but not in other controls including S-anti-gag (α gag).

mer concentrations. S-oligomer, however, is known to be nuclease-resistant and, in fact, there was no detectable effect on TdR uptake in this cell culture assay (Fig. 18).

Radiolabeled Oligomer

To study uptake by cells, we initially employed an end-labeled [^{35}S]phosphorothioate oligomer, which is relatively phosphatase resistant (Eckstein, 1985) Figure 19 shows a profile of cellular uptake of ^{35}S-end-labeled S-dC$_{28}$. The uptake approached a plateau at approximately 1 hr after exposure to the oligomer. Incubation of cells at 4°C for 1 hr gave one-third the number of counts in comparison to that at 37°C (M. Matsukura, un-

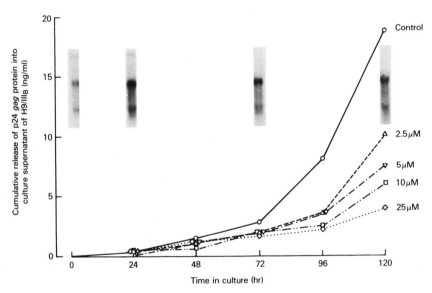

Fig. 15. After 96 hr, dose-dependent inhibitory effect of S-anti-rev_{28} to p24 gag protein production is observed. Prior to the change in p24 gag production, it appears to be a significant change in mRNA profile of HIV-1.

published data) indicating a temperature-dependent uptake mechanism. Because there is uncertainty about the extent of possible degradation of the terminal phosphate in a phosphorothioate end-labeled oligomer, we are now investigating the uptake with internally ^{35}S-labeled oligomer (Marti et al., work in progress). Two sets of preliminary data using terminally labeled or internally labeled oligomers appear to be similar, confirming the observation in Figure 19. In the experiments using end-labeled S-dC$_{28}$, there was significantly less uptake (almost half) of the oligomer by the U937 cell line at 37°C, of which each cell is similar in size to H9 (M. Matsukura, unpublished data). This observation suggests that there could be a cell type-specific difference in uptake profile and, consequently, different extents of sequence-specific anti-HIV activity. To elucidate these crucial problems, further uptake experiments and additional viral expression inhibition assays should be carried out using multiple cell types including normal peripheral blood cells.

Fluorescent-Labeled Oligomer

A fluorescein-conjugated phosphorothioate oligomer was prepared to study the distribution and localization of phosphorothioate oligomer inside of cells. Briefly, within 30 min, an apparent steady-state distribution of the 28-mer phosphorothioate oligomers was reached, in accordance with the results from experiments using radiolabelled oligomer (*vide supra*) in H9 cells. Also there seems to be significant accumulation of the oligomer in the cell nucleus and perinuclear organelles in HeLa cells (Marti et al., work in progress). Moreover, there is an intense fluorescence signal in the nucleolus. These characteristics of phosphorothioate oligomers could be beneficial for the acquisition of antisense activity.

PHARMACOKINETICS AND CLINICAL USAGE

Very little information has been published regarding the pharmacologic behavior of antisense oligomers, in general, and S-oligos in particular. The anti-HIV activity that we and others have found for S-oligos has, however, recently prompted initiation of preliminary pharmacokinetic, distribution, and toxicity studies by the Developmental Therapeutics Program of the National Cancer Institute (J. Covey, personal communication) and others (P. Iversen, personal communication). Although the results and discussion of this work have yet to be published, it is encouraging that Agrawal et al. (1989) found no overt evidence of toxicity in rats given a single intraperitoneal injection of an S-oligo at 150 mg/kg. Obviously, however, *in vivo* experiments are necessary to obtain initial information on possible toxicological profiles of the phosphorothioate oligomers.

The milligram amounts of S-oligos that derive from current automated DNA synthesis equipment are clearly inadequate to supply enough material for typical preclinical/clinical drug studies, which require kilogram quantities. In addition to a potent, sequence-specific oligomer, successful first-generation antisense therapy thus requires major advances in synthesis scale-up and efficiency with regard to the raw materials/solvent that are used. Exploratory investigations have achieved reasonable economies at the 1-g scale (Andrus et al., 1989). Extensions of such efforts are underway to reach highly efficient multiple gram pilot manufacturing runs, per reactor, which can in turn be multiplexed for initial production of large quantities in reasonable periods of time (G. Zon, personal communication). On the other hand, initial costs for such material will be relatively high. Development of efficient drug administration/delivery protocols is also very important. Second-generation antisense drugs and formulations will probably need to incorporate novel features that increase specificity and/or potency. In fact, we found that intercalator-conjugated phosphorothioate oligomers exhibited a 5- to 10-fold enhancement of anti-HIV activity against the viral expression *in vitro* (Matsukura, unpublished data).

In conclusion, we have learned that phosphorothioate oligomers can exhibit anti-HIV

16

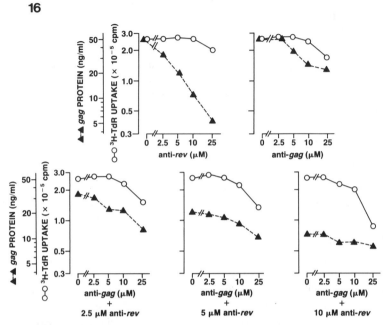

Fig. 16. S-anti-gag shows very marginal inhibitory effect against viral expression. While 25 μM S-anti-rev_{28} profoundly inhibits gag protein production (approx. 90%), S-anti-*gag* shows only 50% inhibition at same concentration. In the combination with S-anti-*gag* and -rev_{28}, no synergistic inhibitory effects are observed.

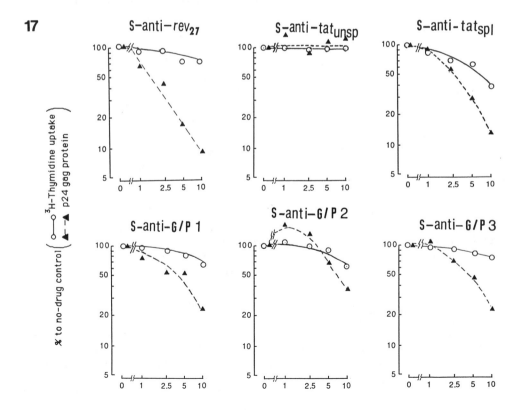

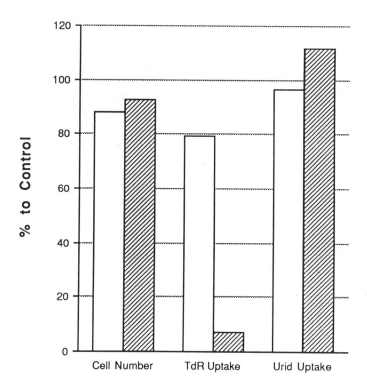

Cell Growth

Fig. 18. While 25 μM S-anti-rev_{28} shows comparable results of toxicity in viable cell number, [^{3}H]thymidine uptake, and [^{3}H]uridine uptake, 25 μM n-anti-rev_{28} shows a reduction in [^{3}H]thymidine uptake compared to the other parameters. The reduction in [^{3}H]thymidine uptake in n-anti-rev_{28} suggests that n-anti-rev_{28} was degraded into monomer nucleotides (unlabeled), and that unlabeled thymidine competed with [^{3}H]thymidine in DNA synthesis. □, S-oligo; ▨, n-oligo.

activities by multiple mechanisms, namely sequence-specific (antisense) and sequence-nonspecific (nonantisense) mechanisms. We have also discussed the importance of well-defined experimental conditions (i.e., distinction between a *de novo* infection assay using uninfected cells and a chronically infected cell assay). To establish *real* sequence specificity, it is necessary to have control sequences (e.g., sense, random, N^3-methyl-thymidine containing antisense, homooligo-

mer, and other antisense constructs), as many as is reasonable or necessary. It is also very critical to distinguish the antisense activity from the toxicities of oligomers, especially in cell proliferation inhibition assays such as those in antioncogene antisense (anti-c-*myc* and etc.) experiments. The field has shown growth, but much research remains to be done. In particular, for the development of S-anti-*rev* as an anti-HIV compound, a detailed profile of possible differences of its anti-

◄ **Fig. 17.** Sequences other than S-anti-rev_{27} (see Table 2) show less inhibitory effect against viral expression. Note the difference of the antisense activities and cytotoxicities between antisense oli-

gonucleotides against unspliced *tat* (including splice junction) and spliced *tat* (mRNA sequence, upstream of initiation site for *tat* mRNA).

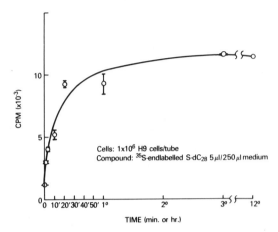

Fig. 19. Accumulation of S-dC$_{28}$ end-labeled with [^{35}S]phosphorothioate appears to reach a plateau in 1 hr. The assay is performed in duplicate except for the sample at 12 hr. H9 cells (1×10^6) are incubated with 250 μl complete medium (see "viral gene expression inhibition assay" in text) containing 5 μl of end-labeled S-dC$_{28}$ in 37°C. Cells were incubated at 37°C for a certain period of time, pelleted, washed twice with 1 ml of phosphate-buffered saline, and the pellet was suspended with scintillation cocktail and the radioactivity was counted.

HIV activity among various viral isolates and cell types should be explored. This caveat also holds for any other intended anti-HIV antisense oligonucleotides.

SUMMARY

In in vitro assay systems, nuclease-resistant phosphorothioate oligodeoxynucleotides (S-oligos) exhibited potent anti-HIV activity with multiple mechanisms. The S-oligos inhibited de novo infection of *uninfected target cells (ATH8)* by HIV in a sequence nonspecific manner, which means that S-oligos with antisense, sense, random, and homo sequences all inhibited de novo infection by HIV. On the other hand, the expression of HIV in *chronically infected cells (H9/III$_B$)* was inhibited by an antisense S-oligo against mRNA *rev* (S-anti-*rev*) but not by other sequences used as controls. Unmodified oligodeoxynucleotides (n-oligos: normal oligodeoxynucleotides) were not active in each of the as-

says we used. The results we obtained clearly demonstrate that S-oligos can inhibit the life cycle of HIV at multiple points. We also demonstrated that S-oligos can enter cells in a dose-dependent manner, and furthermore can enter into the cell nucleus.

S-oligos appear to be a potential candidate for drug treatment against AIDS, by virtue of their unique biological properties. The development of large-scale low cost synthesis and purification is now underway to enable clinical application.

ACKNOWLEDGMENTS

We thank Drs. Gerald Zon, Jack Cohen, Ichiro Matsuda, and Shinji Harada for their support during this project, and Drs. John Dahlberg, William Egan, and Patrick Iversen for helpful discussion.

REFERENCES

Agrawal S, Ikeuchi T, Sun D, Sarin PS, Konopka A, Maizel J, Zamecnik PC (1989): Inhibition of human immunodeficiency virus in early infected and chronically infected cells by antisense oligodeoxynucleotides and their phosphorothioate analogues. Proc. Natl. Acad. Sci. U.S.A. 86:7790–7794.

Agris CH, Blake KR, Miller PS, Reddy MP, Ts'o PO (1986): Inhibition of vesicular stomatitis virus protein synthesis and infection by sequences specific oligodeoxynucleoside methylphosphonates. Biochemistry 25:6268–6270.

Andrus A, Efcavitch JW, McBride LJ, Giusti B (1988): Novel activating and capping reagents for improved hydrogenphosphonate DNA synthesis. Tetrahedron Lett. 29:861–864.

Andrus A, Geiser T, Zon G (1989): Synthesis of antisense phosphorothioate analogs for pharmacokinetic and pre-clinical studies. Nucleosides Nucleotides 8:967–968.

Eckstein F (1985): Investigation of enzyme mechanisms with nucleoside phosphorothioates. Annu. Rev. Biochem. 54:367–402.

Feinberg MB, Jarrett RF, Aldovini A, Gallo RC, Wong-Staal F (1986): HTLV-III expression and production involve complex regulation at the levels of splicing and translation of viral RNA. Cell 46:807–817.

Felber BK, Hadzopoulou-Cladaras M, Cladaras C, Copeland T, Pavlakis GN (1989): rev protein of human immunodeficiency virus type 1 affects the stability and transport of the viral mRNA. Proc. Natl. Acad. Sci. U.S.A. 86:1495–1499.

Goodchild J, Agrawal S, Civeira MP, Sarin PS, Sun D, Zamecnik PC (1988): Inhibition of human immunodeficiency virus replication by antisense oligodeoxynucleotides. Proc. Natl. Acad. Sci. U.S.A. 85:5507–5511.

Gupta KC (1987): Antisense oligonucleotides provide insight into mechanism of translation initiation of two sendai virus mRNAs. J. Biol. Chem. 262:7492–7496.

Hasan N, Somasekhar G, Szybalski W (1988): Antisense RNA does not significantly affect expression of the *gal*K gene of *Escherichia coli* or the *N* gene of coliphage lambda. Gene 72:247–252.

Heikkila R, Schwab G, Wickstrom E, Loke SL, Pluznik DH, Watt R, Neckers LM (1987): A c-*myc* antisense oligodeoxynucleotide inhibits entry into S phase but not progress from G_0 to G_1. Nature (London) 328:445–449.

Izant JT, Weintraub H (1985): Constitutive and conditional suppression of exogenous and endogenous genes by anti-sense RNA. Science 229:345–352.

Kyogoku Y, Lord RC, Rich A (1967): The effect of substituents on the hydrogen bonding of adenine and uracil derivatives. Proc. Natl. Acad. Sci. U.S.A. 57:250–257.

Lemaitre M, Bayard B, Lebleu B (1987): Specific antiviral activity of a poly(L-lysine)-conjugated oligodeoxyribonucleotide sequence complementary to vesicular somatitis virus N protein mRNA initiation site. Proc. Natl. Acad. Sci. U.S.A. 84:648–652.

Loke SL, Stein CA, Zhang XH, Mori K, Nakanishi M, Subasinghe C, Cohen JS, Neckers LM (1989): Characterization of oligonucleotide transport into living cells. Proc. Natl. Acad. Sci. U.S.A. 86:3474–3478.

Majumdar C, Stein CA, Cohen JS, Broder S, Wilson SH (1989): Stepwise mechanism of HIV reverse transcriptase: Primer function of phosphorothioate oligodeoxynucleotide. Biochemistry 28:1340–1346.

Matsukura M, Shinozuka K, Zon G, Mitsuya H, Reitz M, Cohen JS, Broder S (1987): Phosphorothioate analogs of oligodeoxynucleotides: Inhibitors of replication and cytopathic effects of human immunodeficiency virus. Proc. Natl. Acad. Sci. U.S.A. 84:7706–7710.

Matsukura M, Zon G, Shinozuka K, Stein CA, Mitsuya H, Cohen JS, Broder S (1988): Synthesis of phosphorothioate analogues of oligodeoxyribonucleotides and their antiviral activity against human immunodeficiency virus (HIV). *Gene* 72:343–347.

Matsukura M, Zon G, Shinozuka K, Robert-Guroff M, Shimada T, Stein CA, Mitsuya H, Wong-Staal F, Cohen JS, Broder S (1989): Regulation of viral expression of human immunodeficiency virus in

vitro by an antisense phosphorothioate oligodeoxynucleotide against rev(*art/trs*) in chronically infected cells. Proc. Natl. Acad. Sci. U.S.A. 86:4244–4248.

Matsushita S, Robert-Guroff M, Rusche J, Koit A, Hattori T, Hoshino H, Javaherian K, Takatsuki K, Putney S (1988): Characterization of a human immunodeficiency virus neutralizing monoclonal antibody and mapping of the neutralizing epitope. J. Virol. 62:2107–2114.

McManaway ME, Neckers LM, Loke SL, Al-Nasser AA, Redner RL, Shiramizu BT, Goldschmidts WL, Huber BE, Ghatia K, Magrath IT (1990): Tumor-specific inhibition of lymphoma growth by an antisense oligodeoxynucleotide. Lancet 335:808–811.

Mitsuya H, Popovic M, Yarchoan R, Matsushita S, Gallo RC, Broder S (1984): Suramin protection of T cells in vitro against infectivity and cytopathic effect of HTLV-III. Science 226:172–174.

Mitsuya H, Matsukura M, Broder S (1987): Rapid in vitro systems for assessing activity of agents against HTLV-III/LAV. In Broder S (ed): AIDS: Modern Concepts and Therapeutic Challenges. New York and Basel: Marcel Dekker, pp 303–334.

Sadaie MR, Benter T, Wong-Staal F (1988): Site-directed mutagenesis of two trans-regulatory genes (tat-III, trs) of HIV-1. Science 239:910–913.

Sarin PS, Agrawal S, Civeira MP, Goodchild J, Ikeuchi T, Zamecnik PC (1988): Inhibition of acquired immunodeficiency syndrome virus by oligodeoxynucleoside methylphosphonates. Proc. Natl. Acad. Sci. U.S.A. 85:7448–7451.

Shibahara S, Mukai S, Morisawa H, Nakashima H, Kobayashi S, Yamamoto N (1989): Inhibition of human immunodeficiency virus (HIV-1) replication by synthetic oligo-RNA derivatives. Nucl. Acids Res. 17:239–252.

Smith CC, Aurelian L, Reddy MP, Miller PS, Ts'o POP (1986): Antiviral effect of an oligo(nucleotide methylphosphonate) complementary to the splice junction of herpes simplex virus type 1 immediate early pre-mRNAs 4 and 5. Proc. Natl. Acad. Sci. U.S.A. 83:2787–2791.

Sodroski J, Goh WC, Rosen C, Dayton A, Terwilliger E, Haseltine W (1986): A second post-transcriptional *trans*-activator gene required for HTLV-III replication. Nature (London) 321:412–417.

Stec WJ, Zon G, Egan W, Stec B (1984): Automated solid-phase synthesis, separation, and stereochemistry of phosphorothioate analogues of oligodeoxyribonucleotides. J. Am. Chem. Soc. 106:6077–6079.

Stec WS, Zon G, Uznanski B (1985): Reversed-phase high-performance liquid chromatographic separation of diastereomeric phosphorothioate analogues of oligodeoxyribonucleotides and other back-bone-modified congeners of DNA. J. Chromatogr. 326:263–280.

Stein CA, Matsukura M, Subasinghe C, Broder S, Cohen JS (1989): Phosphorothioate oligodeoxynucleotides are potent sequence nonspecific inhibitors of *de novo* infection by HIV. AIDS Res. Hum. Retro. 5:639–646.

Stein CA, Iversen PL, Subasinghe C, Cohen JC, Stec WJ, Zon G (1990): Preparation of ^{35}S labeled polyphosphorothioate oligodeoxyribonucleotides by use of hydrogen-phosphonate chemistry. Anal. Biochem. 188:11–16.

Stephenson ML, Zamecnik PC (1978): Inhibition of Rous sarcoma viral RNA translation by a specific oligodeoxyribonucleotide. Proc. Natl. Acad. Sci. U.S.A. 75:285–288.

Veronese FM, Sarngadharan MG, Rahman R, Markham PD, Popovic M, Bodner AJ, Gallo RC (1985): Monoclonal antibodies specific for p24, the major core protein of human T-cell leukemia virus type III. Proc. Natl. Acad. Sci. U.S.A. 82:5199–5202.

Veronese FM, Copeland TD, DeVico AL, Rahman R, Oroszlan S, Gallo RC, Sarngadharan MG (1986): Characterization of highly immunogenic p66/p51 as the reverse transcriptase of HTLV-III/LAV. Science 231:1289–1291.

Wickstrom E, Simonet W, Medlock K, Ruiz-Robles I (1986): Complementary oligonucleotide probe of vesicular stomatitis virus matrix protein mRNA translation. Biophys. J. 49:15–17.

Wickstrom EL, Bacon TA, Gonzalez A, Freeman DL, Lyman GH, Wickstrom E (1988): Human promyelocytic leukemia HL-60 cell proliferation and c-myc protein expression are inhibited by an antisense pentadecadeoxynucleotide targeted against c-myc mRNA. Proc. Natl. Acad. Sci. U.S.A. 85:1028–1032.

Yakubov LA, Deeva EA, Zarytova VF, Ivanova EM, Ryte AS, Yurchenko LV, Vlassov VV (1989): Mechanism of oligodeoxynucleotide uptake by cells: Involvement of specific receptors? Proc. Natl. Acad. Sci. U.S.A. 86:6454–6458.

Zaia JA, Rossi JJ, Murakawa GJ, Spallone PA, Stephens DA, Kaplan BE, Eritja R, Wallace RB, Cantin EM (1988): Inhibition of human immunodeficiency virus by using an oligonucleotide methylphosphonate targeted to the tat-3 gene. J. Virol. 62:3914–3917.

Zamecnik PC, Stephenson ML (1978): Inhibition of Rous sarcoma virus replication and cell transformation by a specific oligodeoxynucleotide. Proc. Natl. Acad. Sci. U.S.A. 75:280–284.

Zamecnik PC, Goodchild J, Taguchi Y, Sarin PS (1986): Inhibition of replication and expression of human T-cell synthetic oligonucleotides complementary to viral RNA. Proc. Natl. Acad. Sci. U.S.A. 83:4143–4146.

Zerial A, Thuong NT, Helene C (1987): Selective inhibition of the cytopathic effect of type A influenza viruses by oligodoexynucleotides covalently linked to an intercalating agent. Nucl. Acids Res. 15:9909–9919.

Human Immunodeficiency Virus Type-1 Replication Is Reduced by Intracellular Antisense RNA Expression

Georg Sczakiel, Karola Rittner, and Michael Pawlita

Institut für Virusforschung, Deutsches Krebsforschungszentrum, Im Neuenheimer Feld 280, D-6900 Heidelberg, Germany

INTRODUCTION

The acquired immunodeficiency syndrome (AIDS) is a disease associated with infection by specific viruses, the human immunodeficiency virus types 1 or 2 (HIV-1 and HIV-2) (Barre-Sinoussi et al., 1983; Popovic et al., 1984; Salahuddin et al., 1985). The cells in which these viruses predominantly replicate are lymphocytes and macrophages. Both these cell types are of hematopoietic origin, meaning that they belong to a cellular compartment where physiologically individual differentiated cells have a distinct half-life and are continuously replaced from progeny of bone marrow stem cells. This renewal potential of the bone marrow is exploited in clinical medicine for the therapy of malignancies of hematopoietic cells when bone marrow free of malignant cells is transplanted into a patient to repopulate the hematopoietic system after benign or malignant hematopoietic cells have been destroyed by irradiation. The development of molecular biological techniques now has made it thinkable and will perhaps make it feasible in the future to use bone marrow transplantation as a tool for somatic gene therapy (Dzierzak et al., 1988; Friedman, 1989).

If bone marrow stem cells and their differentiated progeny could be rendered resistant to replication of HIV, such cells could be of therapeutic benefit in an infected individual (Baltimore, 1988). One potential way of making cells resistant to HIV replication could be the intracellular expression of antisense RNA targeted against viral sequences. An alternative application of the antisense principle, the use of short synthetic extracellularly added antisense nucleic acids, will not be discussed here. The difference between host-specific and virus-specific nucleotide sequences can be assumed to lead to high selectivity, which is not always the case for the differences between cellular and viral proteins. The use of the antisense principle for the inhibition of viral replication was already shown in a variety of biologically divergent systems with different forms of antisense nucleic acids including the constitutive intracellular antisense RNA expression (Table 1). This form of application of antisense RNAs, several hundreds of base pairs in length, might have certain advantages. Minor mutations in the target sequences that appear frequently in retroviral genomes including HIV-1 (Coffin, 1986; Alizon et al., 1986; Steinhauer and Holland, 1987) can be assumed not to affect antisense RNA target interactions.

In the following we describe the potential applicability of intracellularly transcribed antisense RNA as an anti-HIV-1 agent in two independent experimental tissue culture systems. In the first antisense RNA expression plasmids are comicroinjected with infectious

Prospects for Antisense Nucleic Acid Therapy of Cancer and AIDS, pages 179–193

TABLE 1. Experimental Inhibition of Virus Replication/Gene Expression by Intracellular Antisense RNA Expression

Virus	Host cell	Target sequence	Reference
Rous sarcoma virus (RSV)	Quail cell line R(−)Q	Envelope gene (*env*)	Chang and Stoltzfuss (1985)
Rous sarcoma virus (RSV)	Quail cell line R(−)Q	3'- and 5'-end noncoding regions	Chang and Stoltzfuss (1987)
Rous sarcoma virus carrying neomycin resistance gene sequences	Chicken fibroblasts	Neomycine resistance gene	To et al. (1986)
Bovine papillomavirus type 1	Mouse C127 cell line	BPV1 69% fragment	Bergman et al. (1986)
Simian 40 (SV40)	African green monkey kidney cells (COS1)	Large T-antigen 5'-end	Jennings and Molloy (1987)
Human papillomavirus type 18 (HPV 18)	Human cervical carc. cell line C4-1	HPV 18 open reading frames E6 and E7	von Knebel-Doeberritz et al. (1988)
Human T-lymphotropic virus type 1 (HTLV-1)	Primary human leukocytes	5'-LTR and 5'-*gag* region,	von Rueden and Gilboa (1989)
Human immunodeficiency virus type 1 (HIV-1)	Human colon carcinoma cell line SW480	5' leader/*gag*	Sczakiel et al. (1990)
Human immunodeficiency virus type 1 (HIV-1)	Human T-lymphoid cell line Jurkat	*tat*-exon 1	Rhodes and James (1990)
Human immunodeficiency virus type 1 (HIV-1)	Human T-lymphoid cell line Jurkat	5' leader/*gag*	Sczakiel and Pawlita (1991)
Tobacco mosaic virus	Transgenic tobacco plants	Coat protein	Powell et al. (1989)
Potato virus X (PVX)	Transgenic tobacco plants	Coat protein	Hemenway et al. (1988)
Cucumber mosaic virus	Transgenic tobacco plants	Coat protein	Cuozzo et al. (1988)
Bacteriophage SP	*E. coli*	Coat protein, replicase	Coleman et al. (1985)
Bacteriophages SP, Q beta, GA	*E. coli*	Maturation gene Shine–Dalgarno sequence of maturation gene	Hirashami et al. (1986)

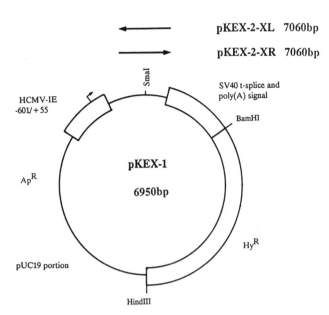

pKEX-2-XL   7060bp

pKEX-2-XR   7060bp

Fig. 1. Schematic representation of the expression vector system used for the intracellular expression of antisense RNA. The arrows above of the *Sma*I site represent pUC131-derived multilinker sites inserted in either orientation in plasmids pKEX-2-XL and pKEX-2-XR, respectively. Abbreviations: HCMV-IE −601/+55, human cytomegalovirus immediate early promoter/enhancer element (positions −601 to +55 in respect to the start of transcription); Ap^R, ampicillin resistance gene; Hy^R, hygromycin B resistance gene, driven by the herpes simplex virus thymidine kinase promoter; SV40, simian virus 40.

proviral HIV-1 DNA into the nuclei of human epithelioid cells. In this rapid and transient assay the inhibition of viral replication can be followed by a decrease in the production of HIV-1-specific antigens in the cell-free culture supernatants (Sczakiel et al., 1990). In the second assay, stably transfected human T-lymphoid cell clones with constitutive antisense RNA expression were used to show inhibition of HIV-1 replication after infection (Sczakiel and Pawlita, 1991).

METHODS AND RESULTS

HIV-1 Antisense RNA Expression Plasmids

The antisense RNA expression plasmids used in this work (Fig. 1) carry the hygromycin B resistance gene (Bernard et al., 1985) in *cis* as a dominant selectable marker gene. This serves for selection of human T lymphoid cells that have integrated the plasmid DNA into their chromosome upon transfection. The expression of HIV-1 antisense RNA is directed by the constitutive immediate early (IE) promoter/enhancer element of the human cytomegalovirus (HCMV) (Boshart et al., 1985) for two reasons. First, the potentially antiviral agent (antisense RNA) should already be present at the time of infection in order to attack more efficiently early steps in the viral replication cycle. Second, possible effects on viral replication by the induction of an inducible promoter, which cannot easily be distinguished from antisense RNA effects, should be avoided. In addition the HCMV-IE promoter/enhancer element is episomally one of the strongest known RNA-

*Pol*II promoters, e.g., in human cells it is two orders of magnitude stronger than the Rous sarcoma virus LTR and three times stronger than the human β-globin promoter (Rittner et al., 1991). Also the integrated HCMV-IE promoter/enhancer element is active in cell types originating from a wide variety of tissues and species.

In this study neither of the widely used retroviral vector systems, which direct integration of expression cassettes into the host cell chromosome, nor self-replicating, T cell-specific vector systems, e.g., such as *Herpes saimirii*-derived ones (Graessmann and Fleckenstein, 1989) was used.

HIV-1 Replication in Cells Comicroinjected With Proviral DNA and Antisense RNA Expression Plasmids (Transient Comicroinjection Assay)

The principle of this assay, schematically depicted in Figure 2, is based on the observation that microinjection of infectious proviral HIV-1 DNA into the nuclei of human fibroblasts and epithelioid cells leads to viral replication and to the production and release of virus particles (Boyd et al., 1988; Sczakiel et al., 1990). Therefore the inhibitory properties of any potential antiviral molecule including proteins or low-molecular-weight compounds can be tested and related to each other in this assay by comicroinjection together with HIV-1 provirus. In this study test plasmids and nondefective proviral HIV-1 DNA (pNL4-3; Adachi et al., 1986) were comicroinjected into SW480 cells (CD4⁻ epithelioid colon carcinoma cells, Leibovitz et al., 1976) at defined amounts and molar ratios. These cells can replicate HIV-1 after microinjection and transfection, respectively, but cannot be infected, probably due to lack of CD4 antigen (Adachi et al., 1986).

Viral replication in microinjected cells was determined after amplification of infectious HIV-1 in cocultivated CD4⁺ MT-4 cells (Harada et al., 1985) by indirect immunofluorescence staining of MT-4 cells with HIV-1 anti-

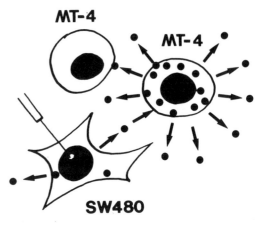

Fig. 2. Principle of the microinjection assay used for testing transiently the antiviral activity of HIV-1 antisense RNA expression plasmids. Human CD4⁻ epithelioid colon carcinoma cells (SW480) are comicroinjected with infectious proviral DNA (10 ng/μl) and a 5-fold molar excess of test plasmids. The HIV-1 virus initially produced in SW480 cells is amplified by cocultivated human CD4⁺ T-lymphoid MT-4 cells that replicate HIV-1 efficiently. The HIV-1 antigen concentrations are measured 4 days after microinjection in cell-free culture supernatants by a polyspecific ELISA.

sera or by HIV-1 antigen ELISA with cell-free coculture supernatants.

Several antisense RNA expression plasmids targeted against various HIV-1 subregions (Table 2) were cloned and tested for their antiviral activity. At a 5-fold molar excess of expression plasmids over proviral HIV-1 DNA inhibition was measured predominantly for antisense RNA expression plasmids whereas sense RNA expression plasmids and unrelated DNAs had no or only minor effects on HIV-1 replication except for two cases (p2s, pSR3), which will be discussed later (Table 3). The first antisense RNA expression plasmid that was found to inhibit HIV-1 replication was p2as (Sczakiel et al., 1990). This plasmid contains the first exon of all known HIV-1 transcripts including the first splice donor site and therefore could interfere with splicing as well as with all known HIV-1-derived RNAs. The most inhibitory plasmids express RNAs complementary to the *gag* p17/p24 coding sequences (pAR3) and the first *tat* and *rev* exons (pAR6). Plasmid pAR3 ex-

TABLE 2. Antisense RNA Expression Plasmids, the Source of the HIV-1 Sequences Used and Their Inhibitory Effects on HIV-1 Replication as Measured in the Comicroinjection Assay[a]

	Plasmid	HIV-1 sequence Length	HIV-1 sequence Subregion	Inhibition
Antisense RNA expression plasmids	p2as	406	5'-leader/*gag* (B:222–629)	+
	pAR3	1000	*gag* p17 and p24 (N:710–1710)	+ + +
	pAR4	91	frame shift (B:1549–1640)	–
	pAR5	662	*vif,vpr* (B:4704–5366)	–
	pAR6	562	*tat,rev,vpr* (B:5366–5928)	+ + +
	pAR7	3109	*env orf* (B:5366–8475)	+ + +
	pAR8	521	*env* (B:7198–7719)	+
	pAR9	330	*tat,rev* exon 2 (N:8190–8520)	–
Control plasmids				
HIV-1 sense RNA	p2s	406	5'-leader/*gag*	+
expression plasmids	pSR3	1000	*gag* p17 and p24	+
	pSR5	662	*vif,vpr*	–
	pSR8	521	*env*	–
	pSR9	330	*tat,rev* exon 2	–
Other plasmids	pBR322		Prokaryotic sequence	–
	bluescript		Prokaryotic sequence	–
	pBS29	410	HIV-1 fragment from p2as cloned into bluescript	–
	pKEX-1		See Figure 1	–
	pKEX-2-XR		See Figure 1	–
	pKEX-2-XR-CAT		CAT coding sequence cloned into pKEX-2-XR	+

[a] –, less than 30% inhibition; +, 30–70% inhibition; + + +, inhibition equal or greater than 70%; B, HIV-1 proviral clone BH10 (Ratner et al., 1985); N, HIV-1 proviral clone pNL4-3 (Adachi et al., 1986).

TABLE 3. Time-Dependent Increase of the Relative HIV-1 Antigen Concentrations in Cell-Free Culture Supernatants of Infected Jurkat Clones Stably Transfected With the Plasmids Indicated in the First Column[a]

Plasmids in stable clones	Number of clones tested	Days post-HIV-1 infection 10	14	17	20
pKEX-1	9	34 ± 25	4430 ± 1520	14800 ± 6060	39000 ± 8470
p2s	6	41 ± 30	4990 ± 1600	18000 ± 3790	40000 ± 16000
p2as (total)	9	13 ± 13	2940 ± 1680	7700 ± 7300	35000 ± 14700
p2as (RNA⁻)	4	20 ± 19	3460 ± 1800	7700 ± 7000	42000 ± 4500
p2as (RNA⁺)	5	5 ± 1	2420 ± 890	7800 ± 5600	28000 ± 15400

[a] The transfection and selection protocol is described in the text. Aliquots of cell cultures were centrifuged and dilutions of the cell-free solutions in phosphate-buffered saline were assayed for HIV-1 antigens by a commercial HIV-1 antigen ELISA and are expressed as relative HIV-1 antigen concentrations (RAC) ± 1 × SD.

presses RNA complementary to a 1-kb region within the *gag* open reading frame and could affect *gag* gene expression. Since the recently identified *cis* elements required for viral packaging (Lever et al., 1989) are potential target sequences of pAR3 as well as p2as, it is possible that antisense RNA transcribed from these two expression plasmids inhibits steps of the viral packaging process. The second most inhibitory antisense RNA expression plasmid (pAR6) targeted to the first *tat* and *rev* exons includes coding sequences for the viral vpu and vpr proteins. Also the recently identified large number of potential protein encoding spliced small HIV-1 RNAs (Salfeld et al., 1990; Schwartz et al., 1990) could be affected.

However, a clear inhibition of HIV-1 replication was also observed with the sense RNA expression plasmid p2s (Table 2). It cannot be excluded that inhibition by p2s is due to negative interference of the transcripts or hypothetical protein products (carboxy-terminally truncated *gag* polypeptides) with steps of viral replication. Negative interference of mutated viral proteins leading to inhibition of HIV-1 replication was already shown (Trono et al., 1989; Mermer et al., 1989; Malim et al., 1989).

In this assay the extent of HIV-1 inhibition did not exceed the range of 75–80%. The reasons for this are unknown. Yet, it seems likely that the low molar excess of antisense RNA expression plasmids over HIV-1 DNA and the fact that proviral DNA is microinjected (only late steps of the retroviral life cycle after synthesis of double-stranded proviral HIV-1 DNA can be affected) contribute to the lack of more complete inhibition.

HIV-1 Infection in Human T Cell Clones With Stable Antisense RNA Expression

In a second independent assay the effects of intracellular antisense RNA on HIV-1 replication were measured in subclones of the human T-lymphoid cell line Jurkat (Schneider et al., 1977; Wendler et al., 1987) that stably express antisense RNA following electrotransfection and hygromycin B selection as described in detail by Döffinger et al. (1988) and Sczakiel and Pawlita (1991). Briefly 2 × 10^7 cells were harvested from a logarithmically growing Jurkat culture, washed with phosphate-buffered saline (PBS), and electroporated in a total volume of 200 μl PBS in the presence of 50 μg linearized plasmid DNA in an electroporation chamber with 4 mm interelectrode distance (Bio-Rad) at a capacitance of 960 μF and an initial voltage of 250 V (gene pulser, Bio-Rad). In contrast to many other human T cell lines Jurkat cells can be used rather easily to establish stably transfected clones. The fact that Jurkat cells do not show an immediate cytotoxic effect a few days after HIV-1 infection as other human T cell lines may do (e.g., MT-4, ATH8, CEM) enables one to analyze HIV-1 infection over a period of several weeks and therefore might be closer to the situation in an infected host. The ratio of cells transfected transiently by the electroporation protocol used in these experiments was in the range of 1.5–2%. The frequencies of stable transfection in selected Jurkat cells were in the range of 4×10^{-6}. In contrast to this, with other human T-lymphoid cell lines [Molt-4 (Minowada et al., 1972) and H9 (Popovic et al., 1984)] either no (Molt-4) or predominantly spontaneous drug-resistant clones (H9) were selected, although transient transfection efficiencies were in the same range as was observed for Jurkat cells.

The status of the transfected recombinant plasmid DNAs in hygromycin B selected Jurkat clones was analyzed by restriction of chromosomal DNA and subsequent Southern hybridization with ^{32}P-labeled plasmid DNA (Fig. 3). The results show that the transfected plasmid DNA usually is integrated into the chromosome as a single copy and without major rearrangements in regard to the expression cassette for antisense RNA. The transcription of HIV-1 antisense RNAs in Jurkat clones was analyzed by Northern hybridization with ^{32}P-labeled double-stranded and single-stranded specific probes (Fig. 3). In cytoplasmic RNA from approximately 65% of the clones with integrated plasmid DNA antisense transcripts were detectable, indicating that antisense RNA is transported into the cytoplasm. In contrast to this there were no HIV-1-specific transcripts detectable in nuclear RNA preparations. The morphology and the proliferation rates of stable Jurkat clones were not apparently different from parental Jurkat cells, indicating that the transfection and selection procedures did not lead to major phenotypic changes.

HIV-1 replication in Jurkat clones stably expressing antisense RNA targeted to the 5′-leader/gag region (p2as). In initial experiments one p2as-Jurkat clone with constitutive antisense RNA expression was infected with HIV-1 and compared with HIV-1-infected parental Jurkat cells (Fig. 4). The HIV-1 antigen production and the percentage of HIV-1 infected cells, monitored by indirect immunofluorescence 10 days postinfection showed a significant reduction of virus replication in the antisense producing p2as clone.

To exclude differences due to clonal variation of parameters not linked to antisense RNA expression, HIV-1 replication was analyzed in infection experiments with a greater number of independent clones established with the plasmid p2as, the corresponding sense RNA expression plasmid p2s, and the original expression vector without any HIV-1 sequence, named pKEX-1 (Fig. 1). In a first set of experiments HIV-1 infection was analyzed only on days 10 and 14 postinfection (Fig. 4). On day 10 there was a 73% reduction of HIV-1 antigens in cell-free supernatants of the average of 10 p2as clones [p2as (total)] in comparison with the average of 9 pKEX-1 clones. On day 14 inhibition was still at 44%. The subgroup of p2as clones with antisense RNA detectable by Northern analysis [p2as (RNA$^+$)] showed an even stronger and longer lasting inhibition, i.e., 88% on day 10 postinfection and 80% on day 14 postinfection.

The time course of HIV-1 production in groups of infected p2as, p2s, and pKEX-1 clones was followed over a period of 3 weeks with measurement of the release of HIV-1-specific antigens into the culture supernatant (Table 3). During the first 4–6 days post-HIV-1 infection of Jurkat cells the amounts of HIV-1-specific antigens in culture super-

natants as well as the ratio of HIV-1-positive cells measured by indirect immunofluorescence were below the sensitivities of the detection methods (data not shown). Then, starting 1 week postinfection there was production of viral antigens and the infection ratio of cultured cells was greater than 1%. At that time and until approximately 3 weeks postinfection there was a reduction of virus replication in antisense RNA-producing cell clones with maximal inhibition of 75%. In the later course of infection, this effect decreased and was not detectable beyond 3 weeks postinfection.

Based on our data with p2as-Jurkat clones as well as with other inhibitory antisense RNA expressing Jurkat clones (Table 4) a general time course of infection could be deduced, schematically depicted in Figure 5. The inhibition of virus replication in antisense RNA-producing cell lines versus control clones at a given time can be expressed as the difference in the relative HIV-1 concentrations. Since the shape of the curves for the time-dependent increase of the HIV-1 antigen concentrations in control and antisense RNA expressing Jurkat cells is the same for both types of clones (e.g., the slopes s1 and s2 are the same), the inhibition $I(t)$ can also be expressed as the time difference between points of the same antigen concentration on both curves (Δt). Hence, improvement of the "antisense RNA effect" is equivalent to an increase of Δt, ideally with Δt reaching infinity. However, it should be noticed that small shifts of the curves along the time axis may result in great differences of virus production in the logarithmic phase of infection.

Increased inhibition by using other target regions. A subset of the antisense RNA expression plasmids that have been shown to lead to stronger HIV-1 inhibition than p2as in the comicroinjection assay was used to establish stably transfected Jurkat clones. At present the course of HIV-1 infection in these cell lines has not been measured with as many independent clones and as many independent infection experiments as were used to analyze p2as transfectants. However, there was an increased transient inhibition (ca. 90%;

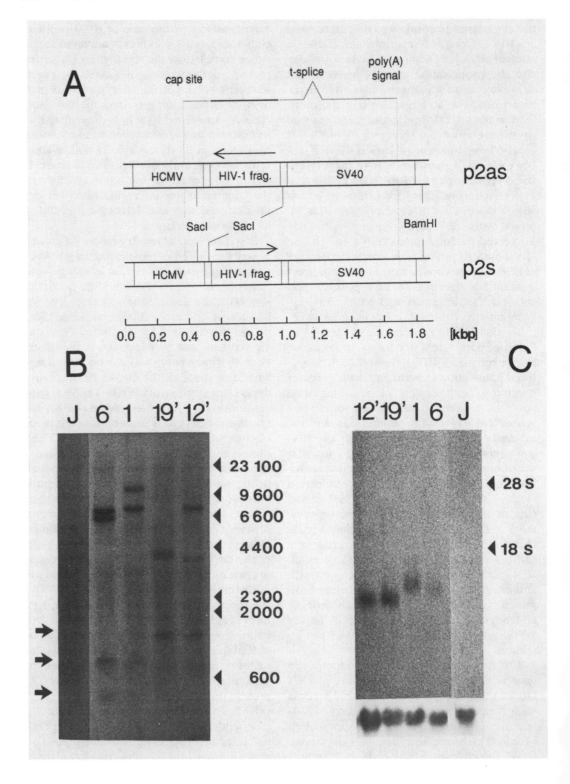

Table 4) for pAR6 clones, demonstrating that an improvement of the antiviral efficacy of intracellular antisense RNA expression is possible by varying the target region. Moreover it should be noticed that at low infective doses (multiplicity of infection (moi) < 0.1) some

---

◄ **Fig. 3.** Chromosomal integration of stably transfected plasmids and expression of HIV-1 antisense RNA and sense RNA in Jurkat clones. (**A**) Schematic representation of the expression cassettes for antisense RNA (p2as) and sense RNA (p2s) transcription. The derived transcripts (above) contain 54 base pairs (bp) of the human cytomegalovirus (HCMV) immediate early promoter/enhancer fragment downstream from the transcriptional start site followed by 409 bp HIV-1 5′-leader/*gag* sequence and the t-splice and polyadenylation signals from simian virus 40. The approximate distances between the restriction sites used in Southern analysis (*Sac*I and *Bam*HI) can be seen with the scale indicated below. (**B**) Autoradiogram from Southern hybridization of *Sac*I/*Bam*HI restricted chromosomal DNA (10 μg) of two p2as clones (no. 1, no. 6), two p2s clones (no. 12′, and no. 19′), and the parental cell line Jurkat (J). DNAs were separated on 0.8% agarose gel, blotted onto gene screen plus membrane, and hybridized with ³²P-labeled plasmid p2s under stringent conditions (hybridization: 50% formamide, 5 × SSC, 42°C; washing: 2 × SSC, 0.1% SDS, 3 × 10 min, 68°C;). In addition to two flanks from the single copy integration in each clone, the small internal *Bam*HI/*Sac*I fragments indicate the difference between p2as and p2s clones. The arrows on the left indicate positions of the 1200 bp *Bam*HI/*Sac*I fragment from p2s clones and the 800 and 400 bp fragments, respectively, in p2as clones. (**C**) Autoradiogram from Northern hybridization of total cytoplasmic RNA (10 μg) isolated from p2as clones (no. 1, no. 6), p2s clones (no. 12′, no. 19′), and parental Jurkat cells. RNAs were separated by formaldehyde/agarose gel electrophoresis, blotted onto gene screen plus membrane, and hybridized with ³²P-labeled plasmid p2s under stringent conditions (hybridization and washing as described above). Hybridization signals at the same positions as shown here were seen with ³²P-labeled strand-specific "riboprobes," additionally distinguishing between the antisense and sense transcripts. As an internal probe for the relative amounts of RNA in each sample, the blot was hybridized with a human β-actin cDNA probe (bottom part of the figure). Reproduced from Sczakiel and Pawlita (1991), with permission of the publisher.

pAR6 clones and also one tested p2as clone displayed a complete inhibition of HIV-1 infection, i.e., no spread of HIV-1 infection in the cell cultures over the observed period of 30 days, whereas in identically treated control clones HIV-1 infection progressed with the kinetics described above.

DISCUSSION

Is the Inhibition of HIV-1 Replication Mediated by Antisense RNA?

Results from both test systems, the transient microinjection assay and the T cell lines with constitutive antisense RNA expression, indicate that antisense RNA is responsible for antiviral effects.

Microinjection assay. (1) The inhibitory effects are specific for the HIV-1 portions of the tested transcripts. For example, the CAT expression plasmid pKEX-2-XR-CAT does not show a significant inhibitory effect. (2) The inhibition is observed with the antisense RNA expression plasmids, not with the corresponding sense RNA expression plasmids with the already discussed exception of the related sense RNA expression plasmids p2s and pSR3. This indicates that the inhibitory effect is caused by the single-stranded (antisense-) transcripts and not by the double-stranded HIV-1 DNA.

T cell Clones. (1) Northern positive stably transfected p2as T cell lines show a clearly stronger inhibition of HIV-1 replication than Northern negative p2as clones, i.e., clones with low abundance or completely absent antisense RNA. (2) The antiviral effects cannot be seen in p2s clones, although sense RNA expression is even stronger than antisense RNA expression in p2as clones.

However, our experiments do not provide direct evidence for inhibitory interactions, i.e., base pairing between antisense RNA and viral nucleic acids.

What Are the Critical Parameters for Efficient Inhibition of HIV-1?

Two parameters appear to have a major influence on the extent of HIV-1 inhibition, namely the selected target region and the

188   Sczakiel et al.

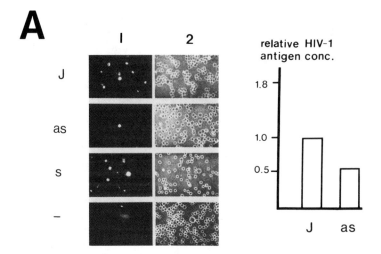

stable Jurkat clones				
		RAC	%IF$^+$	RAC
		10 days	10 days	14 days
transfected plasmid	number of clones tested	p.i.	p.i.	p.i.
pKEX-1	9	45 ± 10	27 ± 12	16800 ± 2500
p2s	7	46 ± 13	26 ± 12	24000 ± 7000
p2as (total)	10	12 ± 10	14 ± 12	9400 ± 5800
p2as(RNA$^-$)	5	19 ± 6	22 ± 20	13500 ± 1400
p2as(RNA$^+$)	5	5 ± 1	5 ± 3	5300 ± 1700

Fig. 4. Inhibition of HIV-1 replication in Jurkat clones stably transfected with the antisense RNA expression plasmid p2as. (**A**) Analysis of HIV-1 infection in parental Jurkat cells (J), one p2as-Jurkat clone (as), one p2s-Jurkat clone (s), and uninfected Jurkat cells serving as a negative control (−) by indirect immunofluorescence staining for HIV-1 antigen-positive cells (column 1: immunofluorescence staining with an AIDS patient serum; column 2: phase contrast) and measurement of HIV-1 antigen concentrations in cell-free supernatants on day 9 postinfection in HIV-1-infected Jurkat cells and the p2as-Jurkat clone tested on the left. (**B**) Summary of the HIV-1 infection data collected with groups of stable Jurkat clones indicated in the first column. The p2as group was subdivided into clones with antisense RNA expression detectable [p2as(RNA$^+$)] and undetectable [p2as(RNA$^-$)] by Northern hybridization. Abbreviations: RAC, relative antigen concentration (±1 × SD); %IF$^+$, percentage of cells positive for HIV-1 antigens detected by indirect immunofluorescence staining; p.i., postinfection.

TABLE 4. Inhibition of HIV-1 Replication in Jurkat Clones Stably Transfected With Various Antisense RNA Expression Plasmids[a]

Plasmid pi	HIV-1 subregion	Number of clones tested	HIV-1 replication on day 10	
			ELISA units (RAC)	Percentage
p2as	5'-leader/*gag*	11	1900 ± 640	35 ± 12
pAR3	*gag* p24	3	1650 ± 100	31 ± 6
pAR6	*tat,rev* first exons	3	<550	<10
pAR7	*env*	3	2150 ± 170	40 ± 8
pKEX-1	—	9	5400 ± 1000	100 ± 19
pKEX-2-XR-CAT	—	6	5400 ± 1890	100 ± 35
p2s	5'-leader/*gag*	8	5600 ± 1200	104 ± 22

[a]RAC, relative antigen concentration.

initial infective dose. The selection of the target sequence determines the nucleotide sequence of the corresponding antisense transcripts and thereby indirectly the secondary and tertiary structures that are assumed to play an important role for intermolecular interactions and hence the antisense effect. In

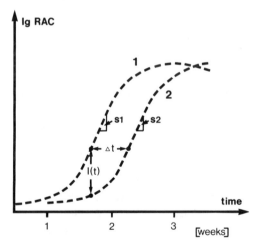

Fig. 5. Schematic representation of the time-dependent increase of the HIV-1 antigen concentrations in cell-free supernatants of Jurkat control clones (curve 1; e.g., pKEX-1 clones) and antisense RNA-producing clones (curve 2; e.g., p2as clones). Abbreviations: 1gRAC, logarithm of the relative antigen concentration (RAC) in the cell-free culture supernatant; $I(t)$, time dependent inhibition of HIV-1 replication in curve 2 versus curve 1; Δt, time difference between points of equal HIV-1 antigen concentrations in curve 1 and curve 2; s1 and s2, slopes of the curves 1 and 2 in the logarithmic phase of antigen production.

addition the selection of the target region may have crucial influence on the selection of the biological step(s) in the course of the viral life cycle, which can be inhibited by antisense RNA and which may lead to an overall reduction of virus replication.

The initial infective HIV-1 dose also determines the course of HIV-1 infection. In the infection experiments performed with Jurkat clones and shown in Table 3 the moi, which had been determined by endpoint dilutions of infective cell-free culture supernatants with MT-4 cells, was in the range of approximately 10 to 25. Lower initial infective HIV-1 doses lead to significantly prolonged reduction of viral replication in antisense RNA-producing Jurkat clones.

In contrast to this, the expression of the CD4 surface antigen, which is the cellular receptor used by the virus (Maddon et al., 1986), does not have a significant influence on HIV-1 replication. For example, CD4 expression in cloned Jurkat cultures ranged between 20 and 90% CD4[+] cells at a given time point (data not shown). However, these different CD4 expression levels did not correlate with HIV-1 infection.

The influence of intracellular antisense RNA levels on HIV-1 replication seems to be critical only at very low levels of antisense RNA expression. On the one hand there is no correlation between HIV-1 infection parameters and the antisense RNA expression levels in p2as (RNA[+]) clones, i.e., within the group of clones with antisense RNA expression detectable by Northern hybridization. However, the group of p2as (RNA$^-$) clones (Northern negative) showed a reduced extent of inhibition. This group of p2as clones may consist of clones without any antisense RNA expression and those with levels of antisense RNA that are undetectable by Northern hybridization. For this reason one can hypothesize that a certain minimal level of intracellular antisense RNA is necessary and sufficient for inhibition of HIV-1 replication in this model system. Therefore a simple increase in intracellular antisense RNA levels above the levels already reached in the p2as (RNA$^+$) group is not expected to improve antiviral effects.

Yet, with higher infective doses the breakthrough of infection cannot be abolished by antisense RNA and the reasons why finally HIV-1 infection breaks through with the same kinetics as in "unprotected" Jurkat clones are unclear. It was found that HCMV directed antisense RNA transcription is not influenced during HIV-1 infection or after microinjection of proviral HIV-1 DNA (data not shown). Additionally, infection experiments with HIV-1 produced in antisense RNA expressing Jurkat clones (pAR6) after break through of infection showed that this HIV-1 is not resistant against the pAR6-derived antisense RNA.

CONCLUSIONS AND PROSPECTS

Our experiments demonstrate that transient and constitutive intracellular expression of antisense RNA in human cells can lead to inhibition of HIV-1 replication. Although at higher infective doses the HIV-1 infection breaks through in antisense RNA producing T cell lines approximately 2 to 3 weeks postinfection, transient inhibition can reach more than 90% (pAR6 clones) depending on the initial infective HIV-1 dose and the antisense RNA target region. This shows the principal applicability of intracellular antisense RNA expression as an inhibitor of HIV-1 replication.

To compare the antiviral effects of intracellular antisense RNA expression with the inhibitory effects of the standard anti-HIV-1 drug 3'-azido-3'-deoxythymidine (AZT), Jurkat cells were infected in the continuous presence of 1 μM AZT. The results show that also in this case HIV-1 infection breaks through beginning at days 8 to 10 postinfection. Since application of AZT against HIV-1 infection in vitro and in vivo has a clear antiviral effect that is not reflected in the Jurkat model system at later stages of infection it might well be that the transient anti-HIV-1 efficacy of intracellular antisense RNA in Jurkat cells leads to underestimation of its antiviral effect. Nevertheless it seems to be possible and promising to further improve the antisense RNA-mediated inhibition of HIV-1 replication in the Jurkat model system. This could be achieved by combinations of different effective antisense RNA sequences on one transcript as was shown in a prokaryotic system (Coleman et al., 1985) or by multiple copies of the same sequence on one transcript. One can also imagine that ribozyme sequences (Uhlenbeck, 1987; Haseloff and Gerlach, 1988; Cotten and Birnstiel, 1989; for review see Cech and Bass, 1986) at certain positions on antisense transcripts support the antiviral activity. Recently it was demonstrated that intracellular expression of an HIV-1-targeted ribozyme sequence leads to a strong but also not complete inhibition of HIV-1 replication on day 7 postinfection of CD4$^+$ HeLa cells (Sarver et al., 1990). However, the further course of infection was not followed.

The promising first results obtained in the Jurkat model system have to be reproduced with further cell culture systems closer to the in vivo situation. This includes primary human HIV-1 permissive cells that could be tested either in vitro or in animal models such as scid/hu mice (Namikawa et al., 1988; for review see McCune, 1990). Unfortunately for the relevant types of human cells one of the major technical problems is the lack of an efficient delivery method for recombinant nucleic acids. So far the use of retroviral expression vector systems in combination with a short-term selection for successfully infected cells seems to be the most promising approach. Alternative delivery systems might also be useful such as improved electroporation protocols and viral vector systems derived from adeno-associated parvoviruses (Lebkowski et al., 1988; Samulski et al., 1989).

In conclusion we think that the intracellular expression of antisense RNA is a useful tool to inhibit HIV-1 replication. Improvements of the antiviral efficacy that lead to even higher extents and longer periods of inhibition should be possible. The major future problems that must be addressed include ways of efficient introduction of recombinant DNA into more relevant model systems, e.g., into primary lymphoid and hematopoietic stem cells. However, in the long-term view one should be able to solve these technical problems, enabling one to proceed further toward the therapeutic application of intracellularly expressed antisense RNA.

SUMMARY

Antisense RNA expressed intracellularly is a potential specific negative regulator for gene expression and replication of cellular and viral nucleic acids. In the preceding it was indicated that transient as well as stable intracellular antisense RNA expression is capable of inhibiting the replication of the human immunodeficiency virus type 1 (HIV-1). In two independent assay systems HIV-1 replication was inhibited by antisense RNA complementary to different target regions. In human T-lymphoid cells with constitutive antisense RNA expression, over 90% of HIV-1 inhibition could be reached transiently over a period of up to 3 weeks. There is experimental evidence that the antisense RNA expression causes the inhibitory effects in a sequence-specific manner, indicating the potential suitability of antisense RNA as an antiviral therapeutic agent. Possible ways for increasing the antiviral efficacy of antisense RNA by optimization of critical biochemical and biological parameters was discussed.

ACKNOWLEDGMENTS

We thank H. zur Hausen for continuous support and encouragement. We also thank R. Gallo, M. Martin, H.-U. Bernard, and M. Boshart for plasmids and G. Moldenhauer and M. Knobloch for CD4 measurements. This work was supported in part by BMFT Grant FKZ II-083-89.

REFERENCES

Adachi A, Gendelman HE, König S, Folks T, Willey R, Rabson A, Martin MA (1986): Production of acquired immunodeficiency syndrome-associated retrovirus in human and nonhuman cells transfected with an infectious molecular clone. J. Virol. 59:284–291.

Alizon M, Wain-Hobson S, Montagnier L, Sonigo P (1986): Genetic variability of the AIDS virus: Nucleotide sequence analysis of two isolates from African patients. Cell 46:63–74.

Baltimore D (1988): Intracellular immunization. Nature (London) 335:395–396.

Barre-Sinoussi F, Cherman JC, Rey R, Nugeryre MT, Chamaret S, Gruest J, Dauget C, Axler-Blin C, Vezinet-Brun F, Rouzioux C, Rosenbaum W, Montagnier L (1983): Isolation of a T-lymphotropic retrovirus from a patient at risk for acquired immunodeficiency syndrome (AIDS). Science 220:868–871.

Bergman P, Ustav M, Moreno-Lopez J, Vennström B, Petterson U (1986): Replication of the bovine papillomavirus type 1 genome: Antisense transcripts prevent episomal replication. Gene 50:185–193.

Bernard H-U, Krämmer G, Röwekamp WG (1985): Construction of a fusion gene that confers resistance against hygromycin B to mammalian cells in culture. Exp. Cell. Res. 158:868–871.

Boshart M, Weber F, Jahn G, Dorsch-Häsler K, Fleckenstein B, Schaffner W (1985): A very strong enhancer is located upstream of an immediate early gene of human cytomegalovirus. Cell 41:521–530.

Boyd AL, Wood TG, Buckley A, Fischinger PJ, Gilden RV, Gonda MA (1988): Microinjection and expression of an infectious proviral clone and subgenomic envelope construct of a human immunodeficiency virus. AIDS Res. Hum. Retroviruses 4:31–41.

Cech TR, Bass BL (1986): Biological catalysis by RNA. Annu. Rev. Biochem. 55:599–629.

Chang LJ, Stoltzfuss CM (1985): Gene expression from both intronless and intron-containing Rous sarcoma virus clones is specifically inhibited by antisense RNA. Mol. Cell. Biol. 5:2341–2348.

Chang LJ, Stoltzfuss CM (1987): Inhibition of Rous sarcoma virus replication by antisense RNA. J. Virol. 61:921–924.

Coffin JM (1986): Genetic variation in AIDS viruses. Cell 46:1–4.

Coleman J, Hirashima A, Inokuchi Y, Green PJ, Inouye M (1985): A novel immune system against bacteriophage infection using complementary RNA (micRNA). Nature (London) 315:601–603.

Cotten M, Birnstiel ML (1989): Ribozyme mediated destruction of RNA in vivo. EMBO J. 8:3861–3866.

Cuozzo M, O'Connell KM, Kaniewski W, Fang R-X, Chua N-H, Tumer NE (1988): Viral protection in transgenic tobacco plants expressing the cucumber mosaic virus coat protein or its antisense RNA. Biotechnology 6:549–557.

Döffinger R, Pawlita M, Sczakiel G (1988): Electrotransfection of human lymphoid and myeloid cell lines. Nucl. Acids Res. 16:11840.

Dzierzak EA, Papayannopoulou T, Mulligan RC (1988): Lineage-specific expression of a human β-globin gene in murine bone marrow transplant recipients reconstituted with retrovirus-transduced stem cells. Nature (London) 331:35–41.

Friedman T (1989): Progress toward human gene therapy. Science 244:1275–1281.

Graessmann R, Fleckenstein B (1989): Selectable recombinant Herpesvirus saimirii is capable of persisting in a human T-cell line. J. Virol. 63:1818–1821.

Harada S, Koyanagi Y, Yamamoto N (1985): Infection of HTLV-III/LAV in HTLV-I-carrying cells MT-2 and MT-4 and application in a plaque assay. Science 229:563–566.

Haseloff J, Gerlach WL (1988): Simple RNA enzymes with new and highly specific endoribonuclease activities. Nature (London) 334:585–591.

Hemenway C, Fang R-X, Kaniewski WK, Chua N-H, Tumer NE (1988): Analysis of the mechanism of protection in transgenic plants expressing the potato virus X coat protein or its antisense RNA. EMBO J. 7:1273–1280.

Hirashami A, Sawaki S, Inokuchi Y, Inouye M (1986): Engineering of the mRNA-interfering complementary RNA immune system against viral infection. Proc. Natl. Acad. Sci. U.S.A. 83:7726–7730.

Jennings PA, Molloy PL (1987): Inhibition of SV40 replicon functions by engineered antisense RNA transcribed by RNA polymerase III. EMBO J. 6:3043–3047.

Lebkowski JS, McNally MM, Okarma TB, Lerch LB (1988): Adeno-associated virus: A vector system for efficient introduction and integration of DNA into a variety of mammalian cell types. Mol. Cell. Biol. 8:3988–3996.

Leibovitz A, Stinson JC, McCombs WB III, McCoy CE, Mazur KC, Mabry ND (1976): Classification of human colorectal adenocarcinoma cell lines. Cancer Res. 36:4562–4569.

Lever A, Göttlinger H, Haseltine W (1989): Identification of a sequence required for efficient packaging of human immunodeficiency virus type 1 RNA into virions. J. Virol. 63:4085–4087.

Maddon PJ, Dalgleish AG, McDougal JS, Clapham PR, Weiss RA, Axel R (1986): The T4 gene encodes the AIDS virus receptor and is expressed in the immune system in the brain. Cell 47:333–348.

Malim MH, Böhnlein S, Hauber J, Cullen BR (1989): Functional dissection of the HIV-1 rev trans-activator—derivation of a trans-dominant repressor of rev function. Cell 58:113–120.

McCune JM (1990): The rational design of animal models for HIV infection. Sem. Virol. 1:229–235.

Mermer B, Felber BK, Campell M, Pavlakis GN (1989): Identification of transdominant HIV-1 rev protein mutants by direct transfer of bacterially produced proteins into human cells. Nucl. Acids Res. 18:2037–2044.

Minowada J, Ohnuma T, Moore GE (1972): Rosette-forming human lymphoid cell lines. I. Establishment and evidence for origin of thymus-derived lymphocytes. J. Natl. Cancer Inst. 49:891–895.

Namikawa R, Kaneshima H, Shultz LD, Lieberman M, Weissman IL, McCune JM (1988): Infection of the SCID-hu mouse by HIV-1. Science 242:1684–1686.

Popovic M, Sarngadharan MG, Read E, Gallo RC (1984): Detection, isolation and continous production of cytopathic retroviruses (HTLV-III) from patients with AIDS and pre-AIDS. Science 224:497–500.

Powell PA, Stark DM, Sanders PR, Beachy RN (1989): Protection against tobacco mosaic virus in transgenic plants that express tobacco mosaic virus antisense RNA. Proc. Natl. Acad. Sci. U.S.A. 86:6949–6952.

Ratner LW, Haseltine W, Patarca R, Livak KJ, Starcich B, Josephs SF, Doran ER, Rafalski JA, Whitehorn EA, Baumeister K, Ivanoff L, Petteway SR Jr, Pearson ML, Lautenberger JA, Papas TS, Ghrayeb J, Chang NT, Gallo RC, Wong-Staal F (1985): Complete nucleotide sequence of the AIDS virus, HTLV-III. Nature (London) 313:277–283.

Rhodes A, James W (1990): Inhibition of human immunodeficiency virus replication in cell culture by endogenously synthesized antisense RNA. J. Gen. Virol. 71:1965–1974.

Rittner K, Stöppler H, Pawlita M, Sczakiel G (1991): Versatile eucaryotic vectors for strong and constitutive transient and stable gene expression. Methods Mol. Cell. Biol. 2:176–181.

Salahuddin SZ, Markham PD, Popovic M, Sarngadhavan MG, Orndorff S, Fladagar A, Patel A, Gold J, Gallo RC (1985): Isolation of infectious human T-cell leukemia/lymphotropic virus type III (HTLV-III) from patients with acquired immunodeficiency syndrome (AIDS) or AIDS related complex (ARC) and from healthy carriers: A study of risk groups and tissue sources. Proc. Natl. Acad. Sci. U.S.A. 82:5530–5534.

Salfeld J, Göttlinger HG, Sia RA, Park RE, Sodroski JE, Haseltine WA (1990): A tripartite HIV-1 tat-env-rev fusion protein. EMBO J.9:965–970.

Samulski RJ, Chang L-S, Shenk T (1989): Helper-free stocks of recombinant adeno-associated virus: Normal integration does not require viral gene expression. J. Virol. 63:3822–3828.

Sarver N, Cantin EM, Chang PS, Zaia JA, Ladne PA, Stephens DA, Rossi JJ (1990): Ribozymes as potential anti-HIV-1 therapeutic agents. Science 247:1222–1225.

Schneider U, Schwenk H-U, Bornkamm G (1977): Characterization of EBV-genome negative "null" and "T" cell lines derived from children with acute lymphoblastic leukemia and leukemic transformed non-Hodgkin lymphoma. Int. J. Cancer 19:621–630.

Schwartz S, Felber BK, Benko DM, Fenyö E-M, Pavlakis GN (1990): Cloning and functional analysis of multiple spliced mRNA species of human immunodeficiency virus type 1. J. Virol. 64:2519–2529.

Sczakiel G, Pawlita M (1991): Inhibition of human immunodeficiency virus type 1 replication in human T-cells stably expressing antisense RNA. J. Virol. 65:468–472.

Sczakiel G, Pawlita M, Kleinheinz A (1990): Specific inhibition of human immunodeficiency virus type 1 replication by RNA transcribed in sense and antisense orientation from the 5′-leader/gag region. Biochem. Biophys. Res. Commun. 169:643–651.

Steinhauer DA, Holland JJ (1987): Rapid evolution of RNA viruses. Annu. Rev. Microbiol. 41:409–433.

To RY, Booth SC, Neiman P (1986): Inhibition of retroviral replication by antisense RNA. Mol. Cell. Biol. 6:4758–4762.

Trono D, Feinberg MB, Baltimore D (1989): HIV-1 gag mutants can dominantly interfere with the replication of the wild type virus. Cell 59:113–120.

Uhlenbeck OC (1987): A small catalytic oligoribonucleotide. Nature (London) 328:596–600.

von Knebel-Doeberritz M, Oltersdorf T, Schwarz E, Gissmann L. (1988): Correlation of modified human papilloma virus early gene expression with altered growth properties in C4-1 cervical carcinoma cells. Cancer Res. 48:3780–3786.

von Rueden T, Gilboa E (1989): Inhibition of human T-cell leukemia virus type I replication in primary human T cells that express antisense RNA. J. Virol. 63:677–682.

Wendler J, Jentsch KD, Schneider J, Hunsmann G (1987): Efficient replication of HTLV-III and STLV-III mac in human Jurkat cells. Med. Microbiol. Immunol. 176:273–280.

Novel Antisense Derivatives: Antisense DNA Intercalators, Cleavers, and Alkylators

Dmitri G. Knorre and Valentina F. Zarytova

Novosibirsk Institute of Bioorganic Chemistry, Siberian Division of the USSR Academy of Sciences, Novosibirsk 90, 630090, USSR

INTRODUCTION

Several essential groups of diseases are caused by expression of either foreign or incorrect genetic information. Thus a crucial event in the course of most viral diseases is multiplication of viral nucleic acids followed by their expression. Uncontrolled cell proliferation during tumor growth is believed to be, to a great extent, due to the expression of oncogenes that appear in cells of a diseased organism as a result of mutations in protooncogenes present in normal cells. Therefore, current antiviral and anticancer chemotherapy depends, in general, on interruption of nucleic acid metabolism. At the same time these drugs must be highly selective in order to discriminate either between viral and host's nucleic acids or between oncogenes and their normal counterparts. Specific targeting of nucleic acids may represent a new general approach to the design of antiviral and anticancer drugs. The desired selectivity may be achieved using the principle of complementarity, which makes it possible to design a complementary chain for any nucleotide sequence of the preselected target. Two main ways to realize this principle are known.

Specific genetic constructs may be prepared by genetic engineering manipulation, which, if transcribed within the cell, produces RNA complementary to the target nucleic acid, thus preventing its translation. Such RNAs are usually referred to as antisense RNAs (Walder, 1988).

The second approach is based upon administration into cells of oligonucleotides, their analogs, or derivatives, complementary to a definite region of the target nucleic acids. The oligonucleotides have a number of important advantages compared to a great variety of other chemical compounds influencing nucleic acids:

1. The mechanisms of their interaction with single-stranded and double-stranded nucleic acids are well known and, therefore, it is relatively easy to design an oligonucleotide specific to a certain region of the target nucleic acid.

2. Numerous oligonucleotides (4^N, where N is the number of nucleotide residues) with different specificities can be prepared using a universal technology.

3. Prior to use the oligonucleotides can be chemically modified in many ways in order to enhance their ability to penetrate into cells, across cell membranes, to form complexes with the selected targets or to damage their sequences.

The first version of this approach was proposed by N. I. Grineva and co-workers in 1967. They suggested that binding of a reactive group to an oligonucleotide may improve selectivity of chemical modification of nucleic acids since the oligonucleotide moiety should deliver this group to the nucleic acid containing the complementary sequence. This method was called complementary ad-

Prospects for Antisense Nucleic Acid Therapy of Cancer and AIDS, pages 195–218
©1991 Wiley-Liss, Inc.

dressed modification (Belikova et al., 1967; Grineva et al., 1968; Zarytova et al., 1968; Grineva and Zarytova, 1968).

A decade later P.C. Zamecnik demonstrated that specific oligonucleotides may be used per se in suppressing multiplication of the Rous sarcoma virus (Zamecnik and Stephenson, 1978). Later this group successfully extended this approach to HIV-I (Zamecnik et al., 1986). In both of the above cases negatively charged hydrophilic oligonucleotides were used. To facilitate penetration via cell membrane. Miller and Ts'o proposed using oligonucleotide derivatives with phosphate residues masked by ethylation (Miller et al., 1971, 1977). A few years later the same authors proved that oligonucleoside methylphosphonates can also be used as hydrophobic oligonucleotide analogs (Miller et al., 1979, 1985; Blake et al., 1985).

A principal advantage of chemical modification of nucleic acids by reactive oligonucleotide derivatives is that they irreversibly crosslink the latter to the target, thus forming a stable duplex. However, stabilization may be achieved by milder means, namely by introduction into the oligonucleotide of aromatic groups capable either of stacking to a terminal base pair of the duplex or of intercalating between its base pairs. This approach was suggested and realized in 1983 by C. Hélène, who attached acridine residues to oligonucleotides (Asseline et al., 1983, 1984a,b).

Obviously some of these methods may be used in combination with the attachment of reactive groups. Thus in 1980 D.G. Knorre and collaborators obtained P-ethylated oligonucleotides carrying reactive alkylating groups. Using these derivatives specific modification of poly(A) tracts was performed in cells (Zarytova et al., 1980; Knorre et al., 1981). Binding of an alkylating group and a duplex-stabilizing phenazinium residue to the opposite ends of oligonucleotide was realized by V.F. Zarytova and co-workers in 1986. They demonstrated significant enhancement of efficiency of complementary addressed alkylation (Zarytova et al., 1986; Vlassov et al., 1986).

The present chapter is devoted to oligonucleotide derivatives carrying reactive groups and the combination of these derivatives with hydrophobic and duplex-stabilizing residues, with emphasis on the recently suggested new types of groups and their combinations. The ways of improving the efficiency and selectivity of interaction with nucleic acids are discussed.

SEQUENCE-SPECIFIC ALKYLATION

Alkylating oligonucleotide derivatives bearing N-2-chloroethyl-N-methylaminophenyl residue (abbreviated as RCl) were the first to be investigated as sequence-specific reagents. Two ways of attachment of this group were proposed. The first one is based on the reaction of an aldehyde ClRHCO with the 3'-terminal cis-diol group of a ribonucleoside residue (Belikova et al., 1967; Grineva et al., 1968; Zarytova et al., 1968; Grineva and Zarytova, 1968; Knorre et al., 1985a).

I

(in this scheme and throughout the paper B stands for an arbitrary base and X for oligonucleotide).

The second variant (Grineva and Lomakina, 1972; Zarytova and Pichko, 1990) is binding of ClRCHR'NH$_2$ to the 5'-terminal phosphate via a phosphoramidate bond:

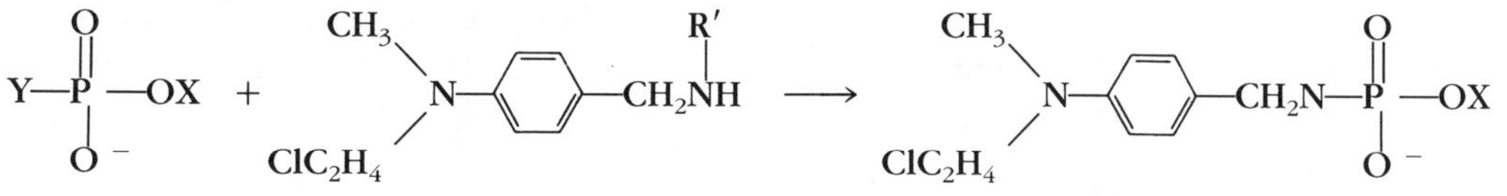

1978). These activating reagents may be used only for deoxyribooligonucleotides, since the internucleotide phosphates of the ribose series are converted into 2',3'-cyclophosphate triesters and are either hydrolyzed or isomerized to 2'-5' internucleotide bonds during the subsequent aqueous treatment. These

(Cl RCH$_2$NR'H)   **II**

R' = H or CH$_3$

where Y is a 3'- or 5'-phosphate-activating residue.

The activation can be performed by both diphenylphosphochloridate (Grineva and Lomakina, 1972), N,N-dicyclohexylcarbodiimide (Budker et al., 1977), and mesitylenecarbonylchloride (Guimautdinova et al., 1977,

by-products are avoided if Y is N-methylimidazole (MeIm), N,N-dimethylaminopyridine, or N-oxide residues, which may be introduced via intermediate activation of terminal phosphate by a mixture of triphenylphosphine and 2,2-dipyridyldisulfide (Zarytova et al., 1987b; Godovikova et al., 1989).

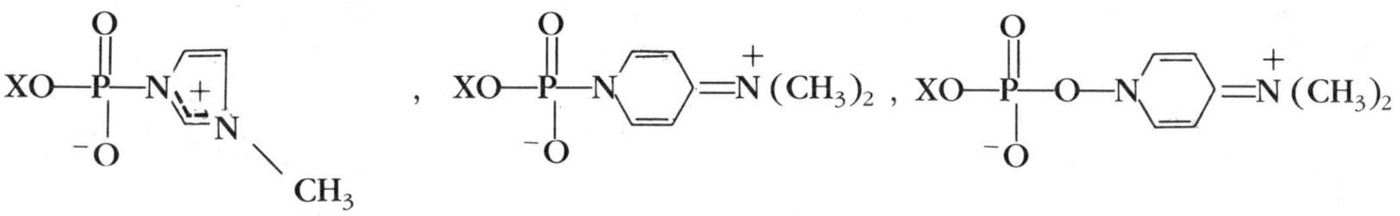

A new approach based on phosphite chemistry has been recently proposed (Levina and Ivanova, 1985). According to this method the

oligonucleotide is successively treated with ethylchlorophosphite and Cl RCH$_2$NH$_2$ in the presence of CCl$_4$.

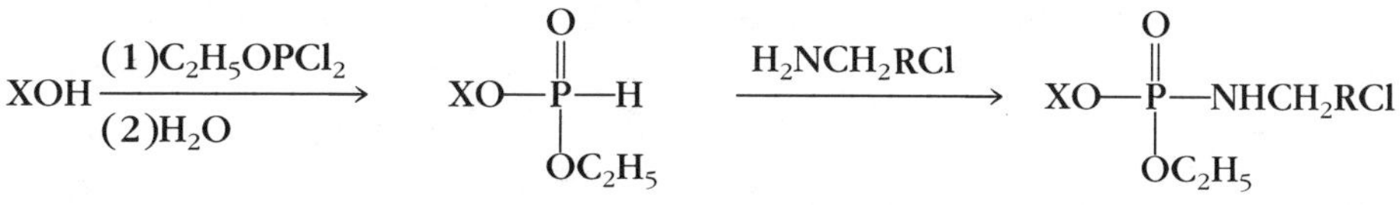

It is known that alkylation with aromatic 2-chloroethylamines proceeds via the formation of a highly reactive ethyleneimonium cation as an intermediate rate-limiting stage

of the overall process (SN$_1$ mechanism) (Vlassov et al., 1970; Grineva et al., 1970, 1977a,b; Benimetskaya et al., 1977).

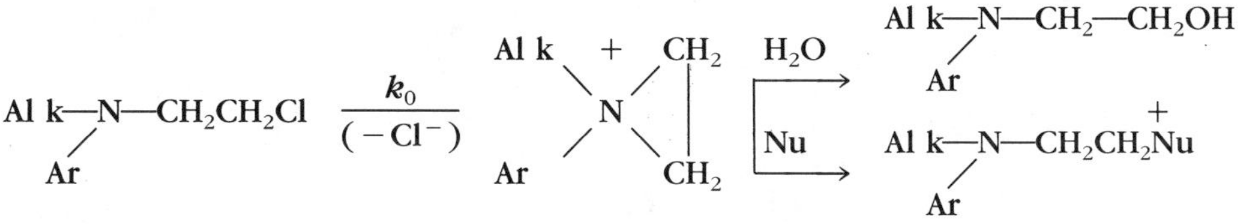

Nu – nucleophile

alkylation product

Ethyleneimonium cation can react with the nucleophiles present in the reaction mixture or be hydolyzed by water. An essential kinetic characteristic of the SN_1 reaction mechanism is that the average reaction time is of the magnitude of $1/k_0$, irrespective of the fate of the intermediate cation. The first experiments on complementary addressed modification were carried out with 16S + 23S rRNA from *E. coli* and alkylating oligonucleotide derivatives (Vasilenko et al., 1973; Grineva and Karpova, 1974, 1975). The results of these experiments revealed for the first time the basic principles of modification in the complex. If alkylating derivatives **I** and **II** are in a complementry complex with the target, 95–99% of the cations formed react with the bases adjacent to the alkylating group. If these reagents are not part of the complexes, the cations are distributed among all nucleophiles of the reaction mixture and the extent of modification of the specific target is several orders of magnitude lower.

Biomolecular modification by alkylating reagents in solution affects the most nucleophilic centers of RNA and DNA, primarily guanine residues. By contrast, the process in the complex fixes the active group at a definite site of the target and any nearby bases are alkylated. It was also shown that the most characteristic feature of modification in the complex is the lowering of maximum extent of modification with temperature (Belikova et al., 1973; Grineva and Karpova, 1974, 1975). Therefore, by raising the temperature one can avoid modification in less stable complexes. On the contrary, by raising reagent concentration, it is possible to attain modification of the sites having low association constants (Grineva and Karpova, 1974).

These criteria are still valid for specifying an intramolecular reaction within a complex prior to the establishing of modification sites. This is important for the *in vivo* experiments, where one may have to deal with a complicated mixture of modified nucleic acids.

At first the specificity of alkylation was checked and modified bases were determined using yeast tRNA (Grineva et al., 1977a) and a 303-nt single-stranded DNA fragment (Vlassov et al., 1986) as targets.

The sites of DNA alkylation by oligonucleotide derivatives were revealed in accordance with the Maxam–Gilbert procedure.

In the present and in all subsequent studies the results obtained for these reagents at the level of cells and living organisms are discussed in another chapter (Vlassov and Yakubov, this volume).

Recently it has appeared possible to modify double-stranded DNA bearing a $(pdC)_n \cdot (dG)_n$ fragment using the Cl-$RCH_2NH(pdC)_n$ alkylating reagent in conditions favoring the formation of triple-stranded complexes (Fedorova et al., 1988; Knorre et al., 1985a,b, 1989a,b).

Recently two novel types of alkylating groups were proposed to be used for site-specific modification of nucleic acids. An $RSCH_3$ fragment was attached to C-5 of deoxyuridine, which was further converted into the corresponding triphosphate and used as a substrate of DNA polymerase I with a template and primer. The $RSCH_3$ group was activated by BrCN:

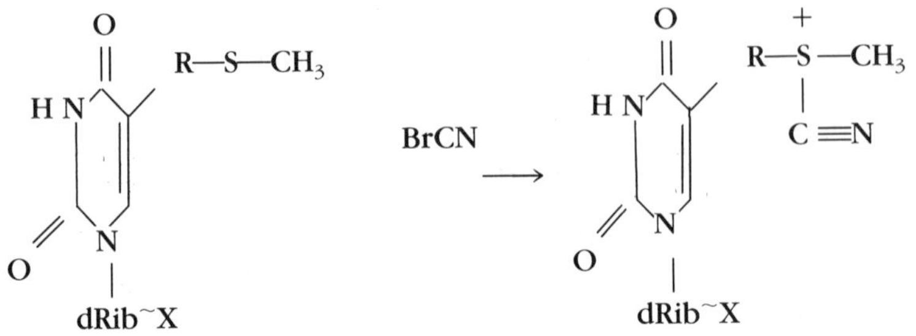

$$R = -CH=CH(CH_2)_2NHCO-(CH_2)_2-$$

Site-specific cleavage of a single-stranded ϕX174 DNA was observed with a 9% yield as a result of reaction with the complementary reagent containing an —SCH$_3$ group. The mechanism involves methyl group transfer from sulfur of the modified dU of the extended primer to N-7 of DNA template guanine (Iverson and Dervan, 1987, 1988).

The oligonucleotides bearing alkylating electrophiles at the C-5 position of dU can function as exceptionally selective cleavage reagents for DNA.

It was also shown that a tetradecadeoxynucleotide with 5-[3-(iodoacetamido)propyl]-2'-deoxyuridine (U*), incorporated instead of T, alkylated a 30-mer containing the sequence complementary to the reactive 14-mer. Analysis of the cleaved products on a sequencing gel indicated that the target strand guanine situated at a distance of 2 base pairs from U is the only site of alkylation (Meyer et al., 1989).

U* =

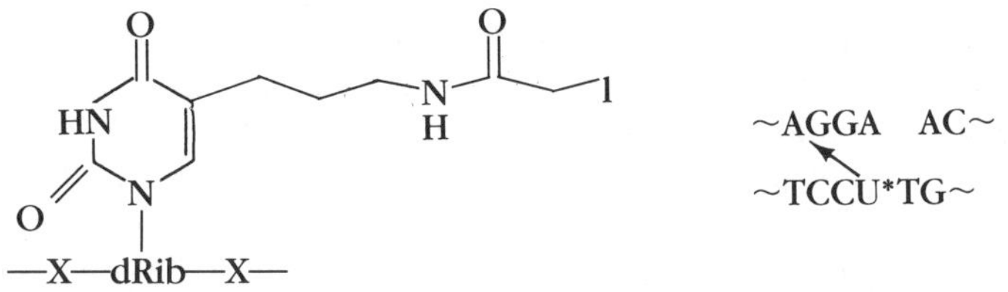

~AGGA AC~
~TCCU*TG~

SEQUENCE-SPECIFIC OXIDATIVE CLEAVAGE OF NUCLEIC ACIDS

It is known that, in the presence of a metal ion, O$_2$ and a reducing agent, several antibiotics (e.g., bleomycin), and substances such as EDTA, porphyrin, and phenanthroline can catalytically split DNA (Knorre et al., 1989a). They have been linked to oligodeoxyribonucleotides to bring about site-specific scission of complementary single-stranded DNA. In the first example EDTA was attached to the 5'-terminal phosphate of an oligonucleotide bearing an NH$_2$-spacer.

Nucleic acids are well known to be damaged by free radicals generated either by X-ray and γ-irradiation or by chelated transition metals in oxidative conditions. Therefore, numerous attempts were made to prepare reagents containing a chelating group tethered to an oligonucleotide, aimed at sequence-specific oxidative cleavage of nucleic acids. The derivatives of EDTA, phenanthroline, and some porphyrins were used as chelating moieties.

An EDTA fragment may be attached to oligonucleotides or their analogs via a phosphodiester or phosphoramide bond.

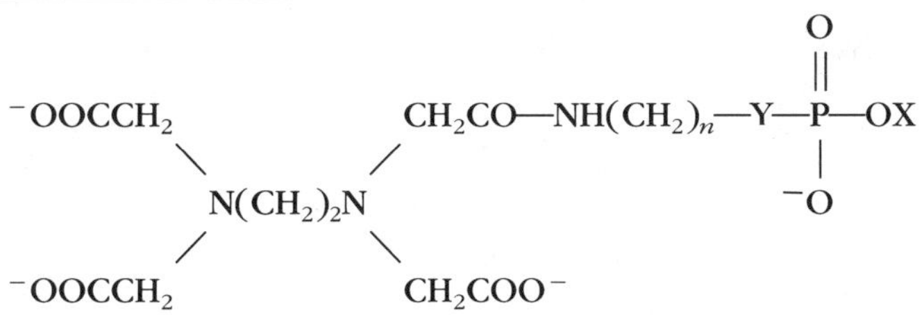

$$^-OOCCH_2 \diagdown \hspace{2cm} CH_2CO-NH(CH_2)_n-Y-\overset{\overset{\displaystyle O}{\|}}{\underset{\underset{\displaystyle ^-O}{|}}{P}}-OX$$

$$N(CH_2)_2N$$

$$^-OOCCH_2 \diagup \hspace{2cm} CH_2COO^-$$

III

$n = 3$, Y = O (Boutorin et al., 1984); $n = 2$, Y = NH (Chu and Orgel, 1985).

Oligonucleotides bearing EDTA residues at heterocyclic bases were synthesized using deoxyuridine carrying EDTA at the C-5 po-

sition, at the appropriate step of phosphoramidite synthesis (Dreyer and Dervan, 1985).

Another way to introduce an EDTA residue into a heterocyclic moiety is transamination of the 4-amino group of Cyt with ethylenediamine, followed by acylation of aminoethyl derivatives of cytosine with EDTA anhydride. It was shown that derivatization at the 5'-position gave higher yield and purer product than that of deoxycytidine (Lin et al., 1989). All oligonucleotide–EDTA hybridization probes are capable of site-specific cleavage of the complementary DNA target. Splitting took place at several (from 5 to 14) points (Dreyer and Dervan, 1985; Chu and Orgel, 1985; Brosalina et al., 1988; Boidot-Forget et al., 1988). Such extended sites of cleavage are likely to be due to diffusion of active oxygen particles (O_2^-, H_2O_2, and $\cdot OH$) formed during interaction of the metal complex with O_2, and to the mobility of the spacer connecting the chelating group with the oligonucleotide.

A characteristic feature of all the above reagents is that in the presence of Fe(II) and DTT, EDTA-derivatized oligomers produce hydroxyl radicals, causing degradation of the sugar–phosphate backbones of both target DNA and RNA. Degradation is specific in the region of the oligomer binding site and is roughly 20-fold more efficient for single-stranded DNA than for RNA. (Lin et al., 1989). A serious disadvantage of EDTA-derivatized reagents is that, although stable *per se* in aqueous solution, they undergo rapid autodegradation in the presence of Fe(II), O_2, and DTT with half lives of about 30 min (Lin et al., 1989).

The second type of oxidative cleavage reagent is the phenanthroline oligonucleotide derivatives (**IV**). 1,10-Phenanthroline was bound to oligonucleotides by coupling of an oligonucleotide imidazolide to 5-glycilamido-1,10-phenanthroline (Chen and Sigman, 1986).

IV

ine same effect can be achieved by alkylation of a terminal thiophosphate of an oligonucleotide with iodoacetamide phenanthroline derivatives (Thuong and Chassignol, 1987; Francois et al., 1989a). The 1,10-phenanthroline derivatives of oligodeoxythymidylates were obtained by condensation of the respective derivative bearing an OH-group with the 3'-phosphate of the protected oligonucleotide (Francois et al., 1988a). The oligonucleotides covalently linked to 1,10-phenanthroline were found to induce cleavage in a complementary polymer, and a 27-mer containing complementary sequence in the presence of copper ion ($CuSO_4$) and reducing agent (mercaptopropionic acid). The sites of cleavage in this reaction were shorter (4–8 nucleosides) than in the case of EDTA oligonucleotide derivatives (Francois et al., 1988a,b, 1989a,b,c). 1,10-phenanthroline–copper attached to deoxyoligonucleotides (21-mer) provided sequence-specific scission of both RNA and DNA phosphodiester backbones with equal reactivity. In both cases the major cutting sites were clustered within six monomer residues of the target. After 2 hr incubation at 37°C, about 20% of the DNA and RNA fragments were converted to oligonucleotides (Chen and Sigman, 1988).

Another type of compound exhibiting cleavage activity toward complementary sequences in the presence of a reducing agent and in aerated solutions is an oligonucleotide covalently linked to a porphyrin (**V**). To prepare them, terminal phosphates of blocked oligonucleotides should be derivatized selectively by phosphotriester condensation with a porhyrin-containing hydroxyl group (Le Doan et al., 1986, 1987b).

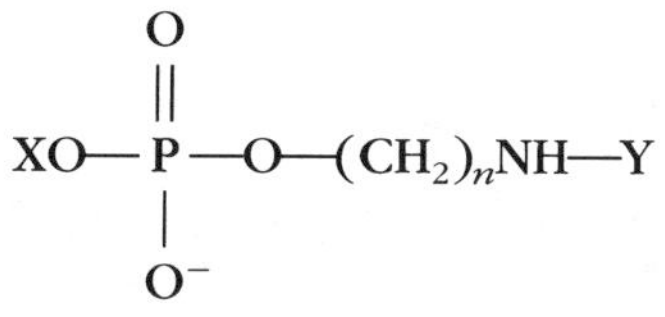

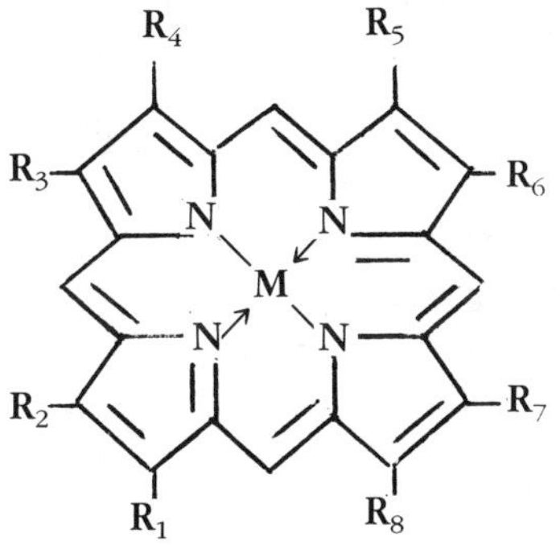

A heptathymidylate covalently linked to a porphyrin (**Va**) can bind complementary sequences, thus inducing local damage to the target DNA. The yield of the total reaction was relatively low, and equaled 1–2% of the original material (Le Doan et al., 1987b).

To obtain compounds like **Vb** and **Vc**, hemin (or deuterohemin) was linked to oligonucleotides via the NH$_2$ group of a 2-aminoethanol spacer by treatment with an anhydride that was prepared by the interaction of hemin (deuterohemin) with acetic acid anhydride in the presence of a nucleophilic catalyst (Ivanova et al., 1988; Frolova et al., 1990).

The yield of the porphyrin oligonucleotide derivative **Vb** was 20% in the presence of *N*-methylimidazole, and 40–60% in the presence of 1-hydroxybenzotriazole. The derivatives obtained were found to react efficiently and in a sequence-specific manner with the DNA target (302-mer). They inhibited the crosslinking reaction, which is sensitive

$$XO-\overset{\overset{O}{\|}}{\underset{\underset{O^-}{|}}{P}}-O-(CH_2)_nNH-Y$$

V

$$Y =$$

Va, Y = methylpyroporphyrin (MPPo) $R_1 = -CH_2CH_2\overset{\overset{O}{\|}}{C}{}^\sim$

   X = (dT)$_7$ $R_2 = R_3 = R_4 = R_5 = R_8 = CH_3$

   $R_6 = R_7 = -C_2H_5$

   $n = 6$, M = Fe, Mn, Co, Cu

Vb, Y = hemin $R_1 = R_8 = -CH_2CH_2C\overset{\diagup O}{\diagdown}$

   X = d(TGACCCTCTCCC)A $R_2 = R_3 = R_5 = R_7 = CH_3$

   $R_4 = R_6 = -CH = CH_2$

   $n = 2$,   M = Fe

Vc, Y = deuterohemin $R_1 = R_8 = -CH_2CH_2C\overset{\diagup O}{\diagdown}$

   X = d(TGACCCTCTTCCCATT) $R_2 = R_3 = R_5 = R_7 = CH_3$

   $R_4 = R_6 = H$

   $n = 3$,   M = Fe

to piperidine treatment. The total yield of this cleavage was 52% for hemin-pN$_{14}$ (**Vb**) and 60% for deuterohemin-pN$_{16}$ (**Vc**). In the latter case 5% of the direct cleavage (without piperidine treatment) was registered.

As in the case of the alkylating reagents (Grineva and Karpova, 1974) the extent of cleavage of nucleic acids by oxidative reagents was shown to depend on oligonucleotide concentration, salt concentration, and temperature. Dissociation of complexes at low salt concentration abolished site-specific cleavage reaction. Piperidine treatment of DNA strongly enhanced the yield of cleavage (Boidot-Forget et al., 1988). The temperature rise appeared to decrease the reaction yield as a result of dissociation of the target−reagent complex (Francois et al. 1988a).

The agents of oxidative cleavage discussed above are supposed to have catalytic effects on the target. However, this effect has not yet been proved.

PHOTOREACTIVE OLIGONUCLEOTIDE DERIVATIVES

The third group of reactive oligonucleotide derivatives is that of reagents carrying photoactivatable groups. These are usually inert in the dark but can be "switched on" by irradiation at a particular wavelength. The first results on site-specific photomodification of nucleic acids were obtained using an N,N-dimethylaminonaphthalene-1-sulfonyl (Dns) group as chromophore, attached to an oligothymidylate. The reagent was prepared according to the scheme:

$$\text{Dns—NH(CH}_2)_n\text{OH} + [(\text{ClPh})\text{pdT}]_9(\text{Lev}) \xrightarrow{\text{TPS, MeIm}}$$

$$\text{Dns—NH(CH}_2)_n\text{O}[(\text{ClPh})\text{pdT}]_9(\text{Lev}) \xrightarrow{25\%\,\text{NH}_3,\ 50°\text{C, 3h}} \text{Dns—NH(CH}_2)_n\text{O(pdT)}_9$$

$$n = 2,3$$

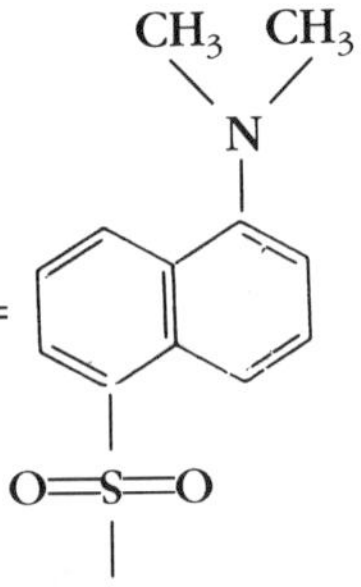

$(n = 2, 3;$ TPS, triisopropylbenzenesulfonyl chloride; Lev = levulinic acid residue).

In the presence of this reagent, polyadenylic acid was split under irradiation, whereas noncomplementary poly(U) and poly(C) remained untouched (Benimetskaya et al., 1983). The authors suggested that the sugar−phosphate backbone was cleaved due to a radiationless transfer of the double-excited chromophore energy to phosphodiester groups. The assumption was based on the observed nonlinear dependence of cleavage extent on the irradiation dose, typical of the two-quantum mechanism (Benimetskaya et al., 1983, 1984, 1988).

Recently a number of oligonucleotides bearing different dye groups were synthesized and were found to allow site-specific photomodification of model oligonucleotide targets.

Alkylation of oligonucleotides bearing a 5'-terminal thiophosphate with p-azidobenzoyl bromide (Praseuth et al., 1987), 3-amino-6-(ω-bromoalkylamino)acridine (Praseuth et al., 1988a), or 3-azido-6(3-bromopropylamino)-acridine (Praseuth et al., 1988b) yielded pho-

toreactive *p*-azidophenacyl (**VI**, 3-azidopro-
flavine (**VII**), or proflavine (**VIII**) derivatives

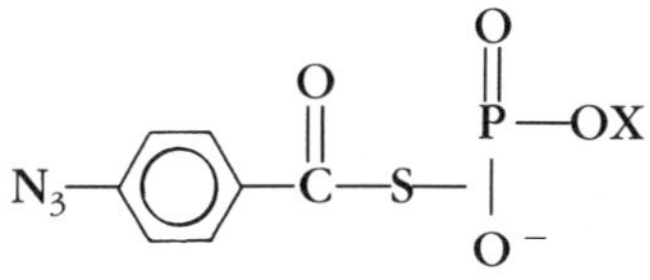

VI

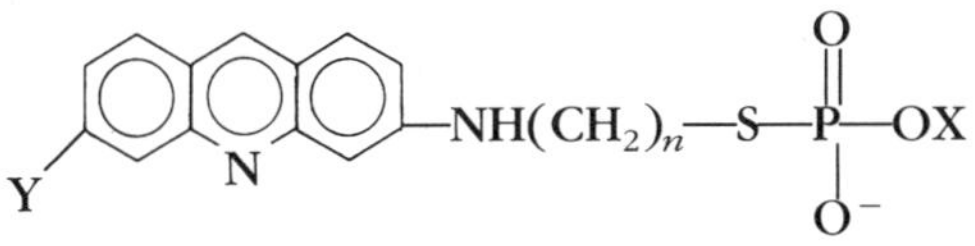

VII   Y = N$_3$

VIII   Y = NH$_2$

where [α]T is the α-anomer of thymidine.

These reagents were used for modification
of a 27-mer carrying a (pdA)$_8$ fragment. It
was shown that β-anomeric derivatives of oligothymidylate reagents performed modification in antiparallel orientation with respect
to the poly(A) tract of a single-stranded DNA
target. In contrast, the modification points by
α-anomeric derivatives made a stretch parallel to the poly(A) tract.

After irradiation and piperidine treatment
of the target complexes with the reagent, both
cleavage and crosslinks were observed (Praseuth et al. 1987, 1988a,b; Le Doan et al.,
1987a,b).

The reaction of derivatized ethidium **IXa**
and **IXb** with the 5′-*N*-methylimidazolide of
an oligonucleotide yielded photoactivable
reagents **Xa** and **Xb**, respectively.

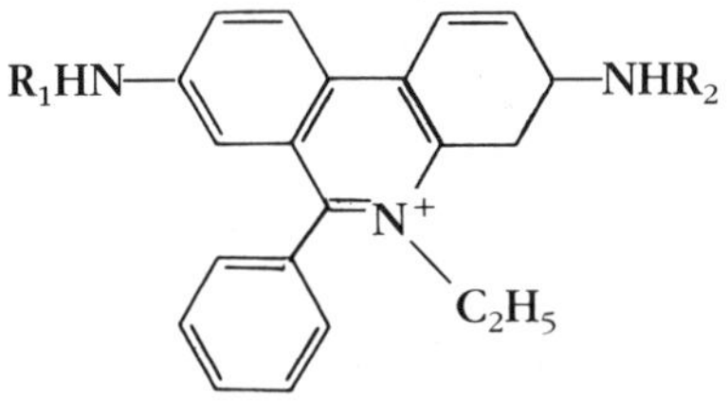

IXa R$_1$ = H;   R$_2$ = —COCH$_2$CH$_2$NH$_2$
IXb R$_1$ = —COCH$_2$CH$_2$NH$_2$;   R$_2$ = H
IXc R$_1$ = R$_2$ = —COCH$_2$CH$_2$NH$_2$
 Xa R$_1$ = H;   R$_2$ = —COCH$_2$CH$_2$Npd(A-A-T-A-C-T-C-T)
 Xb R$_1$ = —COCH$_2$CH$_2$Npd(A-A-T-A-C-T-C-T);   R$_2$ = H

Irradiation induced three types of photo-
damage: direct target cleavage at a specific
site (ca. 12%) and formation of at least two
specific covalent adducts, one of them stable
(20–70%) and the other sensitive (7–27%)
to piperidine treatment. The total yield of
photomodification was 50–80% (Benimetskaya et al., 1988, 1989).

Psoralen derivatives of oligonucleotide XI
were obtained using the phosphoramidite
method. 4′-(Hydroxyethoxymethyl)-4,5′,8-
trimethylpsoralen was phosphorylated with
chloro-[(β-cyanoethoxy)-*N,N*-diisopropyl-
amino]phosphine. The desired psoralenphos-
phoramidite was exploited directly during the
last cycle of solid-phase oligonucleotide synthesis.

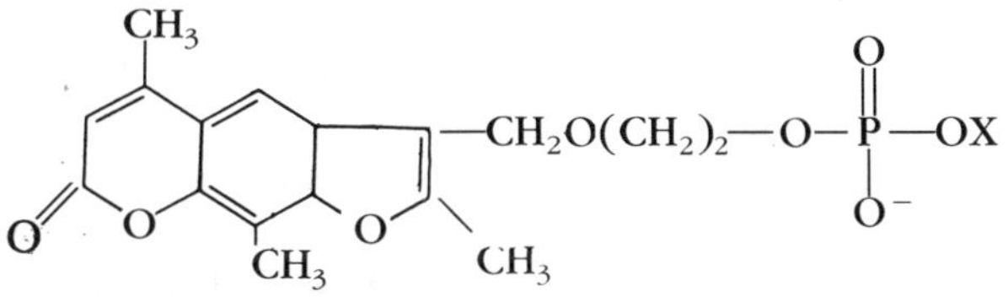

XI

In order to test the photocrosslinking ability of psoralen derivative XI, it was irradiated in the presence of a complementary oligonucleotide. The yield of crosslinking equaled 75% (Pieles and Englisch, 1989).

Photoinduced cleavage of both strands of a double-stranded DNA fragment was realized within the triplex structure formed by ellipticine homopyrimidine oligonucleotide **XII** after irradiation.

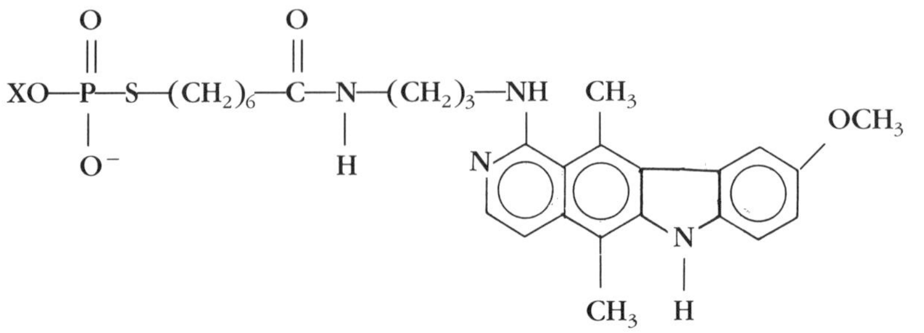

XII

This derivative (**XII**) was synthesized by coupling of a 3′-thiophosphate of an oligonucleotide (11-mer) to 1′-diaminopropyl-5,11-dimethyl-9-methoxyellipticine (Perrouault et al., 1990).

It is known that some metalloporphyrins can act as photosensitizers producing singlet oxygen and therefore triggering DNA photodegradation (Praseuth et al., 1986). Selective photomodification of a 34-mer by a palladium(II)-coproporphyrin derivative of a 17-mer oligodeoxyribonucleotide (**XIII**) was achieved.

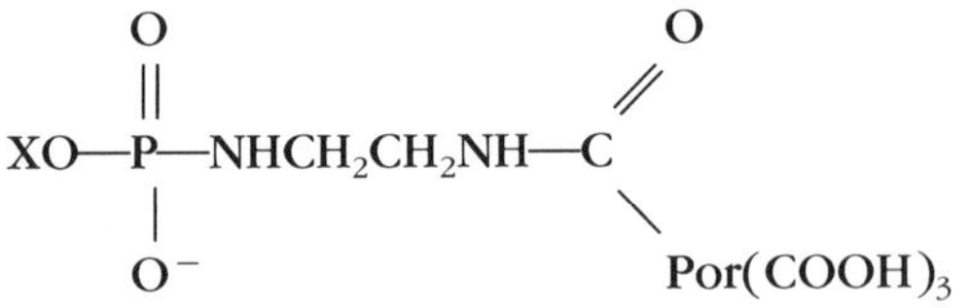

XIII

$$Por(COOH)_4 =$$

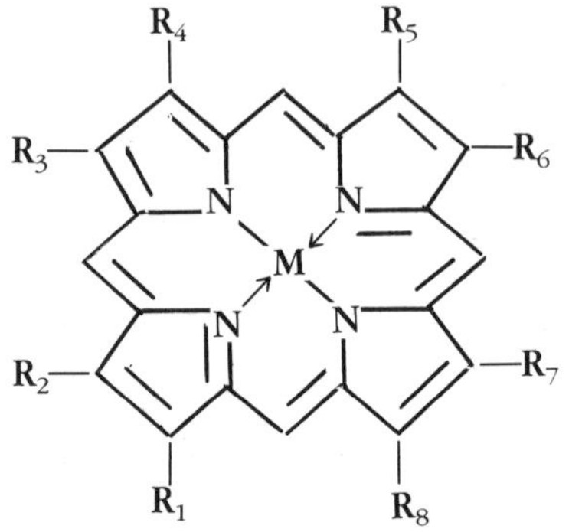

$$X = d(CCGTGGGTCTTGATTAC)$$

$$R_1 = R_3 = R_5 = R_7 = —CH_3$$

$$R_2 = R_4 = R_6 = R_8 = —CH_2CH_2COOH$$

A palladium(II) coproporphyrin was attached to a 5′-terminal phosphate of an oligonucleotide via an amino group spacer upon activation. The extent of modification in complementary complexes was ca. 30%. Further modification was interrupted by self-degradation of the reagent (Fedorova et al., 1990).

OTHER REACTIVE AND CATALYTIC OLIGONUCLEOTIDE DERIVATIVES

The three groups of reagents discussed above, capable of alkylation, oxidative cleavage, and photomodification, do not exhaust chemical and biochemical possibilities of site-specific damage of nucleic acids, although they are the most popular with the investigators in this field.

Attempts to design site-specific analogs of the famous anticancer drug *cis*-PtCl$_2$(NH$_3$)$_2$ have been reported (Vlassov et al., 1983). However, for these derivatives, directly attacking coordinating groups, the level of self-modification seems to be rather high (Vlassov and Kazakov, 1982) and reasonable results may be expected for the derivatives composed of the thymidylate residues with single Ado, Gua, or Cyt. Nevertheless, a heterogeneous oligonucleotide with PtCl$_3^-$ or *trans*-Pt[Cl(NH$_3$)$_2$] attached to the 5'-end *via* a thioethylamido spacer was used to suppress template activity of DNA during DNA polymerase reaction by crosslinking (Chu and Orgel, 1989).

Attachment of addressed oligonucleotides seems to be a promising possibility to enhance specificity of nucleases. A few years ago it was found that oligothymidylates may be covalently linked to RNase A without noticeable loss of nuclease activity (Buneva et al., 1986; Knorre et al., 1986). The most significant success in this field was achieved by P.G. Schultz, who succeeded in specific binding of oligonucleotides to ribonuclease S (Zuckermann and Schultz, 1988) and to staphylococcal nuclease (Corey and Schultz, 1987; Corey et al., 1989a,b; Zuckermann and Schultz, 1989). Highly selective cleavage of both DNA and RNA was demonstrated.

REAGENTS OF ENHANCED EFFICIENCY

It is evident that the efficiency of reactive oligonucleotide derivatives depends on two main factors: the nature of the reactive groups and the stability of the complementary complexes. The latter may be enhanced by attachment to oligonucleotides of polyaromatic heterocyclic residues. The first results were reported nearly a decade ago (Letsinger and Schott, 1981), when the internucleotide phosphate of dTpdT was coupled to phenanthridinium dye.

It was shown that, at low ionic strength of the buffer, T_m of the duplex, formed by poly(A) and modified dTpdT, equaled 45°C, while in 0.1 M salt it was 25°C. The unmodified dinucleotide did not bind poly(A) even at 0°C.

Such stabilization was ascribed to both a strong stacking effect of the dye, intercalating between base pairs, and a favorable electrostatic interaction between the dye cation and the poly(A) polyanion.

Oligonucleotides and their analogs with intercalating acridine (Acr) residues attached via a polymethylene bridge were synthesized by coupling the 3'- or 5'-phosphate of an oligonucleotide with hydroxy groups of acridine derivatives, HO(CH$_2$)$_n$NH(Acr) (Asseline et al., 1983, 1985; Hélène et al., 1985; Thuong and Asseline, 1985; Thuong and Chassignol, 1988; Durand et al., 1989). Later a number of methods for synthesis of Acr-containing oligonucleotides were developed based on the solid phase phosphoramidite technique (Asseline and Thuong, 1988, 1989, 1990; Thuong and Chassignol, 1988).

Oligothymidylate derivatives, AcrNH(CH$_2$)$_5$(pdT)$_n$p(CH$_2$)$_5$NHAcr, were synthesized by linking an oligonucleotide carrying both 3'- and 5'-phosphates with HO(CH$_2$)$_5$NHAcr (Asseline and Thuong, 1988). Acr-derivatives of α-anomers of oligonucleotides were also obtained (Sun et al., 1987; Thuong et al., 1987; Thuong and Chassignol, 1988).

Acr =

It has been shown that Acr attachment to a phosphate group of an oligothymidylate through a polymethylene bridge stabilizes their duplexes with poly(A). Stabilization depends on the length of a polymethylene fragment, with a pentamethylene stretch giving maximal effect. The attachment of a dye residue to the 3'-end of a tetrathymidylate results in a 10 kcal/mol increase in the duplex enthalpy (Asseline et al., 1983, 1984a,b). Quite unexpectedly, the stability of a poly(A) duplex with a 3'-5'-di-Acr derivative of octathymidylate was the same as that with the 3'- or 5'-mono-Acr derivatives. This may be due to the repulsion of two positively charged Acr residues belonging to contiguous oligothymidylate derivatives on the poly(A) template. NMR analysis showed that the Acr residue intercalates mostly between the first and the second base pairs of the duplex (Lancelot et al., 1986, 1988). Further investigations showed that polyaromatic systems stabilize both the duplexes formed by two β-anomeric oligonucleotides, and those composed of an α- and a β-anomeric oligonucleotide (Thuong et al., 1987). Also it was recently shown that acridine residues stabilized the formation of triple-stranded structures (Sun et al., 1989). A detailed survey of acridine derivatives has been described (Cohen, 1989).

Also the duplexes may be stabilized by phenazinium residues. A simple method of introducing an N-2-oxyethylphenazinium residue (Phn) in deblocked oligonucleotides was elaborated. Phenazinium quaternary salts can easily substitute for a hydrogen atom at position 2 for both primary or secondary aliphatic amino groups. This property was exploited to attach Phn to the amino group of a spacer tethered to an oligonucleotide.

Phn =

Unlike Acr-containing oligonucleotides, their phenazinium counterparts have maximal T_m with a spacer having $n = 2$; the attachment of Phn to the 5'-end contributes a stronger stabilizing effect than that to the 3'-end (ΔT equal to 19° and 13°C, respectively).

Simultaneous introduction of two Phn groups (3'- and 5'-end) results in additional stabilization ($\Delta T = 27$°C) (Zarytova et al., 1986, 1989).

A stabilizing effect was also shown for a number of heterocyclic groups such as oxazolopyridocarbazole (Gauthier et al., 1987), 9-aminoellipticine (Vasseur et al., 1988), aflatoxin B (Gopalakrishnan et al., 1989), daunomycin (Godovikova et al., 1990), psoralen (Pieles and Englisch, 1989), and ethidium (Benimetskaya et al., 1989).

All these stabilizing groups appeared to be inert in the dark. However, at least some of them, like Phn (Zarytova et al., 1990a), ethidium (Benimetskaya et al., 1989), and psoralen (Kulka et al., 1989), were found to be photoactivatable.

It seemed then reasonable to try to enhance the efficiency of reactive oligonucleotide derivatives by attaching duplex-stabilizing residues to reagents.

Oligomers **XIV** and **XV** bearing Acr and EDTA or methylpyroporphyrin **XXI** (MPPo) groups were synthesized by coupling acridine-containing oligomers to HO(CH$_2$)$_6$NH-EDTAMe$_3$, followed by alkaline treatment removing the blocking methyl residues, or to HO(CH$_2$)$_6$NH(MPPo).

$$Y - (pdT)_8p - Z$$

XIV a Y = Acr,   Z = EDTA;

  b Y = EDTA, Z = Acr

XV a Y = Acr,   Z = MPPo;

  b Y = MPPo, Z = Acr

They were targeted to a single-stranded DNA fragment, 27 nucleotides in length, bearing an octadeoxyadenylate sequence. The cleavages induced by Fe·EDTA-(CH$_2$)$_6$(pdT)$_8$p(CH$_2$)$_5$-Acr and by (pdT)$_7$p(CH$_2$)$_6$-EDTA·Fe were compared, and revealed a 3- to 7-fold higher efficiency of the Acr-containing reagent at 0–10°C (Boidot-Forget et al., 1988). It should be emphasized that such comparison is not quite correct because there is considerable difference in the points of attachment of the reactive group and Acr residues, as well as in the lengths of addressing oligonucleotide moieties.

The level of cleavage of double-stranded DNA containing a (pdA)$_8$ stretch was significantly lower than that of single-stranded DNA.

Octathymidylate reagents **XVI** and **XVII**, each bearing a stabilizing and a photoreactive

group, were synthesized. Acr was attached to the 5′-phosphate (via a pentamethylene linker) whereas a *p*-azidophenacyl group (N$_3\phi$) was linked to thiophosphate.

XVI   Y − (pdβT)$_8$p − Z

XVII   Y − (pdαT)$_8$p − Z

  a Y = Acr, Z = N$_3\phi$

  b Y = N$_3\phi$, Z = Acr

Oligothymidylate **XVIa**, covalently bound to an acridine derivative and a *p*-azidophenacyl group, was shown to photocrosslink and cleave both sides of the oligo(dA) track of the 27-mer at high salt concentration. This may be due to the formation of both double- and triple-stranded structures (Praseuth et al., 1987, 1988a,b).

Unlike β-anomer **XVI**, α-anomer **XVII** forms parallel stretches with the target, regardless of whether it is covalently linked to an intercalating residue. **XVIIa** was found to cleave at the 3′-side of the target while **XVIIb** cleaved at the 5′-side. Among the three reagents, **XVIa**, **XVIb**, and **XVIIa**, the latter produced the most effective cleavage (Praseuth et al., 1988a,b).

All of the results mentioned above do not allow reliable comparison of the stabilizing effects of the attached groups. Such comparative study has been made with alkylating oligonucleotide derivatives **XVIII** and **XIX** bearing a stabilizing 2-hydroxyethylphenazinium (Phn) residue. The Phn residue can be easily introduced in alkylating reagents in the same way as in parent oligonucleotides bearing a —CH$_2$CH$_2$NH$_2$ spacer.

XVIII   Phn − pd(NpN...N)rN>CHRCl

XIX   Y − pd(NpN...N)p − Z

a Y = Phn, Z = CHRCl; **b** Y = CHRCl, Z = Phn

The kinetics of modification by reagent with and without Phn demonstrates that, in the former case, the reaction rate and maximum extent of modification are higher. It is noteworthy that Phn residue does not change the site of modification of the reagent as was demonstrated on model systems (Zarytova et al., 1986, 1987a, 1989). Figure 1 presents the temperature dependence of maximum extent of modification. The yield of modification is higher for the Phn reagent in a wide range of temperatures. The S-shaped curves obtained look similar to the denaturing profile of complementary complexes, reflecting the interdependence of complex stability and efficiency of target modification. Attachment of a stabilizing group allows the achievement of the same modification level either with a shorter reagent or at higher temperature. Another important consequence of stabilization is the possibility to decrease the concentrations necessary to attain a biological effect. For instance, a hexanucleotide reagent Cl-RCH$_2$NH-pd(TTCCC)A complementary to the DNA fragment corresponding to TBEV RNA (25°C) does not modify the target at 1×10^{-4} M concentration. The analogous reagent bearing a Phn residue allows a 50% modification at 2×10^{-5} M concentration. The possibility to decrease reagent concentration by attaching stabilizing groups seems to be especially important for *in vivo* experiments, as it ensures against nonspecific modification of many other biopolymers of cells.

It is worth mentioning that to achieve the

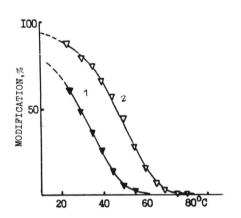

Fig. 1. Temperature dependence of the maximum extent of alkylation of the target pd(AACCTGTTTGGC) with the reagents pd(CCAAAC)A>CHRCl (1) and Phn-NHCH$_2$CH$_2$-NH-pd(CCAAAC)A>CHRCl (2); 1×10^{-5} M target, 2×10^{-5} M reagents (Zarytova et al., 1987a).

stabilization effect, it is not always necessary to introduce a reactive group and a duplex-stabilizing residue into the same oligonucleotide. They may be attached to two different oligonucleotides complementary to contiguous sequences of the target nucleic acid, in this way forming a tandem. Due to strong cooperative interactions between these two oligonucleotide derivatives within their complex with the target, a rather stable nicked duplex containing both the additional residues is formed. The derivative bearing a stabilizing group enhances stability of the complex formed by its reactive counterpart and, consequently, increases the efficiency of the reaction with the target (Zarytova et al., 1988; Kutyavin et al., 1988; Amirkhanov et al., 1989; Amirkhanov and Zarytova, 1990). Therefore the stabilizing derivative may be considered as an effector of sequence-specific modification. Obviously, the nicked structure formed is not as stable as that produced by a single oligonucleotide covering simultaneously both complemenary sequences of the target. However, being built of shorter oligonucleotides, it seems to have some advantages.

The first is an easier availability of short oligonucleotide carriers that may be especially essential for future medical applications of reactive oligonucleotide derivatives.

The second is the possibility to control modification specificity by oligonucleotide effectors. The longer an oligomer is, the greater is the number of imperfect complexes it forms with different nucleic acids and/or with different parts of the same polynucleotide chain. The formation of such complexes may result in modification of undesirable regions of the target or of undesirable targets (e.g., not only viral nucleic acids but also host nucleic acids). This was clearly demonstrated in the case of alkylation of *Escherichia coli* 16 S rRNA with oligonucleotide derivatives complementary to different regions of this RNA (Zenkova et al., 1990a,b). A reagent with a short oligonucleotide moiety forms significantly fewer imperfect and partially paired complexes. At the same time, it may form a greater number of completely paired complexes. However, at low concentrations of the reagent and in the absence of effector these complexes are rather unstable and are unlikely to provide notice-able modification of the target. Effectors al-low discrimination of these multiple com-plexes because they stabilize only those complexes where they form a tandem with the reagent. Consequently the effector will "switch on" only one of numerous recogni-tion sites. Thus, using the effectors, improve both the selectivity and efficiency of se-quence-specific modification.

These possibilities were demonstrated us-ing a 303-nt single-stranded DNA fragment of cDNA of TBEV RNA as a target, alkylating derivatives of oligonucleotides as reagents, and the above described phenazinium deriv-atives as effectors. A scheme of the experi-ments is given in Figure 2. The regions A and B of the target contain the same hexanucleo-tide sequence d(TGGGAA) at positions 144–149 and 261–266. At a $10^{-5} M$ concentra-tion, no modification of the DNA fragment by a $ClRCH_2NH$-pd(TTCCC)A reagent was revealed. The presence of an oligonucleo-tide, complementary to sequence 136–143, carrying one and especially two phenazinium residues allowed noticeable alkylation of re-gion B and forbade modification of region A.

Similar phenazinium derivatives complemen-tary to sequence 252–259 produced just the opposite effect (Fig. 3) (Kutyavin et al., 1988). The same reagent squeezed between two ef-fectors was found to be still more efficient (comp. 38 and 89% respective yields of al-kylation with one and two effectors in iden-tical conditions) (Zarytova et al., 1990b).

HYDROPHOBIC DERIVATIVES AND ANALOGS OF OLIGONUCLEOTIDES CARRYING REACTIVE GROUPS

A key problem of application of oligonu-cleotides and their reactive derivatives to liv-

Fig. 3. Eight percent denaturing polyacrylamide gel of ^{32}P-labeled products of the DNA fragment (Fig. 2) modification by alkylating reagent Cl-RCH_2NH-pd(TTCCCA) (lane 2) and by the same reagent in the presence of effectors: Phn-$NHCH_2CH_2NH$-pd(TTCAAGGC)p-$NHCH_2CH_2NH$-Phn (lane 1) and Phn-$NHCH_2CH_2NH$-pd(GAGCCTGG)p-OCH_2CH_2NH-Phn (lane 3). Lane 4 is A + G reaction. Modifications were carried out at 37°C; time 18 hr; $1 \times 10^{-5} M$ target, $1 \times 10^{-5} M$ oligonucleotide derivatives. (Reproduced from Kutyavin et al., 1988, with permission of the pub-lisher.)

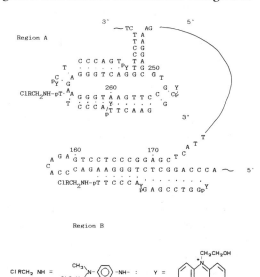

Fig. 2. The structures of the target regions of the DNA fragment (302-mer), reagents, and effectors used.

ing systems is their poor permeability into cells. One of the ways to solve it is using nonionic oligonucleotide derivatives or analogs. At first it was proposed to use P-ethyl esters of oligonucleotides (Miller et al., 1974). Later it was found that similar result may be achieved with oligomers prepared from nucleoside methylphosphonates (Miller et al., 1977, 1979). A common feature of both types of compounds is the chirality of P-atoms of all hydrophobic internucleoside fragments. Since there is no general approach to stereoselective synthesis of these oligomers they are produced as mixtures of diastereomers. Meanwhile the ability of S_p isomers to form complementary complexes is much lower than that of R_p isomers. Therefore, in order to be used as antisense compounds, these hydrophobic oligomers as well as their reactive derivatives must be fractionated either at each step of the oligomer chain growth (Abramova and Lebedev, 1983; Abramova et al., 1985) or at a final stage of the synthesis.

The first hydrophobic reactive oligonucleotide derivative described was [dTp(Et)]$_9$pU>CHRCl carrying an alkylating aromatic N-2-cloroethylamino residue RCl. It was found to modify preferentially poly(A) fragments of polyadenylated mRNAs within Krebs ascites carcinoma cells (Karpova et al., 1980; Knorre et al., 1981).

Oligodeoxyribonucleoside methylphosphonates with modified 5'-phosphates were obtained (Lee et al., 1988a) by coupling of a 5'-terminal phosphate oligomer to the amino group of 4'-(aminoalkyl)-4,5'-8-trimethylpsoralen in the presence of water-soluble carbodiimide. Psoralen-derivatized methylphosphonate oligomers efficiently (up to 98%) crosslink to single-stranded DNA targets under irradiation at 365 nm. Irradiation of the duplex with a less stable G·T base pair or a single mismatch considerably decreases crosslinking. At temperatures above the T_m, there is a drastic drop in crosslinking (Lee et al., 1988a,b). Psoralen-derivatized methylphosphonate oligomers allowed a 50–80% photomodification of rabbit globin mRNA *in vitro*.

The correlation between the quality of DNA modification and substitution of methylphosphonate residues for the phosphate ones was studied. Alkylating derivatives of methylphosphonate analogs of octathymidylate **XX** (a mixture of isomers)

XX   ClRCH$_2$NH-pdTpdT[p(CH$_3$)dT]$_6$

and an octathymidylate having alternating stereoregular methylphosphonate and phosphate residues, XXI-XXIII, were prepared.

XXI–XXIII   Y-pdTp[dTpdTp(R)]$_3$dTp-R'

XXI Y = $-$NHCH$_2$RCl, R' = $-$SH$_3$

a R=CH$_3$, R$_p$ configuration;

b R=CH$_3$, S$_p$ configuration

XXII Y = $-$NHCH$_2$RCl, R' = Phn

a R=CH$_3$, R$_p$ configuration;

b R=CH$_3$, S$_p$ configuration

XXIII Y = Phn, R' = $-$NHCH$_2$RCl

a R=CH$_3$, R$_p$ configuration;

b R=CH$_3$, S$_p$ configuration

(Amirkhanov and Zarytova, 1989a,b,c, 1990).

The highest yield of modification was achieved for the R_p isomers, which was even greater than for the parent oligonucleotide ClRCH$_2$NH(pdT)$_8$ (Fig. 4). The lowest yield was obtained for reagent X whose addressed moiety was a mixture of isomers containing six methylphosphonate residues and for reagent **XXIb** (S$_p$ isomer).

It is essential that the position of modification by reagents **XXI–XXIII** should remain the same as that for the phosphodiester reagents. A phenazinium residue present in reactive derivatives of both R$_p$ and S$_p$ methylphosphonate analogs enhances the efficiency of target alkylation, as in the case of phosphodiester derivatives (Amirkhanov and Zarytova, 1989, 1990; Amirkhanov et al., 1989). The possibility to enhance the reaction efficiency of methylphosphonate analogs by at-

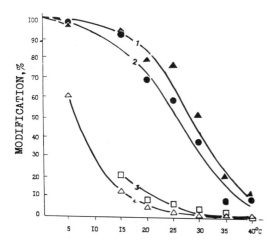

Fig. 4. Temperature dependence of the maximum extent of alkylation of the target pd($C_5A_8C_5$) by reagents: curves 1 and 3, ClRCH$_2$NH-pdTp[dTpdT(R)]$_3$dTp-SCH$_3$, R = CH$_3$, R_p and S_p configuration, respectively; curve 2, ClRCH$_2$NH-pd(T_8); curve 4, ClRCH$_2$NH-pdTpdT[p(CH$_3$)dT]$_6$; 1×10^{-5} M target, 2×10^{-5} M reagents (Amirkhanov and Zarytova, 1989c).

taching phenazinium residues makes the mixtures of diastereoisomers more reliable reactive group carriers.

5′- and cytidine EDTA derivatives of methylphosphonate oligonucleoside were synthesized and their interaction with nucleic acids was characterized. It was shown that derivatization of methylphosphonate oligomers with EDTA reduced the melting temperature by 5°C. The EDTA derivatives obtained cleaved an ssDNA fragment in a site-specific manner (Lin et al., 1989).

Reactive derivatives of methylphosphonate oligonucleotides were also shown to be subject to the influence of contiguous oligonucleotides. With EDTA derivatives and unmodified adjacent oligonucleotides, a 2-fold increase in the level of modification was achieved (Lin et al., 1989). A diphenazinium oligonucleotide effector brought about a 7-fold increase in the efficiency of sequence-specific alkylation directed by an oligonucleotide analog with alternating methylphosphonate and phosphodiester internucleotide fragments (Amirkhanov and Zarytova, 1990).

Another method for enhancement of permeability was imparting of hydrophobicity to oligonucleotides by means of lipophilic residues. Cholesterol, ergosterone, and testosterone residues were attached to blocked oligonucleotides under conditions of phosphotriester method (Ivanova et al., 1988; Boutorin et al., 1989a,b; Zarytova et al., 1988; Zarytova et al., 1990c). Oligonucleotides and phosphorothioate oligonucleotide analogs with cholesteryl groups covalently joined at the terminal internucleoside phosphorus were synthesized on supports by oxidative phosphoramidation with carbon tetrachloride and cholesteroloxycarbonylaminoethylamine (Letsinger et al., 1989). Steroids were also introduced in deblocked oligonucleotides, only this time they carried a spacer with an aliphatic amino group. Terminal phosphates were activated by a mixture of Ph$_3$P and Py$_2$S$_2$ in the presence of a nucleophilic catalyst (Zarytova et al., 1987b). Alkylating residues were coupled to the steroid oligonucleotide derivatives obtained.

XXIV   Y – pd(NpN...N)p – Z

Y = cholesterol, ergosterol,

testosterone residues

Z = NHCH$_2$RCl

The ability of the resultant cholesterol compounds to bind to cells was tested.

They turned out to have enhanced capacity to cross cell membranes. Moreover they affect cell nucleic acids nearly two orders of magnitude more efficiently than the reagents lacking steroid residues. In fact, the level of DNA modification increased from 0.79 to 54 mol of reagent per 10^6 mol of nucleotide residues (Boutorin et al., 1989a,b).

This approach, therefore, makes it possible to avoid the problems of separation of diastereoisomers.

The most essential biological applications of reactive oligonucleotide derivatives are discussed in the chapter by Vlassov and Yakubov (this volume).

SUMMARY

The aim of the present chapter has been to consider the chemical aspects of gene-directed oligonucleotide derivatives that are potential antiviral and antitumor agents. It has dealt with the synthesis of oligonucleotide derivatives carring various reactive groups such as alkylating or photoactivatable chain-cleaving moieties, and combination of these derivatives with hydrophobic and duplex-stabilizing residues. The main criteria of site-specific modification of nucleic acids fragment with oligonucleotide reagents were reviewed. The ways of enhancement of permeability of oligonucleotide derivatives into cells, either by replacing the negatively charged internucleotide phosphates, or by attachment to the oligonucleotide tails of bulky hydrophobic radicals such as steroids, were discussed. Attachment of stabilizing groups to oligonucleotide reagents allows achievement of the same modification level either with a shorter reagent, or at higher temperature, or at lower concentration. To improve both selectivity and efficiency of sequence-specific modification of nucleic acids, targeted tandem short oligonucleotide derivatives (oligonucleotides having reactive groups, and flanked with oligonucleotides bearing polyaromatic residues) may be used.

REFERENCES

Abramova TV, Lebedev AV (1983): Investigation of non-ionic diastereomeric analogs of oligonucleotides. Synthesis and separation of diastereomers of di- and tetrathymidylate ethyl esters. Bioorg. Khim. 6:824–831.

Abramova TV, Knorre DG, Lebedev AV, Pichko NP, Fedorova OS (1985): Investigation of diastereomers of non-ionic oligonucleotide analogues. III. Complementary addressed modification of poly(dA) by alkylating derivatives synthesized from individual diastereomers of triethyl phosphotriesters of tetrathymidylyluridine. Bioorg. Khim. 11:1642–16.

Amirkhanov NV, Zarytova VF (1989a): Reactive oligonucleotide derivatives bearing methylphosphonate groups. I. Synthesis and separation of 5-0-dimethoxytrityl[thymidylyl-(3′-methylphosphonate 5′)-thymidine]-3′-0-β-cyanomethylthiophosphate. Bioorg. Khim. 15:259–266.

Amirkhanov NV, Zarytova VF (1989b): Reactive oligonucleotide derivatives bearing methylphosphonate groups. II. Synthesis of stereoregular octathymidylates containing an alkylating 4-(N-2-chloroethyl-N-methylamino)benzylamine group and alternating methylphosphonate residues. Bioorg. Khim. 15:267–276.

Amirkhanov NV, Zarytova VF (1989c): Reactive oligonucleotide derivatives bearing methylphosphonate groups. III. Affinity modification of nucleic acid target by 4-(N-2-chlorethyl-N-methylamino)benzyl-3′- and 5′-phosphoamide derivatives having stereoregular methylphosphonate residues. Bioorg. Khim. 15:379–386.

Amirkhanov NV, Zarytova VF (1990): Reactive oligonucleotide derivatives bearing methylphosphonate groups. V. Alkylating methylphosphonate octathymidylate derivatives bearing a N-(2-hydroxyethyl)phenazinium residue increase the efficiency of complementary addressed modification of DNA-target. Bioorg. Chem. 16:370–378.

Amirkhanov NV, Zarytova VF, Levina AS, Lokhov SG (1989): Reactive oligonucleotide derivatives bearing methylphosphonate groups. IV. Stabilization of complementary complexes of methylphosphonate octathymidylates by N-(2-hydroxyethyl)-phenazinium residue. Bioorg. Khim. 15:1618–1626.

Asseline U, Thuong NT (1988): Oligothymidylate substitués par un dérivé de l'acridine en position 5′, à la fois en position 5′, à la fois en 5′ et 3′, ou sur un phosphate internucleotique. Nucleosides Nucleotides 7:431–455.

Asseline U, Thuong NT (1989): Solid-phase synthesis of modified oligodeoxyribonucleotides with an acridine derivative or a thiophosphate group at their 3′-end. Tetrahedron Lett. 30:2521–2524.

Asseline U, Thuong NT (1990): New solid-phase method for automated synthesis of oligonucleotides containing an aminoalkyl linker at their 3′-end. Tetrahedron Lett. 31:81–84.

Asseline U, Thuong NT, Hélène C (1983): New substances with high and specific affinity toward nucleic acid sequences: Intercalating agents covalently linked to an oligodeoxynucleotide. C. R. Acad. Sci.Paris, Ser. III 297:369–372.

Asseline U, Toulmé F, Thuong NT, Delarue M, Montenay-Garestier T, Hélène C (1984a): Oligodeoxynucleotides covalently linked to intercalating dyes as base sequence-specific ligands. Influence of dye attachment site. EMBO J. 3:795–800.

Asseline U, Delarue M, Lancelot G, Toulmé F, Thuong NT, Montenay-Garestier T, Hélène C (1984b): Nucleic acid-binding molecules with high affinity and base sequence specificity: Intercalating agents covalently linked to oligodeoxynucleotides. Proc. Natl. Acad. Sci. U.S.A. 81:3297–3301.

Asseline U, Thuong NT, Hélène C (1985): Oligonucleotides covalently linked to intercalating agents. Influence of positively charged substituents on binding to complementary sequences. J. Biol. Chem. 260:8936–8941.

Belikova AM, Zarytova VF, Grineva NI (1967): Synthesis of ribonucleosides and diribonucleoside phosphates containing 2-chloroethylamine and nitrogen mustard residues. Tetrahedron Lett. (37):3557–3562.

Belikova AM, Grineva NI, Knorre DG, Mysina SD (1973): Specific cleavage of DNA on guanine residues by 2',3'-O-[4-(N-2-chloroethyl-N-methylamino)benzylidene] uridine and its 5'-methylphosphate. Dokl. Acad. Nauk SSSR 212:876–879.

Benimetskaya LZ, Grineva NI, Karpova GG, Pichko NP, Chimitova TA (1977): On the mechanism of nucleic acid alkylation within complementary complexes. The kinetics of RNA and DNA alkylation by 2', 3'-0-[4-N-2-chloroethyl-N-methylamino)-benzylidene] oligonucleotides. Bioorg. Khim. 3:903–913.

Benimetskaya LZ, Bulychev NV, Kozionov AL, Lebedev AV, Nesterikhin YuE, Novozhilov SYu, Rautian SG, Stockman MI (1983): Two-quantum selective laser scission of polyadenylic acid in the complementary complex with a dansyl derivative of oligothymidylate. FEBS Lett. 163:144–149.

Benimetskaya LZ, Bulychev NV, Kozionov AL, Lebedev AV, Novozhilov SYu, Stockman MI (1984): Two-quantum selective laser modification of poly- and oligonucleotides in complementary complexes with dansyl derivatives of oligodeoxynucleotides. Nucl. Acids Res., Symp. Ser. No. 14:323–324.

Benimetskaya LZ, Bulychev NV, Kozionov AL, Koshkin AA, Lebedev AV, Novozhilov SYu, Stockman MI (1988): High-efficiency complementary addressed laser modification (cleavage) of oligodeoxynucleotides. Bioorg.Khim. 14:48–57.

Benimetskaya LZ, Bulychev NV, Kozionov AL, Koshkin AA, Lebedev AV, Novozhilov SYu, Stockman MI (1989): Site-specific laser modification (cleavage) of oligodeoxynucleotides. Biopolymers 28:1129–1147.

Blake KR, Murakami A, Spitz SA, Glave SA, Reddy MP, TS'o POP, Miller PS (1985): Hybridization arrest of globin synthesis in rabbit reticulocyte lysates and cells by oligodeoxyribonucleoside methylphosphonates. Biochemistry 24:6139–6145.

Boidot-Forget M, Chassignol M, Takasugi M, Thuong NT, Hélène C (1988): Site-specific cleavage of single-stranded and double-stranded DNA sequences by oligodeoxyribonucleotides covalently linked to an intercalating agent and an EDTA-Fe chelate. Gene 72:361–371.

Boutorin AS, Vlassov VV, Kazakov SA, Kutyavin IV, Podyminogin MA (1984): Complementary addressed reagents carrying EDTA-Fe(II) groups for

directed cleavage of single-stranded nucleic acids. FEBS Lett. 172–43–46.

Boutorin AS, Vlassov VV, Guskova LV, Zarytova VF, Ivanova EM, Kobets ND, Ryte AS, Yurchenko LV (1989a): Synthesis and properties of alkylating oligonucleotide derivatives containing cholesterol or phenazinium groups covalently attached to 3'-termini and their interaction with eukaryotic cells. Mol. Biol. 23:1382–1390.

Boutorin AS, Gus´kova LV, Ivanova EM, Kobets ND, Zarytova VF, Ryte AS, Yurchenko LV, Vlassov VV (1989b): Synthesis of alkyating oligonucleotide derivatives containing cholesterol or phenazinium residues at their 3'-termini and their interaction with DNA within mammalian cells. FEBS Lett. 254:129–132.

Brosalina EB, Vlassov VV, Kazakov SA (1988): Complementary addressed modification of single-stranded DNA with [Fe·EDTA]-oligonucleotide derivatives. Bioorg. Khim. 14:125–128.

Budker VG, Zarytova VF, Knorre DG, Kobets ND, Ryazankina OI (1977): The synthesis of oligonucleotide 5'-triphosphates. Bioorg. Khim. 3:618–625.

Buneva VN, Godovikova TS, Zarytova VF (1986): Modification of RNAse by di- and oligodeoxyribonucleotide 5'-phospho-N-methylimidazolide derivatives. Bioorg. Khim. 12:906–910.

Cazenave C, Loreau N, Thuong NT, Toulmé J-J, Hélène C (1987): Enzymatic amplification of translation inhibition of rabbit β-globin mRNA mediated by antimessenger oligodeoxynucleotides covalently linked to intercalating agents. Nucl. Acids Res. 15:4717–4736.

Chen CB, Sigman DS (1986): Nuclease activity of 1,10-phenanthroline-copper: Sequence-specific targeting. Proc. Natl. Acad. Sci. U.S.A. 83:7147–7151.

Chen CB, Sigman DS (1988): Sequence-specific scission of RNA by 1,10-phenanthroline-copper linked to deoxyoligonucleotides. J. Am. Chem. Soc. 110:6570–6572.

Chu BCF, Orgel LE (1985): Nonenzymatic sequence-specific cleavage of single-stranded DNA. Proc. Natl. Acad. Sci. U.S.A. 82:963–967.

Chu BCF, Orgel LE (1989): Inhibition of DNA synthesis by crosslinking the template to platinum-thiol derivatives of complementary oligodeoxynucleotides. Nucl. Acids Res. 17:4783--4798.

Cohen JS (1989): Oligonucleotides Antisense Inhibitors of Gene Expression. Boca Raton, FL: CRC Press, Chap 2, pp 25–51.

Corey DR, Schultz PG (1987): Generation of a hybrid sequence-specific single-stranded deoxyribonuclease. Science 238:1401–1403.

Corey DR, Pei D, Schultz PG (1989a): Sequence-selective hydrolysis of duplex DNA by an oligonucleotide-directed nuclease. J. Am. Chem. Soc. 111:8523–8525.

Corey DR, Pei D, Schultz PG (1989b): Generation of a catalytic sequence-specific hybrid DNase. Biochemistry 28:8277–8286.

Dreyer GB, Dervan PB (1985): Sequence-specific cleavage of single-stranded DNA: Oligodeoxynucleotide-EDTA FE(II). Proc. Natl. Acad. Sci. U.S.A. 82:968–972.

Durand M, Maurizot JC, Asseline U, Barbier C, Thuong NT, Hélène C (1989): Oligothymidylates covalently linked to an acridine derivative and modified phosphodiester backbone: Circular dichroism studies of their interactions with complementary sequences. Nucl. Acids Res. 17:1823–1837.

Fedorova OS, Knorre DG, Podust LM, Zarytova VF (1988): Complementary addressed modification of double-stranded DNA within a ternary complex. FEBS Lett. 228:273–276.

Fedorova OS, Savitskii AP, Shoikhet KG, Ponomarev GV (1990): Palladium(II)-coproporphyrin I as a photoactivable group in sequence-specific modification of nucleic acids by oligonucleotide derivatives. FEBS Lett. 259:335–337.

Francois J-C, Saison-Behmoaras T, Chassignol M, Thuong NT, Sun J, Hélène C (1988a): Periodic cleavage of poly(dA) by oligothymidylates covalently linked to the 1,10-phenanthroline-copper complex. Biochemistry 27:2272–2276.

Francois J-C, Saison-Behmoaras T, Hélène C (1988b): Sequence-specific recognition of the major groove of DNA by oligodeoxynucleotides via triple helix formation. Footprinting studies. Nucl. Acids Res. 16:11431–11440.

Francois J-C, Saison-Behmoaras T, Chassignol M, Thuong NT, Hélène C (1989a): Sequence-targeted cleavage of single- and double-stranded DNA by oligothymidylates covalently linked to 1,10-phenanthroline. J. Biol. Chem. 264:5891–5898.

Francois J-C, Saison-Behmoaras T, Barbier C. Chassignol M, Thuong NT, Hélène C (1989b): Sequence-specific recognition and cleavage of duplex DNA via triple-helix formation by oligonucleotides covalently linked to a phenanthroline-copper chelate. Proc. Natl. Acad. Sci. U.S.A. 86:9702–9706.

Francois J-C, Saison-Behmoaras T, Thuong NT, Hélène C (1989c): Inhibition of restriction endonuclease cleavage via triple helix formation by homopyrimidine oligonucleotides. Biochemistry 28:9617–9619.

Frolova EI, Ivanova EM, Zarytova VF, Vlassov VV (1990): Porphyrin-linked oligonucleotides: Synthesis and sequence-specific modification of ssDNA. FEBS Lett. 269:101–104.

Gautier C, Morvan F, Rayner B, Huynh-Dinh T, Igolen J, Imbach J-L, Paoletti C, Paoletti J (1987): α-DNA IV: α-anomeric and β-anomeric tetrathymidylates covalently linked to intercalating oxazolopyridocarbazole. Synthesis, physicochemical properties and poly(rA) binding. Nucl. Acids Res. 15:6625–6641.

Godovikova TS, Zarytova VF, Maltzeva TV, Khalimskaya LM (1989): Reactive oligonucleotide derivatives with a zwitterionic terminal phosphate group for affinity reagents and probe construction. Bioorg. Khim. 15:1246–1252.

Godovikova TS, Zarytova VF, Lokhov SG, Maltseva TV, Sergeev DS (1990): Synthesis, structure and properties of daunomycin oligonucleotide derivatives. Bioorg. Khim. 16:1369–1378.

Gopalakrishnan S, Stone MP, Harris TM (1989): Preparation and characterization of an aflatoxin B_1 adduct with the oligodeoxynucleotide d(ATCGAT)$_2$. J. Am. Chem. Soc. 111:7232–7239.

Grineva NI, Karpova GG (1974): Complementary addressed alkylation of ribosomal RNA with alkylating derivatives of oligonucleotides. Mol. Biol. 8:832–844.

Grineva NI, Karpova GG (1975): Alkylation of ribosomal RNA complementary complexes with 2′, 3′-0-[4-2-chloroethyl-N-methylamino)-benzylidene] oligonucleotides. Chemical selectivity and positional specificity. Bioorg. Khim. 1:588–597.

Grineva NI, Lomakina TS (1972): Alkylating derivatives of nucleic acid components. XIV. Synthesis of 5′-phosphoramides of oligonucleotides-derivatives of 4-(N-2-chloroethyl-N-methylamino)benzylamine Zhur. Obsh. Khim. XLII:1630–1634.

Grineva NI, Zarytova VF (1968): Alkylating derivatives of nucleic acid components. III. Reaction of uridilyluridine with 4-dialkylaminobenzaldehydes. Izv. Sib. Otd. Akad. Nauk SSSR, Ser. Khim. Nauk 5:125–129.

Grineva NI, Zarytova VF, Knorre DG (1968): Alkylating derivatives of nucleic acids component. II. The synthesis of uridylyl-(3′ → 5′)-2′,3′-0-[4-N-2-chloroethyl-N-methylamino)-benzylidene] uridine. Izv. Sib. Otd. Akad. Nauk SSSR, Ser. Khim. Nauk 5:118–124.

Grineva NI, Zarytova VF, Knorre DG, Kurbatov VA (1970): Influence of charge of intermediate component on the alkylation of tRNA by 2-chloroethylamines. Dokl. Acad. Nauk SSSR 194:331–334.

Grineva NI, Karpova GG, Kuznetsova LM, Venkstern TV, Bayev AA (1977a): Complementary addressed modification of yeast tRNA$_1^{Val}$ with an alkylating derivative of d(pC-G)-A. The position of the alkylated nucleotides and the course of the alkylation in the complex. Nucl. Acids Res. 5:1609–1631.

Grineva NI, Lomakina TS, Tegeyeva NG, Chimitova TA (1977b): Kinetics the C-Cl ionization in the nucleoside and oligonucleotide 4-(N-2-chloroethyl-N-methylamino)benzylamino)benzyl 5'-phosphoamides. Bioorg. Khim. 3:210–214.

Guimautdinova OI, Grineva NI, Denisova LYa, Lomakina TS, Pustoshilova NM (1977): Aqueous treatment mediated degradation and isomerization of phosphodiester bounds in ribooligonucleotides acylated by mesitoyl chloride. Bioorg. Khim. 3:1633–1640.

Guimautdinova OI, Grineva NI, Karpova GG, Lomakina TS, Shelpakova EI (1978): Preparation of 4-(N-2-chloroethyl-N-methylamino)benzyl ribooligoadenylate 5'-phosphamides and their interaction with E. coli rRNA and DNA. Bioorg. Khim. 4:917–927.

Hélène C, Montenay-Garestier T, Saison T, Takasugi M, Toulmé JJ, Asseline U, Lancelot G, Maurizot JC, Toulmé F, Thuong NT (1985): Oligodeoxynucleotides covalently linked to intercalating agents: A new class of gene regulatory substances. Biochimie 67:777–783.

Ivanova EM, Mamaev SV, Fedorova OC, Frolova EI (1988): Complementary addressed modification of a single-stranded DNA fragment by iron-porphyrin derivative of an oligonucleotide. Bioorg. Khim. 14:551–554.

Iverson BL, Dervan PB (1987): Nonenzymatic sequence-specific cleavage of single-stranded DNA to nucleotide resolution. DNA methyl thioether probes. J. Am. Chem. Soc. 109:1241–1243.

Iverson BL, Dervan PB (1988): Nonenzymatic sequence-specific methyl transfer to single-stranded DNA. Proc. Natl. Acad. Sci. U.S.A. 85:4615–4619.

Karpova GG, Knorre DG, Ryte AS, Stephanovich LE (1980): Selective alkylation of RNA inside the cell with an oligonucleotide ethyl ester derivative bearing a 2-chloroethyl-aminogroup. FEBS Lett. 122:21–24.

Knorre DG, Zarytova VF, Karpova GG, Stephanovich LE (1981): Specific modification of nucleic acids inside the cell with alkylating derivatives of oligonucleotides ethylated at internucleotide phosphates. Nucl. Acids Res., Symp. Ser. No. 9:195–198.

Knorre DG, Vlassov VV, Zarytova VF, Karpova GG (1985a): Nucleotide and oligonucleotide derivatives as enzyme and nucleic acid targeted irreversible inhibitors. Chem. Aspects. Adv. Enzyme Reg. 24:277–299.

Knorre DG, Vlassov VV, Zarytova VF (1985b): Reactive oligonucleotide derivatives and sequence-specific modification of nucleic acids. Biochimie 67:785–789.

Knorre DG, Buneva VN, Baram GI, Godovikova TS, Zarytova VF (1986): Dynamic aspects of affinity labelling as revealed by alkylation and phospho-rylation of pancreatic ribonuclease with reactive deoxyribodinucleotide derivatives. FEBS Lett. 194:64–68.

Knorre DG, Vlassov VV, Zarytova VF, Lebedev AV (1989a): Reactive oligonucleotide derivatives as tools for site specific modification of biopolymers. Sov. Sci. Rev. B. Chem. 12:269–337.

Knorre DG, Vlassov VV, Zarytova VF (1989b): Oligonucleotides linked to reactive groups. In Cohen JS (ed): Oligonucleotides Antisense Inhibitors of Gene Expression. Topics in Molecular and Structural Biology, Vol. 12. Boca Raton, FL: CRC Press, pp 174–196.

Kulka M, Smith CC, Aurelian L, Fishelevich R, Meade K, Miller P, Ts'o POP (1989): Site specificity of the inhibitory effects of oligo(nucleoside methylphosphonate)s complementary to the acceptor splice junction of herpes simplex virus type 1 immediate early mRNA 4. Proc. Natl. Acad. Sci. U.S.A. 86:6868–6872.

Kutyavin IV, Podyminogin MA, Bazhina YuN, Fedorova OS, Knorre DG, Levina AS, Mamayev SV, Zarytova VF. (1988): N-(2-Hydroxyethyl)phenazinium derivatives of oligonucleotides as effectors of the sequence-specific modification of nucleic acids with reactive oligonucleotide derivatives. FEBS Lett. 238:35–38.

Lancelot G, Thuong NT (1986): Nuclear magnetic resonance studies of complex formation between the oligonucleotide d(TATC) covalently linked to an acridine derivative and its complementary sequence d(GATA). Biochemistry 25:5357–5363.

Lancelot G, Guesnet JL, Asseline U, Thuong NT (1988): NMR studies of complex formation between the modified oligonucleotide d(T*TCTGT) covalently linked to an acridine derivative and its complementary sequence d(GCACAGAA). Biochemistry 27:1265–1273.

Le Doan T, Perrouault L, Hélène C (1986): Targeted cleavage of polynucleotides by complementary oligonucleotides covalently linked to iron-porphyrins. Biochemistry 25:6736–6739.

Le Doan T, Perrouault L, Praseuth D, Habhoub N, Decout J-L, Thuong NT, Lhomme J, Hélène C (1987a): Sequence-specific recognition, photo-crosslinking and cleavage of the DNA double helix by an oligo-[α]-thymidylate covalently linked to an azidoproflavine derivative. Nucl. Acids Res. 15:7749–7760.

Le Doan T, Perrouault L, Chassignol M, Thuong NT, Hélène C (1987b): Sequence-targeted chemical modifications of nucleic acids by complementary oligonucleotides covalently linked to porphyrins. Nucl. Acids Res. 15:8643–8659.

Letsinger RL, Zhang G, Sun D, Ikeuchit, Sarin PS (1989): Cholesteryl-conjugated oligonucleotides: Synthesis, properties, and activity as inhibitors of replication of human immunodeficiency virus in cell culture. Proc. Natl. Acad. Sci. U.S.A. 86:6553–6556.

Lee BL, Murakami A, Blake KR, Lin S-B, Miller PS (1988a): Interaction of psoralen-derivatized oligodeoxyribonucleoside methylphosphonates with single-strended DNA. Biochemistry 27:3197–3203.

Lee BL, Blake KR, Miller PS (1988b): Interaction of psoralen-derivatized oligodeoxyribonucleoside methylphosphonates with synthetic DNA containing a promoter for T7 RNA polymerase. Nucl. Acids Res. 16:10681–10697.

Letsinger RL, Schott ME (1981): Selectivity in binding a phenanthridinium-dinucleotide derivative to homopolynucleotides J. Am. Chem. Soc. 103:7394–7396.

Levina AS, Ivanova EM (1985): Use of ethyldichlorophosphite in the synthesis of alkylating phosphoramidates on nonionizable oligonucleotide analogues. Bioorg. Khim. 11:231–238.

Lin S-B, Blake KR, Miller PS, Ts'o POP (1989): Use of EDTA derivatization to characterize interactions between oligodeoxyribonucleoside methylphosphonates and nucleic acids. Biochemistry 28:1054–1061.

Meyer RB Jr, Tabone JC, Hurst GD, Smith TM, Gamper H (1989): Efficient, specific cross-linking and cleavage of DNA by stable, synthetic complementary oligodeoxynucleotides. J. Am. Chem. Soc. 111:8517–8519.

Miller PS, Fang KN, Kondo NS, Ts'o POP (1971): Synthesis and properties of adenine and thymine nucleoside alkyl phosphotriesters, the neutral analogs of dinucleoside monophosphates. J. Am. Chem. Soc. 93:6657–6665.

Miller PS, Barrett JC, Ts'o POP (1974): Synthesis of oligonucleotide ethyl phosphotriesters and their specific complex formation with transfer ribonucleic acid. Biochemistry 13:4887–4896.

Miller PS, Braiterman LT, Ts'o POP (1977): Effects of trinucleotide ethyl phosphotriester G^mp(Et)G^mp(Et)U, on mammalian cells in culture. Biochemistry 16:1988–1996.

Miller PS, Yano J, Yano E, Carrol C, Jayaraman K, Ts'o POP (1979): Nonionic nucleic acid analogues. Synthesis and characterization of dideoxyribonucleoside methylphosphonates Biochemistry 18:5134–5142.

Miller SP, Agris CH, Aurelian L, Blake R, Murakami A, Reddy P, Spitz SA, Ts'o POP (1985): Control of ribonucleic acid function by oligonucleotide methylphosphonates. Biochimie 67:769–776.

Perrouault L, Asseline U, Rivalle C, Thuong NT, Bisagni E, Giovannangeli C, Le Doan T, Hélène C (1990): Sequence-specific artificial photoinduced endonucleases based on triple helix-forming oligonucleotides. Nature (London) 344:358–362.

Pieles U, Englisch U (1989): Psoralen covalently linked to oligodeoxyribonucleotides: synthesis, se-

quence-specific recognition of DNA and photocrosslinking to pyrimidine residues of DNA. Nucl. Acids Res. 17:285–299.

Praseuth D, Gaudemer A, Verlhac J-B, Kraljic I, Sissoeff I, Guille E (1986): Photocleavage of DNA in the presence of sythetic water-soluble porphyrine. Photochem. Photobiol. 44:717–724.

Praseuth D, Chassignol M, Takasugi M, Le Doan T, Thuong NT, Hélène C (1987): Double helices with parallel strands are formed by nuclease-resistant oligo-[α]-deoxynucleotides and oligo-[α]-deoxynucleotides covalently linked to an intercalating agent with complementary oligo-[β]-deoxynucleotides. J. Mol. Biol. 196:939–942.

Praseuth D, Le Doan T, Chassignol M, Decout J-L, Habhoub N, Lhomme J, Thuong NT, Hélène C (1988a): Sequence-targeted photosensitized reactions in nucleic acids by oligo-α-deoxynucleotides and oligo-β-deoxynucleotides covalently linked to proflavin. Biochemistry 27:3031–3038.

Praseuth D, Perrouault L, Le Doan T, Chassignol M, Thuong N, Hélène C (1988b): Sequence-specific binding and photocrosslinking of α and β oligodeoxynucleotide to the major groove of DNA via triple-helix formation. Proc. Natl. Acad. Sci. U.S.A. 85:1349–1353.

Sun J-S, Asseline U, Rouzaud D, Montenay-Garestier T, Thuong NT, Hélène C (1987): Oligo-[α]-deoxynucleotides covalently linked to an intercalating agent. Double helices with parallel strands are formed with complementary oligo-[β]-deoxynucleotides. Nucl. Acids Res. 15:6149–6158.

Sun J-S, Francois J-C, Montenay-Garestier T, Saison-Behmoaras T, Roig V, Thuong NT, Hélène C (1989): Sequence-specific intercalating agents: Intercalation at specific sequences on duplex DNA via major groove recognition by oligonucleotide-intercalator conjugates. Proc. Natl. Acad. Sci. U.S.A. 86:9198–9202.

Thuong NT, Asseline U (1985): Chemical synthesis of natural and modified oligodeoxynucleotides. Biochemie 67:673–684.

Thuong NT, Chassignol M (1987): Synthése et reactivite d'oligothymidylates substitués par un agent intercalant et un groupe thiophosphate, Tetrahedron Lett. 28:4157–4160.

Thuong NT, Chassignol M (1988): Solid phase synthesis of oligo-α- and oligo-β-deoxynucleotides covalently linked to an acridine. Tetrahedron Lett. 29:5905–5908.

Thuong NT, Asseline U, Roig V, Takasugi M, Hélène C (1987): Oligo(α-deoxynucleotide)s covalently linked to intercalating agents: Differential binding to ribo- and deoxyribopolynucleotides and stability towards nuclease digestion. Proc. Natl. Acad. Sci. U.S.A. 84:5129–5133.

Vasilenko SK, Grineva NI, Karpova GG, Kozorovitskyi A, Lomakina TS, Saarma MYu, Tiunov MP (1973):

The formation of complexes of mixed oligonucleotides and their alkylating 5′-derivatives with rRNA and alkylation in these complexes. Dokl. Acad. Nauk SSSR 212:1227–1230.

Vasseur J-J, Gauthier C, Rayner B, Paoletti J, Imbach J-L (1988): Efficient and easy synthesis of octathymidylate covalently linked to intercalating 9-aminoellipticine. Biochem. Biophys. Res. Commun. 152:56–61.

Vlassov VV, Kazakov SA (1982): Complementary addressed modification of poly(I) with cis-hydroxodiamminoplatinum derivatives of oligocytidylates. Bioorg. Khim. 8:499–506.

Vlassov VV, Grineva NI, Zarytova VF, Knorre DG (1970): The effectivity of the acetals of 4-(N-2-chloroethyl-N-methyl-amino)-benzaldehyde derivatives of oligonucleotides in tRNA alkylation. Mol. Biol. 4:201–204.

Vlassov VV, Gorn VV, Ivanova EM, Kazakov SA, Mamaev SV (1983): Complementary addressed modification of oligonucleotide d(pGpGpCpGpGpA) with a platinum derivative of oligonucleotide d(pTpCpCpGpCpCpTpTpT). FEBS Lett. 162:286–289.

Vlassov VV, Zarytova VF, Kutyavin IV, Mamaev SV, Podyminogin MA (1986): Complementary addressed modification and cleavage of a single stranded DNA fragment with alkylating oligonucleotide derivatives. Nucl. Acids Res. 14:4065–4076.

Walder J (1988): Antisense DNA and RNA: Progress and prospects. Genes Dev. 2:502–504.

Zamecnik PC, Stephenson ML (1978): Inhibition of Rous sarcoma virus replication and cell transformation by a specific oligodeoxynucleotide. Proc. Natl. Acad. Sci. U.S.A. 75:280–284.

Zamecnik PC, Goodchild J, Taguchi Y, Sarin PS (1986): Inhibition of replication and expression of human T-cell lymphotropic virus type III in cultured cells by exogenous synthetic oligonucleotides complementary to viral RNA. Proc. Natl. Acad. Sci. U.S.A. 83:4143–4146.

Zarytova VF, Pichko (1990): The alkylating properties of 4-(N-2-chloroethyl-N-methylamino) benzylmethyl-5′-phosphoamide oligonucleotide derivatives. Izv. Sib. Otd. Akad. Nauk SSSR, Ser, Khim. Nauk 3:3–11.

Zarytova VF, Sokolova NI, Grineva NI, Knorre DG, Shabarova ZA (1968): The alkylating derivatives of nucleic acid components. V. Synthesis of 2-chloroethylamino group-containing derivatives of mixed ribo- and deoxyribooligonucleotides. Izv. Sib. Otd. Akad. Nauk SSSR, Ser. Khim. Nauk. 6:102–105.

Zarytova VF, Karpova GG, Knorre DG, Popova VS, Stephanovich LE, Sheshegova EA (1980): Alkylating derivatives of oligonucleotide ethylesters—

complementary-addressed reagents taken up by cells. Dokl. Akad. Nauk SSSR 225:110–113.

Zarytova VF, Kutyavin IV, Silnikov VN, Shishkin GV (1986): Modification of nucleic acids in stabilized complementary complexes: I. Synthesis of alkylating derivatives of oligodeoxyribonucleotides having a 5′-terminal N-(2-hydroxyethyl)-phenazinium residue. Bioorg. Khim. 12:911–920.

Zarytova VF, Kutyavin IV, Podyminogin MA, Silnikov VN, Shishkin GV (1987a): Modification of nucleic acids in stabilized complementary complexes. II. Efficiency and direction of alkylation of dodecadeoxyribonucleotide d(pA-A-C-C-T-G-T-T-T-G-G-C) with 4-(N-2-chloroethyl-N-methylamino)benzylamine derivative of complementary heptanucleotide d(pC-C-A-A-A-C)-A bearing an N-(2-hydroxyethyl) phenazinium residue at its 5′-terminus. Bioorg. Khim. 13:1212–1220.

Zarytova VF, Godovikova TS, Kutyavin IV, Khalimskaya LM (1987b): Reactive oligonucleotide derivatives as affinity reagents and probes in molecular biology. In Bruzik KS, Stec WJ (eds): Biophosphates and Their Analogues—Synthesis, Structure, Metabolism and Activity. Amsterdam: Elsevier, pp 149–164.

Zarytova VF, Kutyavin IV, Levina AS, Mamaev SV, Podyminogin MA (1988): N-(2-Hydroxyethyl)phenazinium derivatives of oligonucleotides as effectors of complementary addressed modification of nucleic acids. Dokl. Akad. Nauk SSSR 302:102–104.

Zarytova VF, Ivanova EM, Kutyavin IV, Sergeev DS, Silnikov VN, Shishkin GV (1989): Modification of nucleic acids in stabilized complementary complexes. III. Synthesis of oligonucleotide derivatives bearing N-(2-hydroxyethyl) phenazinium residues at their 3′-terminus. Izv. Sib. Otd. Akad. Nauk SSSR, Ser. Khim. Nauk 6:3–9.

Zarytova VF, Kutyavin IV, Maltseva TV, Mamaev SV, Maltsev VP (1990a): Photomodification of nucleic acids by N-(2-hydroxyethyl)phenazine derivatives of oligonucleotides. Bioorg. Chem. 16:595–602.

Zarytova VF, Kutyavin IV, Mamaev SV, Podyminogin MA (1990b): Effective and selective modification of a single-stranded DNA fragment with alkylating oligonucleotide derivatives in the presence of N-(2-hydroxyethyl)phenazinium oligonucleotide derivatives as effectors. Bioorg. Khim. 16:1653–1660.

Zarytova VF, Ivanova EM, Chasovskikh MN (1990c): Synthesis and properties of oligonucleotide derivatives containing steroid groups. Bioorg. Khim. 16:610–616.

Zenkova MA, Karpova GG, Levina AS (1990a): The RNase H digestion of 16S rRNA with addressed alkaylating derivatives of oligodeoxyribonucleotides, at the reagents' binding sites. Bioorg. Khim. 16:780–787.

Zenkova MA, Karpova GG, Levina AS, Mamaev SV, Soloviev VV (1990b): Complementary addressed alkylation of the *Escherichia coli* 16S rRNA with 2′,3′-O-[4-N-methyl-N-(2-chloroethyl)-amino] benzylidene derivatives of oligodeoxyribonucleotides. IV. Identification of 16S rRNA binding sites with benzylidene derivatives of d(pACCTTGTT)rA,

d(pTTACGACT)rU d(pTTTGCTCCCC)rA. Bioorg. Khim. 16:788–800.

Zuckermann RN, Schultz PG (1988): A hybrid sequence-selective ribonuclease S. J. Am. Chem. Soc. 110:6592–6594.

Zuckermann RN, Schultz PG (1989): Site-selective cleavage of structured RNA by a staphylococcal nuclease-DNA hybrid. Proc. Natl. Acad. Sci. U.S.A. 86:1766–1770.

Antiparallel Triplex Formation at Physiological pH

Ross H. Durland, Donald J. Kessler, and Michael Hogan

Center for Biotechnology, Baylor College of Medicine, The Woodlands, Texas 77381

INTRODUCTION

DNA triple helices can be formed from synthetic polymers (Felsenfeld and Miles, 1967; Broitman et al., 1987; Letai et al., 1988), by the binding of short single-stranded oligonucleotides to purine-rich DNA segments (Moser and Dervan, 1987; Praseuth et al., 1988; Povsic and Dervan, 1989; Hélène and Toulmé, 1989; Birg et al., 1990), and as the result of internal disproportionation of polypurine arrays in duplex DNA (Hanvey et al., 1988; Kohwi and Kohwi-Shigematsu, 1988). In particular, the Dervan (Moser and Dervan, 1987; Povsic and Dervan, 1989) and Hélène groups (Hélène and Toulmé, 1989; Birg et al., 1990) have shown that by targeting a polypurine tract within a naturally occurring DNA region, synthetic DNA oligonucleotides containing C and T residues can form stable triple helices at acidic pH. As expected from the poly(dT)–poly(dA)–poly(dT) triplex described by Arnott and Selsing (1974) and the poly(dTC)–poly(dGA)–poly(dTC) triplexes that have been well studied by Morgan and colleagues (Lee et al., 1979), formation of those localized triplexes is based upon formation TAT and C+GC triplets of the Hoogsteen type (Saenger, 1984), where C+ refers to cytosines protonated at nitrogen 3. In triple helices with a polypyrimidine third strand, it had been shown that the third stand is bound parallel to the purine-rich strand of the underlying duplex (Moser and Dervan, 1987; Praseuth et al., 1988; Hanvey et al., 1988; Povsic and Dervan, 1989; Hélène and Toulmé, 1989; Birg et al., 1990). The parallel

nature of those C+T triplexes is in good agreement with the parallel structure inferred by Arnott and colleagues for the poly(dT)–poly(dA)–poly(dT) triplex (Arnott and Selsing, 1974). Hereafter, we will refer to parallel or antiparallel triplexes when describing the orientation of the third strand relative to the purine-rich strand of the duplex.

As has been well discussed, the requirement for protonation of third strand C residues limits triplex formation to conditions of acidic pH (approximately 4.5 to 6). As a first step toward triplex formation at physiological pH, Dervan has shown that the use of 5-methylcytosine and 5-bromouridine in the third strand appears to raise the apparent pK for triplex formation (Moser and Dervan, 1987), thereby permitting triplex formation in the pH 6 to 7 range. Clearly, a major goal for the field is to explore other nucleoside modifications or other nucleosides as TFO substituents, in order to permit triplex formation at physiological pH (approximately 7.5).

Our approach to the design of TFOs that bind at physiological pH originated as a program to study the site selectivity of triplex formation by oligonucleotides targeted to a region of the human c-*myc* gene. Initial observations (Cooney et al., 1988) have shown that at physiological pH, a discrete, 27-base G-rich oligonucleotide has the capacity to form a stable triplex with high selectivity at a target site within the c-*myc* control region. The preliminary data suggested that the gross secondary structure of the complex might

Prospects for Antisense Nucleic Acid Therapy of Cancer and AIDS, pages 219–226
©1991 Wiley-Liss, Inc.

resemble the Arnott triplex model, but work below will show that the preferred structure displays significant differences.

The c-*myc* target sequence in those preliminary studies was not a simple homopolymer, but rather one of several DNA elements shown by the Bishop laboratory to be required in *cis* for initiation of mRNA synthesis from the c-*myc* promoter (Hay et al., 1987). In the work of Cooney et al., the TFO of interest was designed so as to bind in the parallel orientation relative to the more purine-rich strand of the duplex target, stabilized by TAT, AAT, and GGC base triplets. G-G hydrogen bonding is well known in polymeric triplexes (Letai et al., 1988), in four-stranded DNA structures (Sundquist and Klug, 1989), and in tRNA molecules (Jack et al., 1976), and has been inferred for internal triplex formation at neutral pH in H-DNA (Kohwi and Kohwi-Shigematsu, 1988). Therefore the utility of GGC triplet as an element in TFO design is not unexpected. AAT triplets are also well conserved as tertiary interactions in tRNA (Jack et al., 1976) and have been shown in homopolymeric triple helices by Broitman et al. (1987). Here, we show that GGC and TAT triplets are themselves sufficient to stabilize triple helix formation upon purine-rich DNA sites. Possible hydrogen bonding schemes for GGC and TAT triplets are displayed in Figure 1.

As mentioned above, the initial work of Cooney et al. presented the triplex as a parallel structure, in agreement with the known parallel structures of C + T triplexes. However, GGC base triplets that stabilize tRNA tertiary structure are known to be antiparallel (Jack et al., 1976). Similarly, the GGC triplets inferred by Kohwi et al. (1988) for H-DNA at physiological pH in the presence of magnesium form an antiparallel structure. Therefore, it is interesting to consider that as a general principle, triple helices containing GGC triplets may in fact be antiparallel.

In this work, we show that G-rich oligonucleotides bind with high affinity and site selectivity to the human c-*myc* target to form colinear, antiparallel triplexes. Binding occurs at physiological conditions of pH, tem-

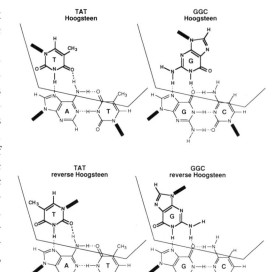

Fig. 1. Possible hydrogen bonding interactions for GGC and TAT triplets. Two possible hydrogen bonding schemes for GGC and TAT triplets, designated Hoogsteen and reverse Hoogsteen, are shown. (Although the terms Hoogsteen and reverse Hoogsteen should properly be applied only to the TAT triplets, we will also use them to describe the analogous GGC triplets for ease of discussion). Bases in the third strand of a triplex bind in the major groove of the duplex (indicated by solid lines). When combined with a syn glycosidic bond orientation, Hoogsteen bonding leads to an antiparallel orientation of a bound TFO, relative to the purine-rich strand of the underlying duplex. An antiglycosidic bond and reverse Hoogsteen bonding will also generate an antiparallel complex. The details of the particular triplets presented here are reasonable, but are as yet unconfirmed by high-resolution structure analysis. Consequently, they should be viewed as a structural hypothesis to explain binding thermodynamics and footprinting analysis. When bound in this way, TAT and GGC triplets are not perfectly isomorphous. However, more detailed modeling suggests that the resulting backbone discontinuity is not severe (not shown).

perature, and divalent ion concentration, suggesting that such compounds may have significant potential as pharmaceutical agents.

RESULTS AND DISCUSSION

We have investigated TFO binding to a target in the human c-*myc* promoter region consisting of base pairs -116 to -142 relative to the principal transcription origin (Fig. 2). As a result of that analysis, we have synthesized a 27-base TFO designed to bind to the c-*myc* target so as to form GGC and TAT triplets in an antiparallel colinear triplex. The resulting TFO molecule (PUGT27a) displays an apparent dissociation constant for its target of approximately $10^{-9}\,M$ in a standard physiological buffer of 5 mM Tris-HCl, pH 7.8, 5 mM MgCl$_2$, 1 mM spermine, and 10% sucrose at 37°C.

Binding data are obtained by titrating trace quantities of radiolabeled DNA with increasing concentrations of TFO, followed by electrophoresis in magnesium-containing polyacrylamide gels. As shown in Figure 3A, triplex formation leads to a reduced mobility of the radiolabeled DNA. The titration is consistent with a two state model in which duplex DNA interacts with free TFO to form a discrete triplex species. Assuming simple equilibrium kinetics apply, the TFO concentration at the midpoint of the transition is an estimate of the apparent dissociation constant for the

complex. In the case of PUGT27a, that concentration is at or below $10^{-9}\,M$.

Binding can also be assessed by titrating the radiolabeled TFO with an unlabeled plasmid DNA digest bearing the target site within one of several fragments. We digested plasmid pMHX DNA (Cooney et al., 1988), containing part of the c-*myc* gene, with *Pst*I and *Taq*I to yield 10 fragments, including a 479 bp fragment that bears the -116 to -142 target site. That fragment array was then purified and mixed with 5'-^{32}P-labeled PUGT27a in the standard binding buffer. In this analysis, stable binding should result in comigration of the labeled TFO with the 479 bp species during electrophoresis. The other nine digestion fragments serve as controls for nonspecific interactions of the TFO with duplex DNA.

As seen in Figure 3B, addition of plasmid to radiolabeled PUGT27a yields detectable binding to only the 479 bp fragment, even under conditions of excess TFO. No stable complexes between PUGT27a and the other nine DNA fragments are observed, indicating that triplex formation is specific to the c-*myc*

C-MYC PROMOTER DOMAIN

```
       -160       -150       -140       -130       -120       -110
        .          .          .          .          .          .
    5'-GGCTGAGTCTCCTCCCCACCTTCCCCACCCTCCCCACCCTCCCCATAAGCG-3'
    3'-CCGACTCAGAGGAGGGGTGGAAGGGGTGGGAGGGGTGGGAGGGGTATTCGC-5'
```

PUGT27p	3'-GGTTGGGGTGGGTGGGGTGGGTGGGGT-5'
PUGT27a	5'-GGTTGGGGTGGGTGGGGTGGGTGGGGT-3'
PU1p	3'-GGAAGGGGTGGGAGGGGTGGGAGGGGT-5'
PU1a	5'-GGAAGGGGTGGGAGGGGTGGGAGGGGT-3'
PU1Ap	3'-GGAAGGGGAGGGAGGGGAGGGAGGGGA-5'
PU1Aa	5'-GGAAGGGGAGGGAGGGGAGGGAGGGGA-3'
PU1ATa	5'-GGTTGGGGAGGGTGGGGAGGGTGGGGA-3'
PU1ACp	3'-GGAACGGGAGGGAGCGGAGGGGAGGCGA-5'
PU1ACa	5'-GGAACGGGAGGGAGCGGAGGGGAGGCGA-3'
PUCT27p	3'-CCTTCCCCTCCCTCCCCTCCCTCCCCT-5'
PUCT27a	5'-CCTTCCCCTCCCTCCCCTCCCTCCCCT-3'
PUGT37p	3'-GTGGTGGGGTGGTTGGGGTGGGTGGGGTGGGTGGGGT-5'
PUGT37a	5'-GTGGTGGGGTGGTTGGGGTGGGTGGGGTGGGTGGGGT-3'
PUGT34p	3'-GTGGTGGGGTGGTTGGGGTGGGTGGGGTGGGTGG-5'
ctrl PUGT37	3'-TGTGGTGGTGGTGGTGGTGGGGGGTGTGGTGGGGGTGG-5'

Fig. 2. Sequences of the human c-*myc* target DNA and the TFOs used in this study. The double-stranded sequence represents the target region for triplex formation, which is a functionally important domain of the regulated human c-*myc* promoter, located -116 to -152 base pairs from the principal transcription start site. Below are the sequences of the TFOs used in this study. TFO names end in the letters p or a, to indicate whether the TFO was designed to bind parallel or antiparallel to the purine-rich strand of the duplex. All oligonucleotides were prepared by standard phosphoramidite synthesis and HPLC purified. In certain instances, the TFOs have been synthesized as the 3' or 5' amine derivatives using solid phase amine modifiers (Glen Research Inc.). Those derivatives are identified by the prefixes 3'A- and 5'A-, where 3'A- refers to 3'-O-CH$_2$-CHOH-CH$_2$-NH$_3^+$ and 5'A- to NH$_3^+$-(CH$_2$)$_6$-OPO$_2$-O-5', respectively.

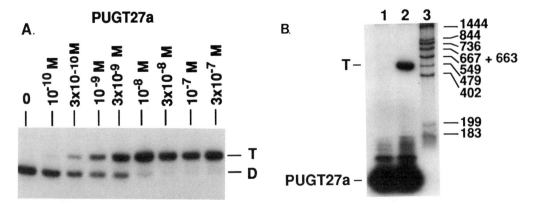

Fig. 3. Electrophoretic analysis of triplex formation on the human c-*myc* promoter region. (**A**) Band shift analysis. Band shift analysis was performed on a cloned 53 bp DNA duplex fragment, consisting of the *Hha*I to *Hinf*I fragment spanning base pairs −111 to −154 (Fig. 2). Band shift analysis was performed by incubating trace quantities of the ^{32}P-labeled 53 bp fragment (approximately 10^{-11} M) with increasing concentration of PUGT27a (measured in strand equivalents) at 37°C in the standard binding buffer: 10 mM Tris-HCl, 5 mM MgCl$_2$, 1 mM spermine, 10% sucrose, pH 7.8. Subsequent to mixing, the solution was incubated for 60 min at 37°C, followed by electrophoresis at room temperature in an 8% polyacrylamide gel buffered with 90 mM Tris-borate pH 7.8, 10 mM MgCl$_2$ (TBM buffer). Duplex and triplex bands have been marked as D and T, respectively. (**B**) Comigration analysis. The plasmid pMHX, which contains the 5′-end of the human c-*myc* gene (Cooney et al., 1988), was digested with *Pst*I and *Taq*I, resulting in 10 fragments of lengths 1444, 844, 736, 667, 663, 549, 479, 402, 199, and 183 base pairs. The triplex target is contained within the 479 base pair fragment. Binding was performed by incubating 10^{-7} M ^{32}P-labeled PUGT27a with 10^{-8} M of the *Pst*I, *Taq*I pMHX digest in the standard binding buffer, as described in (**A**). Plasmic concentration is calculated as moles of binding site equivalents per liter, and TFO concentration as moles of strand equivalents per liter. Triplex formation was visualized by observing the comigration of the ^{32}P-labeled TFO with the 479 bp target duplex (marked T) in a 5% polyacrylamide gel in TBM buffer at room temperature. Free PUGT27a was retained in the gel by means of a 20% polyacrylamide plug at the bottom. In this assay, specificity of triplex formation can be assessed from TFO binding to fragments other than the 479 bp target. As seen, under conditions of excess TFO, the other nine bands of the digest (shown radiolabeled in lane 3 for comparison) show no detectable binding. It is worth noting that the 844 base pair fragment contains the sequence 5′-GGGAGGCGTGGGGGTGGGACGGTGGGG-3′, which is very similar in base composition and purine richness to the triplex target, but shows no binding of PUGT27a under these conditions.

target. It is noteworthy that the 844 bp fragment contains a sequence that is very similar to the c-*myc* target in G content and purine richness (see legend to Fig. 3B). The absence of PUGT27a binding to this fragment is an indication of the sequence specificity of triplex formation.

In order to examine binding in more detail, we have employed DNase I footprinting of triplexes, taking advantage of the reduced DNase I cleavage rates within triple strand structures (Cooney et al., 1988). Buffer conditions for footprinting were essentially as

described for band shift analysis. As shown in Figure 4, PUGT27a protects a region of the duplex target consistent with the formation of a colinear triplex 27 base triplets long. The protected region coincides with the proposed triplex target sequence, and no protection is observed at other sites in this fragment.

Binding of various related TFOs to the c-*myc* target was also examined by DNase I footprinting. Figure 4 shows that PUGT37a, which contains an additional 10 bases at its 3′-end (Fig. 2), protects 10 additional bases

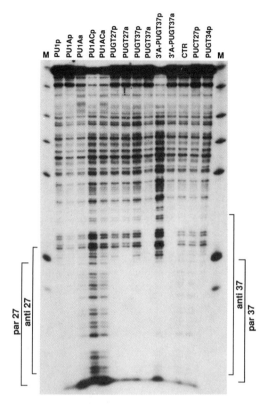

Fig. 4. DNase I footprinting. Triplex formation was assessed at high resolution based upon reduction of DNase I activity at the site of triplex formation (Cooney et al., 1988). DNase I treatment was performed on the *Xma*I to *Sma*I fragment of pMHX (− 99 to − 193 relative P1, 3′-end-labeled at the *Xma*I site) in the standard binding buffer at 20°C, with 10 U/ml of enzyme for 10 min. Following deproteinization, samples were desalted by ethanol precipitation and analyzed by high-resolution electrophoresis. TFO concentration was $5 \times 10^{-7} M$. The sequences of the TFOs indicated above each lane are presented in Figure 2. Brackets mark the proposed binding sites for PUGT27a (anti 27), PUGT27p (par 27), PUGT37a (anti 37), and PUGT37p (par 37). See text for details. As shown, the span of the PUGT37a binding site is approximately 10 base pairs greater than that for PUGT27a. It is also interesting to note that TFOs designed to form C + GC triplets (PUCT27a, PUIACp, PUIACa) do not display detectable binding at this TFO concentration.

of duplex DNA, consistent with the formation of an antiparallel, 37 base triplet structure. The proposed binding sites for PUGT27a and

PUGT37a are indicated in Figure 4 by brackets labeled "anti 27" and "anti 37," respectively, and agree well with the observed footprints. For comparison, we also performed footprint analysis with PUGT27p and PUGT37p, sequence isomers of PUGT27a and PUGT37a designed to form parallel triplexes. These TFOs yield footprints that are clearly distinct from those of the antiparallel isomers. The footprint for PUGT27p is shifted about three base relative to PUGT27a, and is identical to the PUGT37p footprint. Simple models of antiparallel triplex formation suggest a likely explanation for these results. Because the c-*myc* binding site has a pseudo-2-fold axis of symmetry, TFOs designed to bind parallel can potentially bind in the reverse orientation so as to form antiparallel triplexes, while maintaining similar triplet interactions (see Fig. 5B). In this case, the triplex formed by PUGT27p should be shifted over three base pairs at one end (bracket labeled "par 27"), as observed in Figure 4. Antiparallel binding by PUGT37p should lead to a footprint identical to that of PUGT27p (bracket labeled "par 37"), since the additional ten bases at the 3′-end of the TFO would be unable to participate in triplex formation (Fig. 5B). Evidence that PUGT37p does in fact bind antiparallel is provided by the footprint for PUGT34p, which is similar to PUGT37p but lacks three bases at its 5′-end (Fig. 2). Figure 4 shows that PUGT34p protects about three bases fewer than PUGT37p, confirming the position of the 5′-end of these TFOs.

Figure 4 also shows the results of footprint analysis with some additional oligonucleotides whose sequences are given in Figure 2. Of particular interest is PUCT27p, which is designed to permit formation of a parallel C + T-type triplex. No footprint is observed in this experiment, presumably because binding and DNase I digestion were performed at pH 7.8. Additional work indicates that PUCT27p does form a triplex at the c-*myc* site under sufficiently acidic conditions (not shown).

To confirm the antiparallel binding model more directly, we have covalently linked eosin

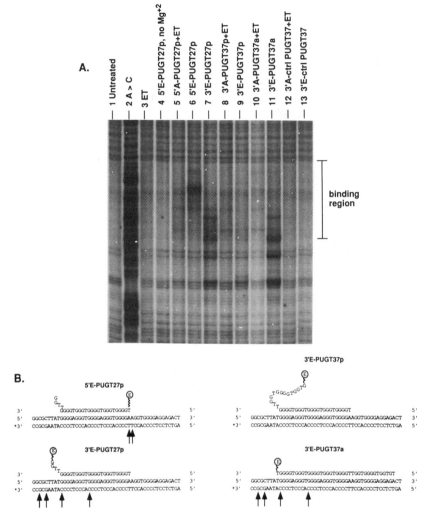

Fig. 5. Eosin photochemical mapping. Eosin isothiocyanate has been affixed through an amine linker to a number of TFOs employing standard aqueous thiourea coupling chemistry. Subsequent to coupling, the dye-oligonucleotide conjugates were purified from free eosin by exclusion chromatography. Photochemistry was performed on the duplex DNA fragment described in Figure 4. The eosin–TFO conjugates (5×10^{-7} M final concentration) were added to labeled DNA fragment in the standard binding buffer at 20°C. Five microliter samples were then irradiated for 5 min at 3 W/cm^2 with the green output of an argon laser, followed by heating to 80°C for 15 min in 10% piperidine to induce hydrolysis of photochemically oxidized bases (Hogan et al., 1987). Binding buffer, eosin isothiocyanate–Tris complex (ET; prepared as in Hogan et al., 1987), or uncomplexed TFO was added as controls at 5×10^{-7} M, then similarly irradiated. Fragmentation patterns were analyzed by electrophoresis on a 7 M urea, 5% polyacrylamide sequencing gel. Eosin linkage to the 5'- or 3'-ends of TFOs is indicated by the prefix 5'E- or 3'E-, respectively. In lane 4, MgCl$_2$ was omitted from the binding buffer to demonstrate the requirement for divalent ions for triplex formation. Lanes 12 and 13 show that an eosin modified TFO with a base composition identical to PUGT37a but with a randomized sequence (ctrl PUGT37; Fig. 2) does not lead to the observed DNA cleavages. (**B**) Proposed structures of eosin-linked TFO complexes. Eosin linked to the 5'- or 3'-end of a TFO is indicated by "E." Sites of cleavage observed in (**A**) are marked by arrows. The labeled end of the duplex is marked with an asterisk.

to either the 5'- or the 3'-terminus of selected TFOs. Eosin is a photochemical reagent which damages DNA by means of singlet oxygen production when irradiated (Hogan et al., 1987). Subsequent piperidine treatment leads to strand cleavage. Eosin is negatively charged and does not itself bind to nucleic acids. Therefore, when covalently linked to a TFO and irradiated with an argon laser, the resulting pattern of DNA cleavage is likely to be a result of the binding characteristics of the TFO rather than that of the dye probe. As expected, we have found that covalent attachment of eosin to TFOs has no measurable effect on their binding affinities (not shown).

Figure 5A shows the pattern of cleavages resulting from eosin linked to the 5'- or 3'-end of selected TFOs. The positions of the cleavages in lanes 6, 7, and 11 are consistent with antiparallel binding and are inconsistent with a significant fraction of parallel triplex. Figure 5B compares the proposed antiparallel triplexes with the observed cleavages. Note that cleavage is observed only when the eosin is covalently linked to the TFO. Addition of free eosin, alone (lane 3) or in combination with an unlinked TFO (lanes 5, 8, 10, 12) does not result in cleavage. Furthermore, eosin covalently linked to a random sequence isomer of PUGT37a (ctrl PUGT37; Fig. 2) does not result in cleavage (lane 13), consistent with the inability of this oligonucleotide to form a triplex (not shown). Finally, magnesium, which is essential for triplex formation (Cooney et al., 1988), is also required for DNA cleavage by eosin-linked TFOs (lane 4).

Detailed examination of the eosin cleavage data indicates that there may be one or more minor antiparallel complexes, differing in the positioning of their TFO termini by nine bases. We suggest that this is a result of the underlying nine base pair repeat of the c-*myc* binding domain (Fig. 2), which permits the formation of a small number of secondary complexes resulting from "slippage." The absolute cleavage efficiency at these various sites is probably not a reliable indicator of the relative proportions of the alternative complexes, since the details of triplex conformation as well as DNA sequence are likely to affect the cleavage rate. Interestingly, no cleavages are observed when eosin is linked to the 3'-end of PUGT37p (Fig. 5A, lane 9). This is consistent with the hypothesis that antiparallel binding of this TFO does not permit the additional 10 bases at the 3'-end to participate in triplex formation. As depicted in Figure 5B, it is likely that this portion of the TFO extends away from the duplex DNA, increasing the distance between the attached eosin and the target, thus reducing the cleavage rate below background levels.

CONCLUSIONS

Data presented here for TFO binding to the human c-*myc* gene promoter lead to the following conclusions. Oligonucleotides designed to form GGC and TAT triplets can bind to polypurine sites in double-stranded DNA to form colinear triplexes with apparent dissociation constants as low as $10^{-9} M$. The third strand in these triplexes is oriented antiparallel to the purine-rich strand of the underlying duplex. Triplex formation is specific for the intended target and occurs at physiological conditions of pH, temperature, and divalent ion concentration. Although the data presented here are confined to studies of the c-*myc* target site, similar studies with TFOs that bind to unrelated targets yield comparable results (not shown).

One hypothesis that guides this work is that reagents that bind with high sequence specificity to DNA under physiological conditions may have useful pharmaceutical properties. The only naturally occurring compounds known to bind DNA in a sequence-specific manner are proteins. However, it is not currently possible to design artificial proteins with predetermined DNA binding characteristics. In contrast, triplex forming oligonucleotides offer the potential for specific DNA-binding reagents that may ultimately be directed to any given target sequence. Such TFOs may potentially be used as agonists or antagonists of regulatory protein binding or as vehicles for targeted DNA damage. Based upon the observation of site-selective triplex formation in several laboratories, it is now clear that triplex formation may potentially lead to pharmaceutically useful, gene-specific ligands that display high affinity and target selectivity at physiological temperature, pH, and ion conditions. The molecules we have described offer one approach to such a drug development program.

SUMMARY

A class of triplex forming oligonucleotides (TFOs) is described that binds to a functionally important target site within the promoter/enhancer region of the human c-*myc* gene. It is shown that binding occurs with high sequence selectivity to the purine-rich target site, with an apparent dissociation constant in the 10^{-9} M range at physiological pH, temperature, and divalent ion concentration. It is also shown that a distinguishing feature of this class of TFO is that it is stabilized by GGC and TAT triplets, with the third strand in an antiparallel orientation relative to the purine-rich strand of the underlying duplex.

ACKNOWLEDGMENTS

This work was supported by grants to M.H. from the Department of the Navy, The National Cancer Institute, The Texas Advanced Technology Program, and Triplex Pharmaceutical Corporation.

REFERENCES

Arnott S, Selsing E (1974): Structures for the polynucleotide complexes poly(dA)·poly(dT) and poly(dT)·poly(dA)·poly(dT). J. Mol. Biol. 88:509–521.

Birg F, Praseuth D, Zerial A, Thuong NT, Asseline U, Le Doan T, Hélène C (1990): Inhibition of Simian virus 40 DNA replication in CV-1 cells by an oligodeoxynucleotide covalently linked to an intercalating agent. Nucl. Acids Res. 18:2901–2908.

Broitman SL, Im DD, Fresco JR (1987): Formation of the triple-stranded polynucleotide helix, poly(A·A·U). Proc. Natl. Acad. Sci. U.S.A. 84:5120–5124.

Cooney M, Czernuszewicz G, Postel EH, Flint SJ, Hogan ME (1988): Site-specific oligonucleotide binding represses transcription of the human c-*myc* gene in vitro. Science 241:456–459.

Felsenfeld G, Miles HT (1967): The physical and chemical properties of nucleic acids. Annu. Rev. Biochem. 36:407–448.

Hanvey JC, Klysik J, Wells RD (1988): Influence of DNA sequence on the formation of non-B right-handed helices in oligopurine·oligopyrimidine inserts in plasmids. J. Biol. Chem. 263:7386–7396.

Hay NJ, Bishop M, Levens D (1987): Regulatory elements that modulate expression of human c-myc. Genes Dev. 1:659–671.

Hélène C, Toulmé JJ (1989): Control of gene expression by oligodeoxynucleotides covalently linked to intercalating agents. In Cohen JS (ed): Oligodeoxynucleotides as Antisense Inhibitors of Gene Expression. Boca Raton, FL: CRC Press.

Hogan ME, Rooney TE, Austin RH (1987): Evidence for kinks in DNA folding in the nucleosome. Nature (London) 328:554–557.

Jack A, Ladner JE, Klug A (1976): Crystallographic refinement of yeast phenylalanine transfer RNA at 2.5 Å resolution. J. Mol. Biol. 108:619–649.

Kohwi Y, Kohwi-Shigematsu T (1988): Magnesium ion-dependent triple-helix structure formed by homopurine-homopyrimidine sequences in supercoiled plasmid DNA. Proc. Natl. Acad. Sci. U.S.A. 85:3781–3785.

Lee JS, Johnson DA, Morgan RA (1979): Complexes formed by (pyrimidine)$_n$·(purine)$_n$ DNAs on lowering the pH are three stranded. Nucl. Acids Res. 6:3073–3091.

Letai AG, Palladino MA, Fromm, E, Rizzo V, Fresco JR (1988): Specificity in formation of triple-stranded nucleic acid helical complexes: Studies with agarose-linked polyribonucleotide affinity columns. Biochemistry 27:9108–9112.

Moser HE, Dervan PB (1987): Sequence-specific cleavage of double helical DNA by triple helix formation. Science 238:645–650.

Povsic TJ, Dervan PB (1989): Triple helix formation by oligonucleotides on DNA extended to the physiological pH range. J. Am. Chem. Soc. 111:3059–3060.

Praseuth DL, et al. (1988): Sequence-specific binding and photocrosslinking of α and β oligodeoxynucleotides to the major groove of DNA via triple-helix formation. Proc. Natl. Acad. Sci. U.S.A. 85:1349–1353.

Saenger W (1984): Principles of Nucleic Acid Structure. New York: Springer-Verlag.

Sundquist WI, Klug A (1989): Telomeric DNA dimerizes by formation of guanine tetrads between hairpin loops. Nature (London) 342:825–829.

Inhibition of DNA/Protein Interactions by Oligonucleotide-Directed DNA Triple Helix Formation: Progress and Prospects

L. James Maher III, Peter B. Dervan, and Barbara Wold

Division of Chemistry and Chemical Engineering (L.J.M., P.B.D.) and Division of Biology (L.J.M., B.W.), California Institute of Technology, Pasadena, California 91125

INTRODUCTION

The transfer of biological information from a stored form (typically double helical DNA) to functional polypeptide via mRNA intermediates comprises the central dogma of molecular biology. As shown in Figure 1, artificial inhibitors of biological functions have traditionally been targeted toward the final protein products of this pathway. In the last 10 years, approaches have been suggested for inhibition of the intermediate information transfer reactions. Such strategies include sense/antisense interactions of oligo- or polynucleotides with mRNA (for reviews see van der Krol et al., 1988; Cohen, 1989; Zon, 1988; Uhlmann and Peyman, 1990; Goodchild, 1990), with the possible inclusion of RNA-catalyzed mRNA cleavage (Haselhoff and Gerlach, 1988; Cameron and Jennings, 1989; Rossi and Sarver, 1990). Here we review the experimental evidence that sequence-specific recognition of double helical DNA by oligonucleotides can occur near physiological conditions (Moser and Dervan, 1987; Le Doan et al., 1987). This highly specific mode of DNA recognition may offer new strategies for inhibiting information transfer reactions involving double helical DNA in vitro and in vivo (Fig. 1).

This review is organized in four sections. First, a brief summary of polymeric nucleic acid triple helices will be presented to provide historical context. Second, experimental evidence documenting double helical DNA recognition by oligonucleotides will be reviewed. Third, the ability of bound oligonucleotides to block DNA recognition by sequence-specific binding proteins will be discussed, and potential applications of this new technology will be suggested. Fourth, we present our experimental approach to the analysis of transcription repression by triple helical complexes. Prospects for artifical transcriptional repression based on triple helix complex formation will then be addressed, with discussion of the significant obstacles that must be overcome if this approach is to be practically implemented in vivo. The fascinating subject of B-form to H-form transitions in double helical DNA (*intra*molecular triple helix formation) is beyond the scope of this presentation (but see Htun and Dahlberg, 1988 and references therein).

HISTORICAL CONTEXT: POLYNUCLEOTIDE TRIPLE HELICAL STRUCTURES

Triple helical polynucleotide complexes were first described in 1957 (Felsenfeld et al., 1957). Ribopolymers poly(U) and poly(A) were shown to form a 2:1 complex, stabilized

Prospects for Antisense Nucleic Acid Therapy of Cancer and AIDS, pages 227–242
©1991 Wiley-Liss, Inc.

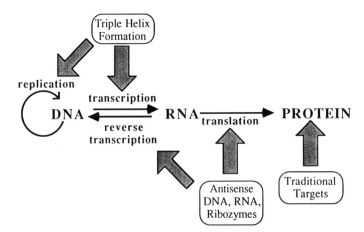

Fig. 1. The central dogma of molecular biology and information transfer targets. The biological transfer and amplification of information from its stored form as double helical DNA to messenger RNA, and then to protein are shown schematically. Proteins represent traditional pharmacological targets, while informational molecules such as mRNA are potential targets for antisense agents. Information transfer reactions involving double helical DNA are potential targets for triple helix-forming reagents. Targeting double helical DNA might be the most direct approach for inhibition of information transfer.

in the presence of magnesium ions. A variety of triple helical polynucleotide complexes were subsequently observed (Michelson et al., 1967; Felsenfeld and Miles, 1967). Guanine oligoribonucleotides form a triple helical complex with two strands of poly(C) under mildly acidic conditions. Half of the pyrimidines in such complexes appear to be protonated (Lipsett, 1963; Lipsett, 1964; Howard et al., 1964). Under similar conditions and in the presence of magnesium ions, poly(U-C) or poly(dT-dC) binds to poly(dT-dC)·poly(dG-dA) (Morgan and Wells, 1968; Lee et al., 1979). X-ray fiber diffraction patterns obtained from [poly(A)·2poly(U)] and [poly(dA)·2poly(dT)] were consistent with an A form (RNA-like) double helical conformation of the Watson−Crick base-paired strands. On this basis, models were put forward with the third strand bound parallel to the purine strand of the double helix in the major groove (Arnott and Selsing, 1974). The proposed triple helix structure would therefor arise from the formation of base triplets (T-A-T and C+G-C) due to Hoogsteen hydrogen bonding of the second pyrimidine strand to the purine strand of the double helix (Hoogsteen, 1959) (Fig. 2, left). Thus, each purine base in a homopurine/homopyrimidine double helical DNA sequence offers two potential sequence-specific hydrogen bonds for oligonucleotide-directed triple helical recognition (Moser and Dervan, 1987; Praseuth et al., 1988). Recent NMR observations have confirmed this model, including direct observation of cytosine protonation in the Hoogsteen strand of triple helical complexes under acidic conditions (Rajagopal and Feigon, 1989; de los Santos et al., 1989; Sklenar and Feigon, 1990; Umemoto et al., 1990).

In addition to triple helix structures based on pyrimidine:purine:pyrimidine complexes (Fig. 2, left), the existence of other triple helical motifs derived from other interactions (A-A-T/U, G-G-C, I-A-U/T, I-G-C) have also been suggested (Letai et al., 1988; Broitman et al., 1987). Recently, detailed examination of strand polarities in triple helical complexes involving this purine motif (predominantly G-G-C but including A-A-T or T-A-T base triplets) has suggested the recognition scheme shown in Figure 2 (right) where the oligopurine strand binds in the major groove, antiparallel to the target purine strand

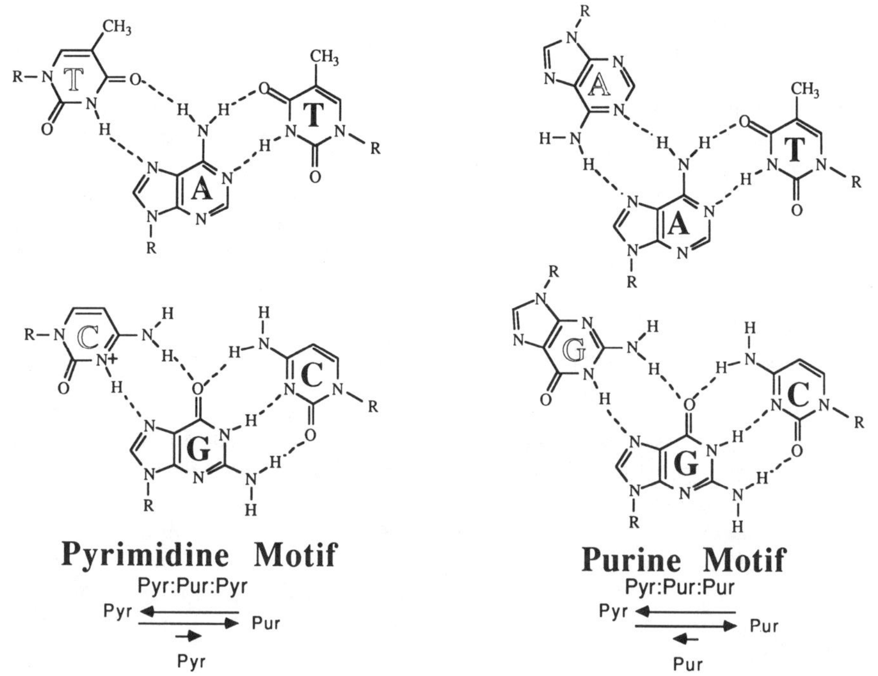

Pyrimidine Motif
Pyr:Pur:Pyr

Purine Motif
Pyr:Pur:Pur

Fig. 2. Left: Triple helix base recognition in the pyrimidine motif. Watson–Crick base pairs (A-T, G-C, dark print) of the DNA double helix are recognized by T and C+ (open print), respectively, bound by Hoogsteen hydrogen bonding in the major groove. The protonation of cytosine N3 is required in this motif. The relative polarities of nucleic acid strands in such a complex are shown below. Right: Postulated triple helix base recognition in the purine motif. Watson–Crick base pairs (A-T, G-C, dark print) of the DNA double helix are recognized by A and G (open print), respectively, bound in the major groove. Nucleic acid strand polarities are shown below. T-A-T base triplets appear to be permitted in the purine motif under some conditions.

(Beal and Dervan, 1991). Noteworthy is the contrast between symmetric and asymmetric radial distributions of backbone groups (R) in the purine and pyrimidine motifs, revealing that there are at least two distinct structural classes of triple helices for double helical DNA recognition (Fig. 2).

OLIGONUCLEOTIDE-DIRECTED RECOGNITION OF DNA BY TRIPLE HELIX FORMATION

Here we present a brief summary of experimental methods for detection of triple helix formation. We then address factors that influence triple helix stability and specificity including oligonucleotide length and sequence, salt, pH effects, and covalent oligonucleotide modifications. Because of the diverse conditions under which triple helical complexes have been observed, direct comparison of experimental results from different studies is sometimes difficult. Also, in contrast to the large and sophisticated body of knowledge available for double-helical DNA, characterization of triple helices is at a much earlier stage. Within these limitations, however, we believe the most coherent view emerges from the expectation that triple helix formation on a double-stranded DNA polymer will prove to be fundamentally analogous with double helix formation on a single-stranded DNA polymer. Thus, examination of transition temperatures for complex dissociation (T_m), dependence of mispairing tolerance on hybridization stringency, and sequence-dependent stabilities for complexes of a given length all are expected to reveal fundamental physicochemical principles underpinning oligonucleotide-directed triple helix formation. Eventually a detailed understanding of triple helix structure and energetics should permit, as has been the case for double helix formation, prediction of the sequence-dependent stabilities of base-paired complexes on the basis of derived nearest-neighbor interaction energies (Breslauer et al., 1986).

Sequence-specific recognition of double helical DNA by oligodeoxyribonucleotide-directed triple helix formation may be detected by at least three experimental approaches. The affinity cleavage technique has been employed using EDTA·Fe(II) [the ethylenediamine tetraacetic acid chelate of Fe(II)] in the presence of reducing agents (Moser and Dervan, 1987), or photocrosslinking agents that cause DNA cleavage upon alkaline work-up (Le Doan et al., 1987; Praseuth et al., 1988; Perrouault et al., 1990). Strong binding of an oligonucleotide to a double helical DNA target may also be detected by electrophoretic gel mobility shift experiments (Lyamichev et al., 1988; Cooney et al., 1988). Finally, the specificity of ligand–DNA interactions is often sensitively monitored by footprinting (for review see Dervan, 1986). Oligonucleotide binding to double helical DNA has been observed in this manner using enzymatic or chemical nucleases, restriction methylases, chemical methylating agents, or ultraviolet light (Povsic and Dervan, unpublished observations; Cooney et al., 1988; Maher et al., 1989; Lyamichev et al., 1990).

Many factors have been found to influence the stability of triple helical complexes. Oligonucleotide-directed triple helix formation in the pyrimidine motif is stabilized by increasing oligonucleotide length (between 11 and 15 nucleotides) as expected by analogy with double helical DNA (Moser and Dervan, 1987). The stability of triple helical complexes containing cytosine residues is also dependent on solution pH, presumably due to stabilization by cytosine N3 protonation

(Fig. 2). This effect is expected to increase for pyrimidine oligonucleotides of increasing cytosine content (Lyamichev et al., 1988). Gel mobility shift experiments with cytosine-containing oligonucleotides in the pyrimidine motif also required acidic conditions (pH ≤ 6) to support complex lifetimes amenable to this technique (Lyamichev et al., 1988).

With the goal of increasing the pH range available for oligonucleotide-directed triple helix formation within the pyrimidine motif, the 5-substituted pyrimidine analogs 5-bromouracil and 5-methylcytosine were investigated (Povsic and Dervan, 1989). Using sequence-specific affinity cleavage of double helical plasmid DNA as an assay system, this study demonstrated that the substituted pyrimidines enhanced binding affinity (5-bromouracil) or enhanced both binding affinity and permissive pH range above pH 7.0 (5-methylcytosine). When both analogs were simultaneously incorporated into oligonucleotides, the stability of the resulting complex (33% cytosine content in pyrimidine oligonucleotide strand) was extended to pH 7.4, a value that may be relevant to physiological conditions. In contrast to the pyrimidine motif with mixed thymine and cytosine oligonucleotides, triple helix formation in the purine motif is not highly pH dependent (Cooney et al., 1988; Beal and Dervan, 1991).

Other solution components also strongly influence triple helix stability. Complexes in both pyrimidine and purine motifs are stabilized by multivalent cations such as magnesium, spermine, spermidine, and Co(III)-hexamine (Moser and Dervan, 1987; Lyamichev et al., 1988; Cooney et al., 1988). This dependence is often interpreted as a manifestation of the ability of such condensed ions to efficiently neutralize phosphate backbone repulsions in DNA triple helices. With a view toward triple helix formation in vivo, it is important to note that the strong stabilization of triple helices by multivalent cations may also pertain in the cell where polyamines are known to exist in millimolar concentration.

In addition to stabilization of triple helical complexes by incorporation of modified bases (above), oligonucleotide conjugation to intercalators has also been examined (Sun et al., 1989). This study demonstrated significant stabilization of a triple helical complex by conjugation of an acridine intercalator at the 5′-terminus of a homopyrimidine oligonucleotide targeted to an 11-bp homopurine/homopyrimidine site in a 32-bp synthetic double helix. The resulting complex protected an additional 2 bp of the double helix in footprinting experiments. Acridine conjugation was shown not to decrease the specificity of oligonucleotide binding, therefore suggesting an attractive strategy for modulating the stability of short triple helices. When added to SV40-infected CV-1 cells, octathymidylates covalently linked to acridine were reported to inhibit viral cytopathic effects at 15–30 μM (Birg et al., 1990). This result appeared to be distinct from effects of the toxic acridine moiety tested alone. Although the oligonucleotide–intercalator conjugate can be shown to form a triple helix in vitro with a short restriction fragment containing the SV40 origin, the actual basis for the observed biological effects remains to be established.

The specificity of oligonucleotide-directed DNA triple helix formation has been addressed in several experiments. For example, under relatively stringent conditions, single base mismatches in the pyrimidine motif are highly destabilizing (Moser and Dervan, 1987). However, in analogy with oligonucleotide binding to single-stranded DNA, increased mismatch tolerance is expected to arise under less stringent conditions [lower temperature, increased multivalent cation concentration, increased oligonucleotide length, or lower pH (pyrimidine motif)]. We expect that this will have important consequences for design of oligonucleotides intended to bind a specific double helical DNA target under physiological conditions. For example, because the physiological milieu defines the relevant stringency in such experiments, overspecification of target length, or excess target affinity is expected to increase the probability of relatively stable interactions with unintended, partially identical, target sites. Opti-

mum triple helical complexes would thus be of sufficient stability to persist at the relevant physiological stringency, while one or more base mismatches would result in sufficient instability to effectively eliminate binding.

A formal analysis of double helical DNA target site size suggests that oligonucleotides of length 16–17 nucleotides should specify unique sites in the 3×10^9 bp haploid human genome (Dervan, 1986). Because of the overrepresentation of homopurine/homopyrimidine sequences in eukaryotic DNA, unique target sites of this type might, in fact, be more abundant (Wells et al., 1988). The ability to specify target sequences greater than 15 bp in length has been exploited to allow chemical recognition and up to 25% double strand cleavage at a single 18 bp site in the 48.5 kbp phage λ genome (Strobel et al., 1988). This result paved the way for extension of this technique to eukaryotic chromosomes. Thus, using a similar procedure wherein a pyrimidine oligonucleotide was conjugated to EDTA·Fe(II) at both termini, Strobel and Dervan (1990) achieved 6% double strand cleavage of an engineered homopurine target site inserted into chromosome III (340 kbp) of S. cerevisiae.

There are also prospects for increasing target specificity by requiring the simultaneous cooperative binding of two or more adjacent oligonucleotides. Two recent experimental results provide support for this idea. In these studies, pairs of nonidentical pyrimidine oligonucleotides were aligned to form contiguous complexes in the major groove of double helical DNA by triple helix formation. In the first case, positive cooperative interaction between the end-to-end oligonucleotides was revealed by the enhancement of affinity cleavage due to one oligonucleotide when the adjacent binding site was occupied (Strobel and Dervan; 1989). In the second case, the 5′-phosphate terminus of one oligonucleotide was activated with cyanogen bromide, imidazole, and nickel chloride. The adjacent 3′-hydroxyl terminus of the second bound oligonucleotide was observed to attack the activated phosphate, resulting in nonenzymatic ligation of the oligonucleotides on the double helical template (Luebke and Dervan, 1989).

INHIBITION OF DNA BINDING PROTEINS

The affinity cleavage experiments described above demonstrated that oligopyrimidines exhibit sequence specificity in binding, but such studies do not address the kinetic characteristics of complex formation and persistence. In particular, slow dissociation rates may be important if such complexes are to be capable of sequestering specific double helical DNA sequences from DNA binding proteins or altering DNA tertiary structure. A series of experiments was performed using a plasmid construct containing a 21 bp triple helix target site as shown in Figure 3 (Maher et al., 1989). The target site for oligonucleotide binding was designed to overlap sequences recognized by restriction endonuclease *Ava*I, restriction methylase M.*Taq*I, and eukaryotic transcription factor Sp1. A minimum triple helix complex lifetime was estimated by an affinity cleavage competition assay (25 mM Tris-acetate, pH 6.8, 100 mM sodium chloride, 1 mM spermine tetrahydrochloride, 22°C). It was observed that detectable plasmid cleavage by 1 μM oligonucleotide-EDTA·Fe(II) was eliminated if a 20-fold excess of non-EDTA oligonucleotide was included in the binding reaction prior to initiation of the cleavage reaction with reducing agent. However, once bound to double helical DNA in the absence of non-EDTA oligonucleotide, equilibration with excess non-EDTA oligonucleotide for at least 2 hr caused no measurable reduction in target cleavage by oligonucleotide-EDTA·Fe(II). Thus, rapid exchange between bound and free oligonucleotides does not appear to occur under these experimental conditions.

Experiments were performed to determine whether oligonucleotides could successfully compete with a restriction endonuclease for binding to double helical DNA. The experimental approach is shown schematically in Figure 4. The triple helical complex overlaps one of the two 6-bp *Ava*I en-

donuclease recognition sites in the test plasmid, while the second *Ava*I site serves as an internal positive control for endonuclease activity. At 37°C in 25 mM Tris-acetate, pH 6.8, 70 mM sodium chloride, 20 mM magnesium chloride, 400 μM spermine, 100 μg/ml bovine serum albumin, 10 mM 2-mercaptoethanol, 80 μM oligonucleotide (20-mer), when added simultaneously with the endonuclease, selectively eliminated endonuclease cleavage at the triple helix target site. Multivalent cations (magnesium or spermine) were included because they strongly stabilize triple helix formation. Substitution of 5-methylcytosine for cytosine reduced the oligonucleotide concentration required for complete endonuclease protection by more than an order of magnitude. Subsequent experiments indicated that restriction inhibition was equally efficient using supercoiled plasmid targets. Similar site-specific inhibition was subsequently reported for endonucleases *Ksp* 632-I and *Eco*RI (François et al., 1989; Hanvey et al., 1989).

Fig. 3. Experimental triple helical complex that ➤ overlaps recognition sites for DNA binding proteins. The homopurine/homopyrimidine target site was constructed by site-directed mutagenesis of a 6.4-kbp plasmid containing the murine metallothionein I promoter (Maher et al., 1989). Binding of a homopyrimidine oligonucleotide (length 20 or 21 nt) results in a triple helical complex that overlaps recognition sites for Ava I restriction endonuclease (top), M.*Taq*I restriction methylase (top), and transcription factor Sp1 (bottom).

Fig. 4. Schematic diagram of restriction of endonuclease protection induced by oligonucleotide-directed triple helix formation. A DNA sequence contains two endonuclease recognition sites. In one case (right), one of the recognition sites overlaps a homopurine/homopyrimidine triple helix target site. In a control experiment (left), no triple helix target site exists. In the presence of a specific oligopyrimidine, triple helix formation occurs only where a target site is present (right). Upon exposure to endonuclease, site-specific endonuclease protection is observed (right), indicating the formation and stability of the triple helical complex over the period of endonuclease exposure.

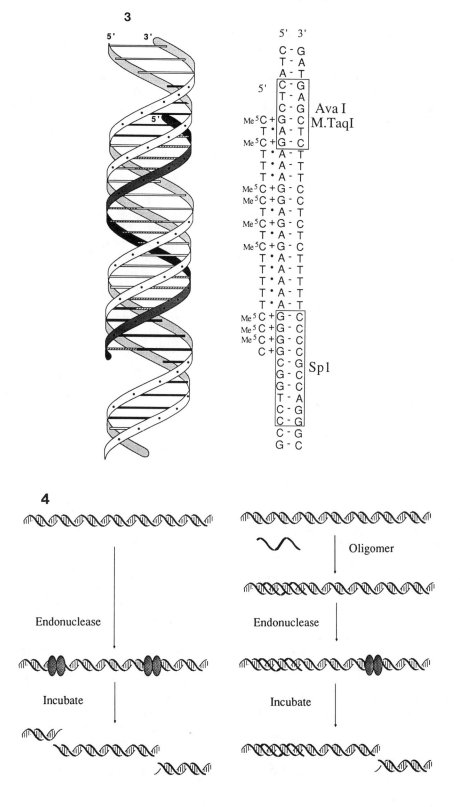

The restriction protection method described above serves as a rapid and convenient nonisotopic assay allowing kinetic and thermodynamic analyses of oligonucleotide-directed triple helix formation (Maher et al., 1990). After exposure to oligonucleotides for varying times and under different reaction conditions, supercoiled plasmid DNA is then challenged briefly (10 min) with restriction endonuclease. Following agarose gel electrophoresis, the distribution of restriction fragments can easily be quantitated to yield binding data. This is, in effect, a low resolution footprinting process. Using this assay, oligonucleotide association kinetics have been determined at 37°C under various reaction conditions (Maher et al., 1990). The results of one such study using the buffer system described above are shown in Figure 5. These data indicate that in the presence of 10- to 100-fold excess of oligonucleotide over plasmid site, the oligonucleotide binding reaction is pseudo-first-order in oligonucleotide concentration, as expected. Under these buffer conditions the triple helix product is sufficiently stable that the reaction is directly analogous to R_0t analyses of the association of DNA strands to excess complementary RNA, and allows estimation of the rate constant for oligonucleotide binding. This kinetic analysis of oligonucleotide-directed DNA triple helix formation revealed several interesting properties. Perhaps the most notable of these is the relatively slow rate of the oligopyrimidine (thymine, 5-methylcytosine) binding compared with DNA double helix formation. The measured pseudo-first-order binding constant is 86 ± 9 liters (mol nucleotide)$^{-1}$ sec^{-1} under the conditions reported above for restriction endonuclease inhibition, and is a further 7-fold slower for the purine motif (Maher et al., 1990; Maher, Dervan, and Wold, unpublished observations). Thus, at 100 nM pyrimidine oligonucleotide (100-fold molar excess over double helical target located in a 6.6-kbp plasmid) the binding reaction requires 2–3 hr for completion. Under these conditions, pyrimidine motif complexes decay with a half-life of approximately 9 hr, leading to a dissociation constant of about 10 nM. A key conclusion from these experiments is that the high oligonucleotide concentration ($\geqslant 10$ nM) necessary for endonuclease protection in initial assays involving simultaneous addition of oligonucleotide and endonuclease reflected the requirement to drive an inherently slow oligonucleotide binding reaction to compete with the protein association rate. A second notable characteristic of triple helical complexes is their sensitivity to monovalent cation concentration. In the presence of stabilizing multivalent cations, increasing monovalent cation concentration from 0.07 to 0.6 M strongly *destabilizes* pyrimidine motif triple helices.

Oligonucleotide-directed triple helix formation can inhibit double helical DNA recognition by other kinds of processing enzymes. For example, restriction endonuclease protection studies were extended to the inhibition of M.*Taq*I, a restriction methylase that catalyzes methyl transfer from S-adenosylmethionine to the N6 position of adenine in the double helical DNA sequence 5'-TCGA-3'. When methylated, this recognition sequence is protected from *Taq*I endonuclease cleavage. One M.*Taq*I recognition site overlaps the experimental triple helix target site as shown in Figure 3. The results of such a methylation protection assay are schematized in Figure 6. Whereas restriction endonuclease protection by oligonucleotide binding produced cleavage inhibition only at the triple helix target site, methylation protection followed by endonuclease treatment (at elevated temperature to destabilize the triple helix) yields cleavage only at the triple helix target site. Nearly complete methylation protection was observed when methylation with M.*Taq*I was performed in the presence of 40 μM oligonucleotide at 44°C (Maher et al., 1989). In addition to establishing the generality of protein competition strategies, this result provided an approach to highly specific enzymatic DNA modification. This oligonucleotide-mediated methylation protection procedure is therefore analogous to the "Achilles-heel" technique proposed by Koob et al. (1988) where a sequence-specific DNA binding protein was employed to induce methylation protection, followed by removal of the protein and endonuclease treatment. Substitution of triple helix formation for protein binding as the methylase inhibitor will likely permit a substantial generalization of this approach. This strategy offers the potential for efficient enzymatic cleavage at single sites in chromosomes (Strobel and Dervan, 1991). Such techniques may become useful for the isolation of selected regions of chromosomal DNA without extensive screening of genomic libraries.

Sp1 is a transcriptional activating protein that binds in the DNA major groove at the consensus sequence 5' G(T)GGGCGGPuPuPy 3' in many eukaryotic promoters (Jones et al., 1985). As shown in Figure 3, an Sp1 recognition site overlaps one terminus of the triple helix target site. The effect of triple helix formation on transcription factor binding was measured by oligonucleotide binding to the target site followed by incubation of the complex with recombinant Sp1 protein. The result is shown schematically in Figure 7. As monitored by DNase I footprinting, triple helix formation resulted in DNase I protection at the homopurine/homopyrimidine target site, and the footprint due to Sp1 binding was prevented. Binding of Sp1 to a control sequence containing an identical Sp1 site but lacking a triple helix target site was unaffected by oligonucleotide. Thus, a eukaryotic transcription factor may be inhibited by sequestering part of its recognition sequence in an intermolecular triple helix.

PROSPECTS FOR ARTIFICIAL GENE-SPECIFIC REPRESSORS

The idea that natural gene regulation might occur via intermolecular triple helix formation is not new. The binding of repressor RNA molecules to double helical DNA has been proposed as a possible repression mechanism both before and after elucidation of the role of repressor proteins (Miller and Sobell, 1966; Britten and Davidson, 1969; Minton, 1985). RNA binding to double helical DNA has also been suggested as a possible regulator of transcription termination by RNA polymerases II

and III in eukaryotes, as well as transcription initiation (Minton, 1985). Oligopyrimidines (pyrimidine motif), oligopurines (purine motif) or their analogs, appear to offer an approach to promoter-specific repression, and may model naturally occurring but as yet unrecognized processes. The steric inhibition of protein/DNA interactions by triple helical complexes (above) suggests that direct competition for binding sites could provide a repression mechanism. This simple competition model might be viewed in analogy to prokaryotic and eukaryotic repressors whose mechanisms may include (but are seldom limited to) occlusion of binding sites for transcription activators, basal factors, or RNA polymerase itself. However, as described below, a number of other fascinating mechanisms may be involved in repression mediated by triple helical complexes.

Encouraging evidence that addresses the goal of transcriptional repression by oligonucleotide binding to double helical DNA has been presented by Hogan and co-workers (Cooney et al., 1988). These studies indicate that a G-rich oligonucleotide (27 nt in length) forms a triple helical complex with a G-rich sequence just upstream of the P1 transcription start site of the human c-*myc* gene. Although it was unclear at the time of the initial report, it appears in retrospect that the resulting triple helical complex likely corresponds to the antiparallel purine motif (Fig. 2). Importantly, formation of an intermolecular triple helix by G-rich oligopurine binding in the human c-*myc* promoter appeared to inhibit transcription from this gene in vitro (Cooney et al., 1988), although the molecular mechanism and overall specificity of this effect remain to be elucidated.

Our research groups are interested in understanding the molecular mechanisms whereby triple helix formation mediates repression of eukaryotic promoters. A detailed understanding of these processes may be of value both for the design of artificial repressors, and in the analysis of natural regulation modes involving multistranded nucleic acid structures.

In order to test some of these possibilities,

5

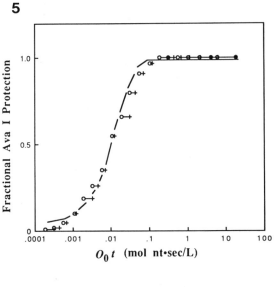

6

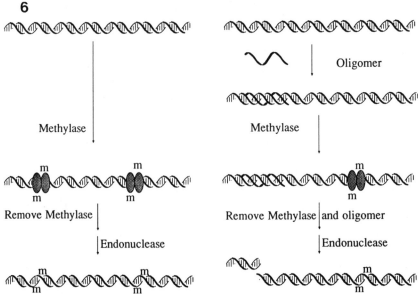

Fig. 5. Oligonucleotide-directed triple helix formation kinetics. Various concentrations of a specific pyrimidine oligonucleotide are incubated with a double helical plasmid containing a triple helix target site. At various times, the extent of triple helix formation is determined by the endonuclease protection assay outlined in Fig. 4. The extent of triple helix formation is plotted in terms of the variable O_0t (oligonucleotide concentration in mole nt/liter multiplied by binding time in seconds). Under conditions of at least 10-fold excess oligonucleotide relative to double helical target site, the resulting curve corresponds well to an association reaction that is pseudo-first-order in oligonucleotide concentration (Maher et al., 1990).

Fig. 6. Schematic diagram of restriction methylase protection induced by oligonucleotide-directed triple helix formation. A double helical DNA sequence contains two sites recognized by a restriction methylase/restriction endonuclease combination, such that methylation of the recognition site prevents endonuclease cleavage. In one case (right), one of the recognition sites over- ➤

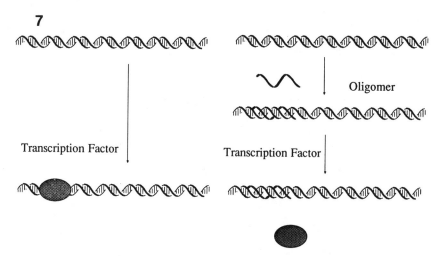

Fig. 7. Schematic diagram of inhibition of transcription factor binding by oligonucleotide-directed triple helix formation. A double helical DNA sequence contains a transcription factor recognition site. In one case (right), the site overlaps a homopurine/homopyrimidine triple helix target site. In a control experiment (left), no triple helix target site exists. When bound to the double helical target, a specific pyrimidine oligonucleotide prevents transcription factor binding (right). Normal transcription factor binding occurs if the oligonucleotide is unable to form a complex with the double helical DNA (left).

we have employed a eukaryotic cell-free transcription system. Transcription of experimental DNA templates bearing triple helical complexes at particular promoter locations can be directly analyzed in such a system; the simple addition of multiple templates provides internal standards for promoter activity. In addition, this cell-free system focuses on macromolecular interactions apart from the requirement for oligonucleotide uptake across membrane barriers. For this purpose we have used nuclear extracts from cultured *Drosophila* cells (Parker and Topol, 1984). Such extracts are particularly useful as they provide a naturally Sp1-deficient environment in which to recapitulate factor-dependent RNA polymerase II transcription by the addition of purified recombinant Sp1 (Kadonaga et al., 1987).

A series of recombinant promoter templates has been created, based on the adenovirus E4 and *Drosophila fushi tarazu* (*ftz*) basal promoters (for examples see Fig. 8). Sp1 binding sites (10 bp) and one or more homopurine triple helix target sites (21 bp) were interspersed in overlapping or nonoverlapping configurations upstream of the conserved TATA sequence element. Supercoiled plasmids bearing these promoters were incubated with specific or nonspecific oligonucleotides under conditions that favor triple helix formation, and were then incubated in nuclear extract with or without added recombinant Sp1. Internal standard templates lacking triple helix sites or Sp1 sites were included in all cases.

A promoter carrying an array of five upstream Sp1 sites was activated by recombi-

laps a homopurine/homopyrimidine triple helix target site. In a control experiment (left), no triple helix target site exists. In the first step of the reaction sequence, a specific pyrimidine oligonucleotide binds to the triple helix target site, blocking methylation at the overlapping methylase recognition site (right). No methylation protec-

tion occurs in the control experiment (left). After removing the methylase, the triple helix is then destabilized (for example, by temperature jump), and restriction endonuclease is added. Enzymatic cleavage of the double helical DNA occurs only where methylation protection had been induced (right).

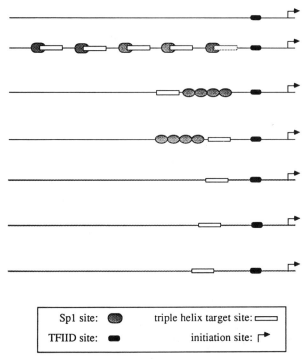

Fig. 8. Several recombinant eukaryotic promoters tested in a cell-free transcription system. Black ovals and arrows indicate TFIID binding sites and transcription initiation sites, respectively. Open rectangles indicate 21-base pair homopurine triplex helix target sites and stippled ovals indicate Sp1 sites. See text for a summary of their activities in the presence of an oligonucleotide that binds the target sites compared to a control, non-binding oligonucleotide.

nant Sp1. We hypothesized that occupation of five triple helix sites (overlapping the Sp1 sites) would abolish this Sp1 activation. To test this hypothesis, we assembled triple helical complexes on the template, and then measured specific transcription from this template in the presence or absence of Sp1. Interestingly, triple helix formation substantially inhibited the basal transcription level observed in the absence of Sp1 (relative to an internal standard). Addition of Sp1 did not relieve the triple-helix repression of the promoter. This observation indicates that triple helical complexes may act as repressors through mechanisms that affect the basal transcription factors, apart from inhibition of upstream transcription activators such as Sp1. To further investigate this point, templates containing single and multiple triple helix sites at various distances from the basal pro-

moter are being tested. Results with some of these constructs raise the possibility that mechanisms other than direct steric occlusion of DNA binding proteins may be responsible for repression.

Although our studies do not yet enable us to define the molecular mechanisms of repression in this model system, it is interesting to consider general mechanisms that seem plausible, based on our present understanding of transcription. Four triple helix repression models are presented in Figure 9. In the first model (Fig. 9A), a triple helix target site overlaps the DNA binding element for an upstream activating factor. If the rate of RNA polymerase initiation from such a promoter is highly dependent on the upstream factor, triple helical complexes that block factor binding will repress the promoter. In the second model (Fig. 9B), the triple helix

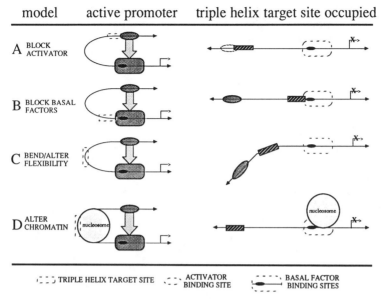

Fig. 9. Possible models by which triple helical complexes modulate transcription initiation in eukaryotic promoters. (**A**) Triple helical complexes overlapping an upstream, positively acting factor binding site repress transcription by steric block of factor binding. (**B**) Triple helix formation causes steric interference with assembly of the transcription initiation complex at or near the TFIID binding site. (**C**) Triple helical complexes alter DNA flexibility or induce local bends, preventing proper DNA tertiary structure organization in the initiation complex. (**D**) Triple helical complexes prevent normal chromatin assembly, leading to nucleosome repositioning. It should be noted that, under certain circumstances, some of these models might also predict transcription activation by triple helical complexes.

target site is located in the promoter region proximal to the TATA element. Triple helical complexes in this position might repress transcription by interference with assembly of the basal transcription apparatus. Alternatively, a less direct mode of triple helix repression, consistent with some cell-free transcription experiments (above), postulates that triple helical complexes induce DNA bends or regions of increased helix stiffness. Such alterations in promoter DNA physical properties could prevent tertiary structure changes (DNA bending or looping) required for transcription (Fig. 9C). Finally, a fourth model for triple helix repression requires consideration of chromatin structure. If triple helical complexes affect the distribution of nucleosomes across the promoter, such complexes could indirectly attenuate transcription by fostering a repressive chromatin configuration (Fig. 9D). Analysis of repres-

sion patterns using a variety of artificial promoter constructs should help to further define the mechanisms of triple helix inhibition; these results may be relevant to artificial repression strategies. Moreover, triple helical complexes may serve to probe and define as yet unrecognized cases of natural regulation that are based on related structures (Davis et al., 1989).

Determination of the effects of local triple helical complexes on other information transfer reactions involving double helical DNA will also be important. Potential target reactions include transcription elongation and termination, DNA replication, recombination, and transposition. Further consideration must also be given to the effects of triple helical complexes on patterns of chromatin assembly and maintenance, including perturbations of nucleosome stability and positioning. For example, under certain conditions triple helical complexes may activate rather than repress transcription by altering nucleosome positioning or chromatin structure, and exposing the promoter to key activating factors.

At present, the potential for artificial repressors based on oligonucleotide-directed triple helix formation is substantially constrained. First, tight binding to double helical DNA in either pyrimidine or purine motifs is typically limited to homopurine/homopyrimidine sites, and such sites do not often overlap important regulatory elements. However, if found to apply in vivo, early evidence from cell-free transcription studies (above) suggesting longer range triple helix effects may extend target sequence flexibility. Ultimately, a more general solution to DNA recognition will undoubtedly be required to facilitate the extension of this technique to targeting any sequence of interest whether studied in vitro or in vivo. Promising recent results suggest that at least some mixed sequence target sites may be recognized in the pyrimidine motif by oligonucleotides in which G residues in the Hoogsteen strand form fairly stable base triplets with T-A base pairs, where the T residues interrupt a homopurine site (Griffin and Dervan, 1989). Moreover, adjacent or nearly adjacent homopurine sequences on opposite DNA strands can be recognized by synthetic pyrimidine oligonucleotides linked at their 3′ termini to accommodate parallel binding in the major groove to both purine strands (Horne and Dervan, 1990). Chemically designed base analogs that recognize mixed sequence DNA may also contribute to improved target flexibility. Reagent stability and cell permeability may be serious limitations to oligonucleotide-based chemotherapy, problems well recognized by researchers in the field of antisense inhibition of RNA targets. Thus, modified oligonucleotides may be required to achieve satisfactory cellular uptake from the circulation and resistance to rapid enzymatic degradation. Such modifications must foster cellular uptake (possibly into particular target cell populations), increase biological half-life, and stabilize the desired specific interaction with target macromolecules in vivo. Thus, although tantalizing observations provide a theoretical foundation for the experimental or therapeutic inhibition of gene expression by triple helix formation, a tremendous effort will be required before such prospects can be realistically and critically evaluated.

SUMMARY

Pyrimidine oligonucleotides exhibit sequence-specific binding to purine sequences in double helical DNA. Molecular recognition occurs through triple helix formation, mediated by Hoogsteen hydrogen bonding of the pyrimidine strand parallel to the double helical DNA purine strand. A second recognition motif involves purine oligonucleotide binding antiparallel to the double helical DNA purine strand. Experiments documenting these interactions were reviewed. Oligonucleotide-directed DNA triple helix formation has been shown to inhibit the recognition of double helical DNA by sequence-specific binding proteins. This observation provides the potential for new approaches to both the analysis of nucleic acids, and the study of nucleic acid/protein interactions. Details of both triple helix motifs are being investigated in our laboratories with the goal of understanding molecular mechanisms whereby such complexes affect protein/DNA interactions involved in eukaryotic transcription.

ACKNOWLEDGMENTS

Supported by an American Cancer Society Postdoctoral Fellowship to L.J.M., and by the Beckman Institute of the California Institute of Technology.

REFERENCES

Arnott S, Selsing E (1974): Structures for the polynucleotide complexes poly(dA)·poly(dT) and poly(dT)·poly(dA)·poly(dT). J. Mol. Biol. 88:509–521.

Beal PA, Dervan PB (1991): Second structural motif for recognition of DNA by oligonucleotide-directed triple helix formation. Science 251:1360–1363.

Birg F, Praseuth D, Zerial A, Thuong NT, Asseline U, Le Doan T, Hélène C (1990): Inhibition of simian virus 40 DNA replication in CV-1 cells by an oligodeoxynucleotide covalently linked to an intercalating agent. Nucl. Acids Res. 18:2901–2907.

Breslauer KJ, Frank R, Blocker H, Marky LA (1986): Predicting DNA duplex stability from the base sequence. Proc. Natl. Acad. Sci. U.S.A. 83:3746–3750.

Britten RJ, Davidson EH (1969): Gene regulation for higher cells: A theory. Science 165:349–357.

Broitman SL, Im DD, Fresco JR (1987): Formation of the triple-stranded polynucleotide helix, poly(A·A·U). Proc. Natl. Acad. Sci. U.S.A. 84:5120–5124.

Cameron FH, Jennings PA (1989): Specific gene suppression by engineered ribozymes in monkey cells. Proc. Natl. Acad. Sci. U.S.A. 86:9139–9143.

Cohen JS (ed) (1989): Oligodeoxynucleotides: Antisense inhibitors of gene expression. Boca Raton, FL: CRC Press.

Cooney M, Czernuszewicz G, Postel EH, Flint SJ, Hogan ME (1988): Site-specific oligonucleotide binding represses transcription of the human c-*myc* gene in vitro. Science 241:456–459.

Davis TL, Firulli AB, Kinniburgh AJ (1989): Ribonucleoprotein and protein factors bind to an H-DNA-forming c-myc DNA element: Possible regulators of the c-myc gene. Proc. Natl. Acad. Sci. U.S.A. 86:9682–9686.

de los Santos C, Rosen M, Patel D (1989): NMR Studies of DNA $(R+)_n \cdot (Y-)_n \cdot (Y+)_n$ triple helices in solution: Imino and amino proton markers of T·A·T and C·G·C$^+$ base triple formation. Biochemistry 28:7282–7289.

Dervan PB (1986): Design of sequence-specific DNA-binding molecules. Science 232:464–471.

Felsenfeld G, Miles HT (1967): The physical and chemical properties of nucleic acids. Annu. Rev. Biochem. 36:407–448.

Felsenfeld G, Davies DR, Rich A (1957): Formation of a three-stranded polynucleotide molecule. J. Am. Chem. Soc. 79:2023–2024.

François J-C, Saison-Behmoaras T, Thuong NT, Hélène C (1989): Inhibition of restriction endonuclease cleavage via triple helix formation by homopyrimidine oligonucleotides. Biochemistry 28:9617–9619.

Goodchild J (1990): Conjugates of oligonucleotides and modified oligonucleotides: A review of their syntheses and properties. Bioconjugate Chem. 1:165–187.

Griffin LC, Dervan PB (1989): Recognition of thymine·adenine base pairs by guanine in a pyrimidine triple helix motif. Science 245:967–971.

Hanvey JC, Shimizu M, Wells RD (1989): Site-specific inhibition of EcoRI restriction/modification en-

zymes by a DNA triple helix. Nucl. Acids Res. 18:157–161.

Haseloff J, Gerlach WL (1988): Simple RNA enzymes with new and highly specific endoribonuclease activities. Nature (London) 334:585–591.

Hoogsteen K (1959): The structure of crystals containing a hydrogen-bonded complex of 1-methylthymine and 9-methyladenine. Acta Crystallogr. 12:822–823.

Horne DA, Dervan PB (1990): Recognition of mixed-sequence duplex DNA by alternate-strand triple-helix formation. J. Am. Chem. Soc. 112:2435–2437.

Howard FB, Frazier J, Lipsett MN, Miles HT (1964): Infrared demonstration of 2- and 3-Stranded helix formation between Poly(C) and guanine mononucleotides and oligonucleotides. Biochem. Biophys. Res. Commun. 17:93–102.

Htun H, Dahlberg JE (1988): Topology and formation of triple-stranded H-DNA. Science 241:1791–1796.

Jones KA, Yamamoto KR, Tjian R (1985): Two distinct transcription factors bind to the HSV thymidine kinase promoter in vitro. Cell 42:559–572.

Kadonaga JT, Carner KR, Masiarz FR, Tjian R (1987): Isolation of cDNA encoding transcription factor Sp1 and functional analysis of the DNA binding domain. Cell 51:1079–1090.

Koob M, Grimes E, Szybalski W (1988): Conferring operator specificity on restriction endonucleases. Science 241:1084–1086.

Le Doan T, Perrouault L, Praseuth D, Habhoub N, Decout J-L, Thuong NT, Lhomme J, Hélène C (1987): Sequence-specific recognition, photocrosslinking and cleavage of the DNA double helix by an oligo-[α]-thymidylate covalently linked to an azidoproflavine derivative. Nucl. Acids Res. 15:7749–7760.

Lee JS, Johnson DA, Morgan AR (1979): Complexes formed by (pyrimidine)$_n$·(purine)$_n$ DNAs on lowering the pH are three-stranded. Nucl. Acids Res. 6:3073–3091.

Letai AG, Palladino MA, Fromme E, Rizzo V, Fresco JR (1988): Specificity in formation of triple-stranded nucleic acid helical complexes: Studies with agarose-linked polyribonucleotide affinity columns. Biochemistry 27:9108–9112.

Lipsett MN (1963): The interaction of poly(C) and guanine trinucleotides. Biochem. Biophys. Res. Commun. 11:224–228.

Lipsett MN (1964): Complex formation between polycytidylic acid and guanine oligonucleotides. J. Biol. Chem. 239:1256–1260.

Luebke KJ, Dervan PB (1989): Nonenzymatic ligation of oligodeoxynucleotides on a duplex DNA template by triple helix formation. J. Am. Chem. Soc. 111:8733–8735.

Lyamichev VI, Mirkin SM, Frank-Kamenetskii MD,

Cantor CR (1988): A stable complex between homopyrimidine oligomers and the homologous regions of duplex DNAs. Nucl. Acids Res. 2165–2178.

Lyamichev VI, Frank-Kamenetskii MD, Soyfer VN (1990): Protection against UV-induced pyrimidine dimerization in DNA by triplex formation. Nature (London) 344:568–570.

Maher LJ, Wold B, Dervan PB (1989): Inhibition of DNA binding proteins by oligonucleotide-directed triple helix formation. Science 245:725–730.

Maher LJ, Wold B, Dervan PB (1990): Kinetic analysis of oligodeoxyribonucleotide-directed triple helix formation. Biochemistry 29:8820–8826.

Michelson AM, Massoulié J, Guschlbauer W (1967): Synthetic polynucleotides. Prog. Nucl. Acids Res. Mol. Biol. 6:83–141.

Miller JH, Sobell HM (1966): A molecular model for gene repression. Proc. Natl. Acad. Sci. U.S.A. 555:1201–1205.

Minton K (1985): The triple helix: A potential mechanism for gene regulation. J. Exp. Pathol. 2:135–148.

Morgan AR, Wells RD (1968): Specificity of the three-stranded complex formation between double-stranded DNA and single-stranded RNA containing repeating nucleotide sequences. J. Mol. Biol. 37:63–80.

Moser HE, Dervan PB (1987): Sequence-specific cleavage of double helical DNA by triple helix formation. Science 238:645–650.

Parker CS, Topol J (1984): A *Drosophila* RNA polymerase II transcription factor contains a promoter-specific DNA-binding activity. Cell 36:357–369.

Perrouault L, Asseline U, Rivalle C, Thuong NT, Bisagni E, Giovannangeli C, Le Doan T, and Hélène C (1990): Sequence-specific artificial photoinduced endonucleases based on triple helix forming oligonucleotides. Nature (London) 344:358–360.

Povsic TJ, Dervan PB (1989): Triple helix formation by oligonucleotide on DNA extended to the physiological pH range. J. Am. Chem. Soc. 111:3059–3061.

Praseuth D, Perrouault L, Le Doan T, Chassignol M, Thuong N, Hélène C (1988): Sequence-specific binding and photocrosslinking of α and β oligodeoxynucleotides to the major groove of DNA via triple-helix formation. Proc. Natl. Acad. Sci. U.S.A. 85:1349–1353.

Rajagopal P, Feigon J (1989): Triple-strand formation in the homopurine:homopyrimidine DNA oligonucleotides d(A-G)$_4$ and d(T-C)$_4$. Nature (London) 239:637–640.

Rossi JJ, Sarver N (1990): RNA enzymes (ribozymes) as antiviral therapeutic agents. Trends. Biot. 8:179–183.

Sklenar V, Feigon J (1990): Formation of a stable triple helix from a single DNA strand. Nature (London) 345:836–838.

Strobel SA, Dervan PB (1989): Cooperative site-specific binding of oligonucleotides to duplex DNA. J. Am. Chem. Soc. 111:7286–7287.

Strobel SA, Dervan PB (1990): Site-specific cleavage of a yeast chromosome by oligonucleotide-directed triple helix formation. Science 249:73–75.

Strobel SA, Dervan PB (1991): Single-site enzymatic cleavage of yeast genomic DNA mediated by triple helix formation. Nature (London) 350:172–174.

Strobel SA, Moser HE, Dervan PB (1988): Double-strand cleavage of genomic DNA at a single site by triple-helix formation. J. Am. Chem. Soc. 110:7927–7929.

Sun J-S, François J-C, Montenay-Garestier T, Saison-Behmoaras T, Roig V, Thuong NT, Hélène C (1989): Sequence-specific intercalating agents: Intercalation at specific sequences on duplex DNA via major groove recognition by oligonucleotide-intercalator conjugates. Proc. Natl. Acad. Sci. U.S.A. 86:9198–9202.

Uhlmann E, Peyman A (1990): Antisense oligonucleotides: A new therapeutic principle. Chem. Rev. 90:543–584.

Umemoto K, Sarma MH, Gupta G, Luo J, Sarma RH (1990): Structure and stability of a DNA triple helix in solution: NMR studies on d(T)6·d(A)6·d(T)6 and its complex with a minor groove binding drug. J. Am. Chem. Soc. 112:4539–4545.

van der Krol AR, Mol JNM, Stuitje AR (1988): Modulation of eukaryotic gene expression by complementary RNA or DNA sequences. BioTechniques 6:958–976.

Wells RD, Collier DA, Hanvey JC, Shimizu M, Wohlrab F (1988): The chemistry and biology of unusual DNA structures adopted by oligopurine·oligopyrimidine sequences. FASEB J. 2:2939–2949.

Zon G (1988): Oligonucleotide analogues as potential chemotherapeutic agents. Pharm. Res. 5:539–549.

Oligonucleotides in Cells and in Organisms: Pharmacological Considerations

Valentin V. Vlassov and Leonid A. Yakubov

Institute of Bioorganic Chemistry, Siberian Division of the Academy of Sciences of the USSR, Novosibirsk 630090, USSR

INTRODUCTION

There is little doubt that the principle of genetic targeting can be used for the development of future therapeutics, oligonucleotide derivatives, that will allow correction of biochemical disorders by modulating expression of specific genes, and fighting viral and other infectious diseases by jamming or damaging foreign genetic programs of the infectious agents. Due to outstanding discriminating ability, oligonucleotides can recognize a single mRNA species within a population of cellular RNAs and modulate its translation. Being the most accurate weapon, oligonucleotide derivatives can attack genetic programs of infectious agents without affecting the host cell programs. Therefore they may become true magic bullets of the future chemotherapy envisioned by Paul Ehrlich. A great advantage of oligonucleotide derivatives is the simplicity and general character of the oligonucleotide target nucleic acid recognition residing upon the Watson–Crick base pairing, which allows one to predict with confidence the interaction of an oligonucleotide with given nucleic acids. This advantage opens the possibility of easy rational design of oligonucleotides targeted to any nucleic acid and an opportunity to affect any function of an organism provided the responsible genetic program is known and sequenced.

In recent years, considerable success has been achieved in the development of oligonucleotide derivatives (for reviews see Zon, 1988; Stein and Cohen, 1988; Caruthers, 1989; Knorre et al., 1988, 1989; Uhlmann and Peyman, 1990; Goodchild, 1989, 1990; Knorre and Zarytova, this volume). The experiments with isolated nucleic acids and cell cultures demonstrated that oligonucleotide derivatives can affect functions of specific nucleic acids in vitro and inside cells (see reviews by Degols et al., 1991; Hélène and Toulmé, 1990; Cohen, 1989). However, many problems related to future pharmacological applications of oligonucleotide derivatives remain to be solved. These are efficient delivery of oligonucleotide derivatives to certain types of cells and into cells, and optimal targeting of oligonucleotides within specific nucleic acids. Besides there are problems related to the applications at the organism level, such as evaluation of toxicity, immunogenicity, and mutagenicity of the compounds, and investigation of their pharmacokinetics, including absorption, distribution, metabolism, and excretion. In the present chapter, we describe the results of biological experiments aimed at answering some of the questions related

Prospects for Antisense Nucleic Acid Therapy of Cancer and AIDS, pages 243–266
©1991 Wiley-Liss, Inc.

to the potential therapeutic use of oligonucleotides.

INTERACTION OF OLIGONUCLEOTIDES WITH CELLS AND THEIR EFFECTS ON THE FUNCTIONING OF NUCLEIC ACIDS

Mechanism of Cellular Uptake of Oligonucleotides

One of the major problems to be solved in order to develop oligonucleotide derivatives of therapeutic value is the delivery of compounds into cells to the target nucleic acids. It was believed that oligonucleotides cannot cross lipid membranes of cells because they are polyanionic molecules. Considerable efforts have been made to design nonionic membrane-permeable analogs (Miller, 1989; Miller and Ts'o, 1987 and references therein) and to develop special delivery techniques (for review see Degols et al., 1991). Nevertheless, numerous studies demonstrated the possibility of affecting translation of specific cellular mRNAs and multiplication of viruses with complementary oligonucleotides, which is possible only when the oligonucleotides can reach target RNAs inside cells. These findings stimulated the investigations of oligonucleotide interactions with mammalian cells aimed at learning its mechanism, which could be used for a more efficient delivery of the compounds into cells.

It was found that incubation of cells with radiolabeled oligonucleotides resulted in binding of the radioactive material to the cells and its accumulation in the cell nuclei (Vlassov et al., 1986a; Zamecnik et al., 1986; Goodchild et al., 1988a). Autoradiography of cells incubated with ^{3}H-labeled oligonucleotides (Goodchild et al., 1988a) revealed the label in both cytoplasm and nuclei of the cells. These experiments provided physical evidence that the compounds were really taken up by the cells, although in such experiments it was difficult to distinguish among the compounds taken up, those absorbed at the cell surface, or metabolized label. Direct evidence for the intracellular localization of the oligonucleotides taken up was obtained in experiments with reactive oligonucleotide derivatives (Karpova et al., 1980; Vlassov et al., 1986b). Phosphodiester oligonucleotides equipped with an alkylating group, an aromatic nitrogen mustard residue, at the 5'-terminal phosphate, were used. Their general structure was $ClRCH_2NHp(N \ldots N)$; chemical formulas of reagents, methods of the synthesis, and properties of the compounds can be found in the chapter by Knorre and Zarytova (this volume). The derivatives of oligothymidylate were known to form complementary complexes with poly(A) under the conditions used and logically they were expected to react with the poly(A) sequences of cellular RNAs when entering the cells. Indeed, it was found that the derivatives were taken up by cells and reacted with cellular biopolymers. The reaction revealed the expected temperature sensitivity, suggesting complex formation to be a necessary condition for the reaction to occur. Analysis of the modified nucleic acids showed that the level of modification of the poly(A) sequences by far exceeded that of the heterogeneous sequences, suggesting the reaction proceeded via formation of a complementary complex. It should be mentioned that, in these experiments, a reaction with cellular DNA was also registered. Analysis of the specificity of this reaction showed that at least a part of the process occurred in a sequence-specific way.

Detailed studies of the mechanism of the oligonucleotide–cell interaction were performed later with fluorescently labeled (Loke et al., 1989; Neckers, 1989) and radiolabeled oligonucleotide derivatives (Yakubov et al., 1989). As physical labels, radioisotopes seem preferable because fluorescent labels may interact specifically and nonspecifically with cellular constituents and modify oligonucleotide properties. For example, the conjugation of oligonucleotides to acridine facilitates the entering of the compounds into trypanosomes (Verspieren et al., 1987).

It was found that oligonucleotides are taken up by cells in a saturable manner compatible with endocytosis. The results of a typical experiment, shown in Figure 1, demonstrate

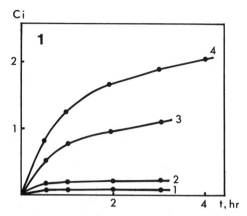

Fig. 1. Time course of binding of oligodeoxynucleotide $(pT)_9$ to mouse fibroblasts L929. Cells were incubated at 37°C with different oligonucleotide concentrations in the media (3, 15, 75, and 150 μM, curves 1–4 respectively). C_i, average concentration in cells is shown as uptake by cell per cell volume, μM.

that under physiological conditions the maximal binding is achieved within 2 hr incubation. The process slows down both with a decrease in temperature and in the presence of the known inhibitors of endocytosis, deoxyglucose, cytochalasin B, and sodium azide.

These data (Goodchild et al., 1988a; Loke et al., 1989; Yakubov et al., 1989) suggest the compounds to be taken up by the endocytosis mechanism. In the cells incubated with labeled oligonucleotides, 20% of the material taken up was bound in the cell nuclei and 50% was associated with the fractions containing mitochondria, lysosomes, and other vesicular structures. This finding is also consistent with an endocytosis mechanism of uptake.

Different cell lines bind somewhat different amounts of oligonucleotides (Yakubov et al., 1989; Ceruzzi and Draper, 1989). It was noticed that the efficiency of oligomer internalization depends on the conditions of cell growth. Thus, an increase in the cell monolayer density from 8×10^4 to 5×10^5 cells/cm^2 resulted in a 3-fold decrease in the maximal binding of the oligomer (per cell).

The investigation of the concentration dependence of oligonucleotide uptake demonstrated that the kinetic curves are similar in a broad concentration range. The dependence of the limit binding level on the oligonucleotide concentration is linear at high concentrations of the compound (Fig. 2). The efficiency of the uptake is quite low at these

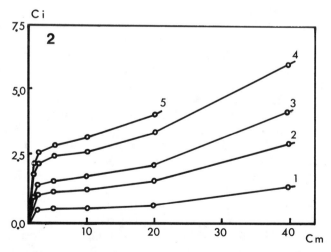

Fig. 2. Dependence of $ClRCH_2NH(pT)_{16}$ uptake by mouse fibroblasts on the concentration of the derivative in the medium. C_i, concentration of the derivative in the cells, μM; C_m, concentration of the derivative in the medium, μM. Curves 1–5 correspond to different times of incubation: 10 min, 30 min, 1 hr, 2 hr, and 4 hr.

concentrations, similar to that of polyvinyl-pyrrolidone, a compound known to be taken up by fluid phase pinocytosis, and average oligonucleotide concentrations achieved in the cells were less than one-tenth of that in the medium. Therefore at high oligonucleotide concentration, the uptake resembles a fluid phase endocytosis of the compounds. A similar low uptake efficiency at such oligonucleotide concentrations was observed for other cells and other oligonucleotide derivatives (Goodchild et al., 1988a; Zamecnik et al., 1986; Wickstrom et al., 1988; Loke et al., 1989). At low concentrations ($<0.5 \mu M$), the uptake efficiency is considerably higher and the average concentration of the oligonucleotide derivatives in the cells exceeds their concentration in the medium. The increased efficiency of endocytosis can be explained by absorption of a limited amount of oligonucleotide on the cell surface, which increases the efficiency of the process. At low oligonucleotide concentration, this absorptive endocytosis plays a major role in the uptake of the oligonucleotides.

The dependence of the binding plateau level on the oligonucleotide concentration (Fig. 2) suggests a complicated nature of the oligonucleotide–cell interaction. This dependence can be explained by the existence of two processes: endocytosis and release of oligonucleotides from cells, that are in equilibrium by the second hour of incubation. The release of the bound oligonucleotides from cells was observed experimentally (Yakubov et al., 1989). It was found that about 70% of the bound labeled material can be liberated from the cells. The rest of the material seems to be bound irreversibly. Electrophoretic analyses of the radiolabeled material isolated from the cells incubated with ^{32}P-labeled HORCH$_2$NHpTTCCTTCCCGTCTTCC (unreactive derivative prepared by hydrolysis of the corresponding alkylating reagent) and of the material from the medium have revealed degradation of the derivatives in the cells after 2–4 hr incubation. The released material is more heavily degraded than that retained by the cells. This may suggest the degradation products to be in those cellular compart-ments that communicate more easily with the external medium. Therefore one may suggest that there are probably two pathways operating in the cellular uptake of oligonucleotides, one of them providing irreversible internalization in an environment with a low nuclease activity.

The results described and the fact that unlabeled oligonucleotides and polynucleotides block the accumulation of the labeled oligonucleotide derivatives in cells in a concentration-dependent manner suggest the existence of a limited number of the nucleic acid binding receptors on the cell surface. Indeed, specific proteins binding oligo- and polynucleotides were detected on the cells. An oligonucleotide-binding 80-kDa protein was isolated by chromatography on oligo(dT) cellulose from surface-iodinated HL-60 cells (Loke et al., 1989). In experiments with oligonucleotide derivatives bearing reactive groups ClR, capable of alkylating nucleophilic centers in both nucleic acids and proteins, the reagents were found to modify a 80-kDa protein in L929 mouse fibroblasts. The reaction was specific according to the competition data: polynucleotides such as tRNA and single-stranded and double-stranded DNA inhibited the modification while heparin and chondroitin sulfates A and B did not affect the alkylation, even when they were present in great excess over the reagent. The modification kinetics were found to plateau at reagent concentrations above 1 μM, similar to the concentration-dependence curve for the oligonucleotide binding process shown in Figure 2. A similar saturable nature of these processes indicates that the protein may participate directly in the specific cell-surface binding of oligonucleotides. A rough dissociation constant was estimated to be 0.2 μM. The number of oligonucleotide-binding proteins was approximately 10^5 per cell. Since the mean time of receptor recycling is known to be about 10 min, such number of receptors is enough to deliver the experimentally determined amount of oligonucleotides into cells.

Competition experiments (Loke et al., 1989) showed that the receptors bind polynucleotides, oligonucleotides, oligonucleotide phosphorothioates, and mononucleotides, but neither nucleosides nor deoxyribose. As expected, methylphosphonate oligonucleotides do not compete with normal oligonucleotides for the receptors. Therefore, the proteins appear to be specific receptors for binding nucleic acids. Their biological role is still unclear. It should be noted that the existence of the nucleic acid binding receptors was favored by a number of other experiments. Thus, in experiments with perfused rat liver it was found that liver cells bind DNA. Since this process was saturable and temperature dependent, the binding was suggested to be receptor mediated. Since the binding was trypsin sensitive, the putative receptor was assumed to be a protein (Emlen et al., 1988). Similarly, the existence of nucleic acid-binding receptors was favored by the results of investigations of DNA binding to human platelets (Dorsch, 1981) and to human leukocytes (Ohlbaum et al., 1979; Bennett et al., 1985) as well as by studies on poly(I)·poly(C) binding to lymphocytes (Diamantstein and Blitstein-Willinger, 1978).

The oligonucleotide binding protein seems to be ubiquitous. In experiments with different cell lines (COS-1 and Vero cells, L-671 mouse myoblasts, mouse hepatocytes, mouse and mink fibroblasts, Ag 17.1, and CHO hamster fibroblasts) the alkylating oligonucleotide derivatives were found to react with 78- and 80-kDa proteins in most of the cells tested (Vlassov et al., 1991a). In COS-1 and Ag 17.1 cells, two additional proteins (82 and 84 kDa) were labeled. They were the only ones attacked in the case of mouse hepatocytes. Moreover, similar proteins were identified in experiments with *Drosophila melanogaster* (Yakubov et al., 1991). When insects were injected with the ^{32}P-labeled alkylating oligonucleotide derivatives $ClRCH_2NHp(TpT)_8$, the modification of a few proteins in different organs was observed. The modified protein pattern was organ and stage specific. Only one protein was labeled in bulbus ejakulatorius; in ovaries another protein was modified. The intensity of its labeling increased with fly maturation. The modification of the

insect proteins was saturable at submicromolar concentrations of the oligonucleotide derivatives (average concentrations in flies) and was suppressed by a coinjection of excess of oligothymidylates or DNA. These results fit the hypothesis about the ubiquitous nature of the nucleic acid-binding receptors.

When considering the problem of oligonucleotide delivery to the target nucleic acids of infectious agents, one should bear in mind that sometimes these nucleic acids are protected inside the cells of infectious agents. The properties of these cells are usually different from those of mammalian cells. Thus the bacterial cell wall seems to be completely impermeable for oligonucleotides due to the polysaccharide cover. In experiments with a mutant *Escherichia coli* strain whose cell wall contains negligible quantities of polysaccharide it was found that oligonucleotide methylphosphonates can enter the mutant cells, although this ability rapidly decreased with the increase in their length (Jayaraman et al., 1981). It was found that oligonucleotide derivatives conjugated to acridine can enter trypanosomes (Verspieren et al., 1987). We investigated the interaction of oligonucleotide derivatives with mycoplasmas and found that the latter are penetrated very easily (Vlassov et al., 1991d). No experiments testing the ability of oligonucleotides to reach nucleic acids in viral particles have been reported yet.

Delivery of Oligonucleotides Into Cells: Oligonucleotide Analogs and Conjugates

A finding that a natural mechanism of oligonucleotide uptake involves adsorptive endocytosis promoted the development of oligonucleotide derivatives bearing groups anchoring them to cell membranes and increasing the endocytosis efficiency. The nonspecific adsorptive endocytosis is not limited by the amount of cell surface receptors and allows higher capacity of binding. Adsorption of oligonucleotides on the cell surface can be achieved simply by their coupling to lipophilic groups, such as cholesterol, used for anchoring of various substances to lipid membranes. The results of the experiments with oligonucleotides bearing 3′-cholesterol groups (Boutorin et al., 1989) (Fig. 3) show that the derivatives bind to the cells very efficiently. Their ability to be taken up by cells has been demonstrated in the experiments with alkylating oligonucleotide derivatives, where the latter were found to react with cellular biopolymers, including DNA. The extent of DNA modification by the cholesterol-linked alkylating oligonucleotide derivatives considerably exceeded that observed with unmodified alkylating oligonucleotide derivatives. The enhanced uptake by cells was also observed for oligonucleotide derivatives bearing octyl, dodecyl, and octadecyl residues, resulting in a 3-, 4-, and 10-fold increase, respectively, compared to the unmodified oligonucleotides (Abramova et al., 1990).

In accordance with the cell-binding data, the lipophilic groups linked to oligonucleotides provided a higher antiviral activity compared to the unmodified oligonucleotides. The conjugation of cholesterol residues to anti-

sense oligonucleotides significantly increased their ability to suppress proliferation of HIV-1 in cell culture (Letsinger et al., 1989; Abramova et al., 1990). An antisense oligonucleotide carrying an undecanol group at the 5′-end was reported to be a considerably more potent inhibitor of the influenza virus polymerase biosynthesis and multiplication of the virus in cell culture than its unmodified counterpart (Kabanov et al., 1990).

To enhance uptake with a hydrophilic substituent, the 3′-ends of oligonucleotides were coupled to ε-amino groups of lysine residues of a known polycationic drug carrier, poly(L-lysine), which is efficiently transported into cells by nonspecific adsorptive endocytosis. The experiments with fluorescently labeled oligonucleotide derivatives conjugated to the carrier showed that internalization of the conjugate follows a classical endocytotic pathway. In the cells, the oligonucleotides are cleaved from polylysine. The uptake of the oligonucleotides conjugated to polylysine was found to be faster and more efficient than that of the parent oligonucleotides (Leonetti et al., 1990a). The polylysine conju-

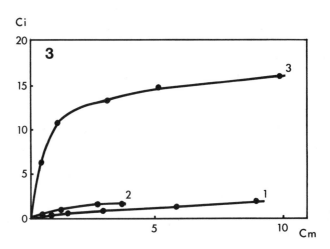

Fig. 3. Dependence of $(pT)_8$ derivatives uptake by ascites carcinoma Krebs 2 cells on concentration of the derivatives in the medium. 1, derivative bearing ClRCH$_2$NH group at its 5′-end; 2, derivative 1 with phenazinium group at its 3′-end; 3, derivative 1 with cholesterol group at its 3′-end. C_i, concentration of the derivative in the cells, μM; C_m, concentration of the derivative in the medium, μM. (Reproduced from Boutorin et al., 1989, with permission of the publisher.)

gated antisense oligonucleotides, complementary to vesicular stomatitis virus (VSV) protein N mRNA, inhibited the synthesis of VSV proteins and exerted antiviral activity in cell culture experiments at 0.1 μM concentrations. At such low concentration, the unmodified oligonucleotides were completely inefficient (Lemaitre et al., 1987; Degols et al., 1989). Poly(L-lysine)-conjugated antisense oligonucleotides were reported to protect cells against HIV-1 infection much more efficiently than the unmodified oligonucleotides and their methylphosphonate analogs (Stevenson and Iversen, 1989).

Nonionic oligonucleotide analogs are able to pass across cell membranes easily (Miller, 1989), to interact with target nucleic acids (Karpova et al., 1980) and to affect their functions (for review see Miller, 1989; Miller and Ts'O, 1987). The main problem with these derivatives is that the present technology does not allow stereospecific synthesis of the compounds (for review see Caruthers, 1989; Zon, 1988; Miller, 1989).

Delivery of Oligonucleotides Into Cells: Membrane Carriers

An alternative approach to the delivery of oligonucleotide derivatives into cells consists in using membrane carriers, liposomes, and viral envelopes. These vehicles can deliver into cells any derivatives irrespective of their ability to enter cells via natural uptake mechanisms. It is important that oligonucleotides being incorporated in the carriers are protected from the enzymes present in the blood stream. The liposomes may be taken up by endocytosis and release the encapsulated material in the cell. Liposomes were used to enhance the efficiency of delivery of phosphorothioate antisense oligonucleotide analogs, targeted to c-*myc* mRNA, into HL-60 cells (Loke et al., 1988). It was found that when incorporated into liposomes, the oligonucleotides more actively reduced the c-*myc* protein expression in the cells.

Liposomes are naturally targeted to the reticuloendothelial system, liver, and spleen. However, coupling of liposomes to various ligands, such as antibodies or protein A, allows their targeting to specific cell populations. Oligonucleotides complementary to mRNA encoding protein N of vesicular stomatitis virus were encapsulated into the protein A-bearing liposomes and delivered into the L929 mouse fibroblasts pretreated with H-2K-specific antibodies that are targeted to the mouse major histocompatibility complex-encoded H-2K protein expressed on the L cells. Due to the protein A–antibody interaction, the liposomes bound to the target cells and were efficiently internalized by endocytosis. The treated cells became less permissive for multiplication of the virus. The encapsulated oligonucleotides were found to be active in amounts of two orders of magnitude lower than the unencapsulated ones (Leonetti et al., 1990b).

Reconstituted envelopes of viruses that penetrate into cells via fusion of virion membranes with cell membranes are efficient vehicles for the delivery of oligonucleotides in cell cytoplasm. Most detailed experimental studies were done with reconstituted envelopes of Sendai virus (RSVE). An efficient technique providing a 25–30% yield of entrapment of oligonucleotide derivatives in these carriers has been developed (Vlassov et al., 1989). It is based on repeated freezing and thawing of empty RSVE in the solution of a substance to be incorporated. Since the procedure is performed fast and at low temperature, it can be used for the incorporation of labile reactive oligonucleotide derivatives. The RSVE loaded with the oligonucleotide derivatives bind to cells with an efficiency similar to that of the virus, which is about 60% (Fig. 4). The loaded vesicles are stable in the course of 2–3 hr incubation with cells; neither leakage nor substantial oligonucleotide degradation is observed in the carriers. Experiments with the Krebs 2 ascites carcinoma cells and RSVE loaded with the radiolabeled alkylating derivative $ClRCH_2NH(pT)_{16}$ showed that the derivative was delivered into cell cytoplasm efficiently and reacted with the poly(A) sequences of cellular RNAs (Table 1). These results provide evidence that, using

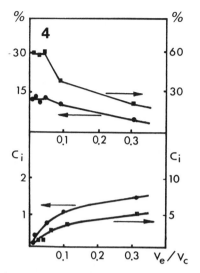

Fig. 4. Delivery of ^{32}P-ClRCH$_2$NH(pT)$_{16}$ loaded into reconstituted Sendai virus envelopes (RSVE) into mice fibroblasts (●, left ordinate) and ascites carcinoma Krebs 2 cells (■, right ordinate). Upper box: ordinate, binding of the RSVE to cells, percent of the added material; abscissa, relative amount of RSVE and cells in the system; V_e, total volume of RSVE; V_c, total volume of the cells. Lower box: ordinate, concentration of the oligonucleotide in the cells; abscissa, the same as in the upper box (Vlassov et al., 1989).

RSVE, one may achieve efficient transfer of oligonucleotide derivatives into cells. Loading of oligonucleotide derivatives targeted to the genomic RNA of tick-borne encephalitis virus into RSVE enhanced their ability to suppress multiplication of the virus in cell culture (Pogodina et al., 1988).

Another type of natural carrier is human erythrocyte membrane preparations, "erythrocyte ghosts." They can be prepared by lysis of erythrocytes by a hypoosmotic shock, and filled by incubation with the necessary substances in a salt solution under conditions of recovery of erythrocyte membrane. A modified variant of the incorporation procedure (Vlassov et al., 1989) is mild enough for charging of the ghosts with reactive oligonucleotide derivatives. In the presence of the UV-inactivated Sendai virus, which is used as a fusion agent, the erythrocyte ghosts efficiently bind to cells and deliver the oligonucleotide derivatives to the cytoplasm. In experiments with the erythrocyte ghosts loaded with an alkylating derivative of oligothymidylate, the reagent appeared to be delivered to the target poly(A) sequences of cellular RNAs much more efficiently than by natural uptake (Table 1).

TABLE 1. Alkylation of RNA in Ascites Carcinoma Krebs II Cells by Oligonucleotide Derivatives Delivered by Membrane Carriers[a]

Reagent and type of vehicle	Average concentration in the medium (μM)	Mean concentration within cells (μM)	Binding to cells (%)	Label in polymer RNA, percent of cell-bound reagent	Poly(A)$^+$ RNA modification, moles of reagent per mole of nucleotide
Free					
ClRCH$_2$NH(pT)$_{16}$	0.2	0.12	0.3	0.02	0.3×10^{-5}
In RSVE					
ClRCH$_2$NH(pT)$_{16}$	0.01	0.4	25	—	1.3×10^{-5}
	0.04	1.7	25	—	8×10^{-5}
ClRCH$_2$NH(pQ)[b]	0.02	1.0	28	—	7.5×10^{-7}
In erythrocyte ghosts					
ClRCH$_2$NH(pT)$_{16}$	0.04	2.1	30	9	2×10^{-4}
ClrCH$_2$NH(pQ)	0.04	2.7	40	11	3.3×10^{-6}

[a]^{32}P-labeled alkylating oligonucleotide derivatives were added to cell culture media. After 1.5 hr incubation of the derivatives with cells, binding of radioactivity and its incorporation in cellular RNAs were determined (Vlassov et al., 1989).
[b](pQ), oligonucleotide pTTCCTTCCCGTCTTCC.

Stability of Oligonucleotide Derivatives in Cells

Potential chemotherapeutic applications of oligonucleotides require analogs that are resistant to nucleic acid-degrading enzymes present in cells and in bloodstream. Although oligonucleotides are stable in usual serum-free culture media, such as Dulbecco's modified essential medium, they are short-living inside cells and in serum. Thus in bovine calf serum, complete hydrolysis of oligonucleotides occurs within 15 min (Wickstrom, 1986). In different types of cells half-lives of oligonucleotides were estimated to be from 10 to 30 min to a few hours (Cazenave et al., 1987; Babkin et al., 1988; Vlassov et al., 1986a; Wickstrom et al., 1988; Yakubov et al., 1989). Judging from the gel patterns of degradation products, a 3'-exonuclease activity is responsible for most of the oligonucleotide degradation in serum.

Bulky groups at the oligonucleotide termini and modifications of terminal phosphodiester groups were found to protect the compounds from degradation. Introduction of a single methylphosphonate linkage at the oligonucleotide end is sufficient to retard the degradation by both 5'- and 3'-exonuclease activities. Two adjacent internal methylphosphonates in oligonucleotide block exonuclease hydrolysis much more efficiently than one (Quartin et al., 1989; Furdon et al., 1989; Goodchild et al., 1988b). Covalent attachment of an acridine residue to the 3'-phosphate and a methylthiophosphate group at the 5'-end phosphate increases their half-life time in oocytes from 10 to 40 min (Cazenave et al., 1987). Blocking of the 5'-end of oligonucleotides by coupling the ClRCH$_2$NH group does not allow dephosphorylation of oligonucleotides in the presence of Krebs 2 ascites carcinoma cells and mice fibroblasts, and inside cells. Coupling of cholesterol, phenazinium, or ethidium residues to the 3'-terminus of oligonucleotides protects them from exonucleolytic degradation and makes them considerably more stable in the medium with cells and inside cells: 80–90% of the compounds remained intact after 24 hr incubation with cells under conditions, where unmodified oligonucleotides were completely degraded (Abramova et al., 1991; Koshkin et al., 1991). These modifications were found to be protective even in mycoplasmas that are known for a high content of nucleases and an ability to rapidly metabolise exogenous nucleic acids. In mycoplasmas, the half-life times of oligonucleotides bearing a ClRCH$_2$NH group at the 5'-phosphate and a cholesterol or phenazinium residue at the 3'-end were more than 10 times as long as of unmodified oligonucleotides (Vlassov et al., 1991d). One should take into account that conjugation of different groups can change properties of oligonucleotides as substrates for nucleases as well as localization of the compounds on and inside the cells. Thus, cholesterol-linked oligonucleotides are likely to be anchored to the cell membranes, their occupied ends becoming less accessible to nucleases due to steric reasons.

Ribose-modified oligonucleotides showed a good resistance to enzymatic hydrolysis. Oligo-[α]-deoxynucleotides are resistant to different nucleases (Imbach et al., 1989) and hydrolytic enzymes inside *Xenopus* oocytes (half-life time >8 hr compared to 10 min for normal oligonucleotides) (Cazenave et al., 1987).

Phosphate-modified oligonucleotides, such as phosphorothioate oligonucleotide analogs, and nonionic methylphosphonates and ethyl phosphotriesters of oligonucleotides, are highly resistant to nucleolytic degradation. In the cells, degradation of these analogs is slow. It starts with deethylation for ethyl phosphotriesters and with deglycosylation for methylphosphonates (Miller, 1989). The problem of sensitivity to nucleolytic enzymes can be solved completely by development of oligonucleotide analogs with nonnatural internucleoside linkages (Miller, 1989; Caruthers, 1989).

Interaction of Oligonucleotides With Cellular Nucleic Acids

Taken up by a cell, oligonucleotides must find their way to target nucleic acids in cytoplasm and in cell nucleus. The discussed results of sequence-specific alkylation of poly(A) sequences of cellular RNAs and the studies on mRNA translation arrest by complementary oligonucleotides and oligonucleotide derivatives inside the cell provide evidence that RNAs in the cytoplasm are readily accessible for oligonucleotides that entered the cell and can hybridize to them.

Cellular DNA is a very attractive target for the oligonucleotide derivatives for the achievement of permanent arrest of the expression of certain genes. One may expect the oligonucleotides taken up by cells to have easy access to chromatin through the nuclear pores. However, it seems very improbable that the oligonucleotides could interact with cellular DNA, which is known to be heavily structured and shielded by histones and a great number of other proteins. Tight packaging of the DNA in chromatin and protein cover is likely to interfere with the formation of both known types of complexes between oligonucleotides and DNA: specific duplexes with partially unwound DNA and triple-stranded complexes. Although in the course of transcription and replication DNA double helices undergo local opening, their transiently unwound regions are unlikely to be as accessible as free single-stranded DNA because they are known to be covered with specific single-stranded DNA-binding proteins. The exception may be the sites of initiation of transcription and replication where the oligonucleotide derivatives may be taken up within the transcription initiation complexes as primers (Boutorin et al., 1987) and may interact with the DNA sequences where the replication or transcription has started, resulting in the inhibition of the corresponding processes. Against these expectations, experiments with oligonucleotides and their alkylating derivatives have revealed that the compounds can interact with DNA in chromatin and in cell nuclei (Vlassov et al., 1990). It was found that the oligonucleotides, $(pT)_{16}$, $(pA)_{16}$, $(pApC)_6$, and the corresponding alkylating derivatives bearing a $CLRCH_2NH$ group at the 5'-terminal phosphates formed complexes with human and rodent chromatin and that the oligonucleotide derivatives reacted with both DNA and chromatin

proteins. It is noteworthy that, under the same conditions, the double-stranded DNA isolated from the same cells interacted with neither the oligonucleotides nor the derivatives. The interaction with chromatin was specific, because the reaction of the oligonucleotide derivatives with chromatin DNA was inhibited by the excess of parent oligonucleotides and was not influenced by oligonucleotides of random sequence. In experiments with an alkylating oligonucleotide derivative targeted to a sequence in the human Alu repeat, it was found that the reaction did take place within the target Alu sequences of the chromatin isolated from HeLa cells. In experiments with oligothymidylate derivatives, the reactivity of cellular DNA toward the reagents was found to be apparently independent of the state of the cells. Similar reactivity was observed in rapidly proliferating liver cells of mice after partial hepatectomy and in the transcriptionally silent chicken erythrocytes. Therefore it was concluded that the chromatin DNA must have either partially unwound sequences or sites readily opening in the presence of complementary oligonucleotides. The latter possibility may be ascribed to torsional stress produced by negative supercoiling of some regions in DNA. DNA from isolated mice methaphase chromosomes was also reported to interact with oligonucleotide derivatives, although to a smaller extent than chromatin DNA (Belyaev et al., 1986).

Suppression of Specific Gene Expression and Antiviral Activity of Oligonucleotide Derivatives: Experiments With Cell Cultures

In theory, oligonucleotide derivatives can be used for modulation of gene expression at both transcriptional and translational levels. Besides the ability to bind to the unwound DNA regions, some oligonucleotides form specific triple-stranded complexes with poly(Pu)·poly(Pu) sequences and affect binding to DNA of specific proteins (Dervan, 1989). This opens the possibility to control the transcription process. Suppression of c-*myc* mRNA synthesis by triple-stranded complex formation in an in vitro system was reported (Cooney et al., 1988) and it was found that oligothymidylates covalently linked to an acridine derivative inhibit replication of SV40 virus DNA in cells, apparently due to the interaction with the AT-rich region within the viral origin of replication (Birg et al., 1990). The possibility of forming triple-stranded complexes with DNA is limited to a few classes of purine–pyrimidine-rich sequences (Dervan, 1989); therefore it is necessary to design oligonucleotide analogs bearing new heterocyclic bases so as to target them to heterogeneous nucleotide sequences in DNA.

Oligonucleotides can arrest specific mRNA translation in vitro and inside cells (for reviews see Paoletti, 1988; Goodchild, 1990; Degols et al., 1991; Hélène and Toulmé, 1990). In fact, they are already used as tools for the investigation of the role of specific genes. As the present state of the art does not allow accurate prediction of RNA secondary and tertiary structure, the information about RNA sequences that are available for oligonucleotide binding is empirical. In many cases oligonucleotides complementary to specific sequences of mRNAs turned out to be inefficient inhibitors, apparently due to insufficient complex formation. In mRNA and pre-mRNA, the available targets are (1) the 5'-end sequences in the region of binding of initiation factors and ribosome, and (2) acceptor and donor splicing sites that must be available to oligonucleotide binding due to a splicing mechanism involving the formation of complementary complexes at these sites. The arrest of translation of different cellular and viral mRNAs was achieved with oligonucleotides and their analogs complementary to the 5'-terminal sequences and to the sequences around the initiating codon of the messengers (for reviews see Hélène and Toulmé, 1990; Vlassov and Yurchenko 1990). The apparent mechanism of inhibition is physical interference of the tethered oligonucleotides with the binding of the initiation factors and ribosome due to unfolding of the biologically active conformation of mRNA needed for the interaction with the ribosome and the factors. The most efficient inhibition was achieved

when a cassette of oligonucleotides formed a contiguous duplex structure with the juxtaposed sequences of the messenger (Maher and Dolnick, 1988; Sankar et al., 1989; Vlassov et al., 1991b). Oligonucleotide binding to the coding regions of mRNA does not affect translation since the elongating ribosome is capable of opening the RNA secondary structure and of displacing the oligonucleotides from the messenger. However arrest is observed in those cases, when the complex formed between the oligonucleotide derivative and the mRNA is a substrate for RNase H (when the enzyme is active enough in the cell) (for review see Walder and Walder, 1988; Cazenave et al., 1989; Dash et al., 1987). Although RNase H is normally found in the cell nucleus, some of its activity is traced in the cytoplasm. The existence of the RNase H activity associated with the HIV-1 reverse transcriptase has been reported recently (Dudding et al., 1990). This enzyme hydrolyzes RNA bound to deoxyribooligonucleotides and some analogs (phosphorothioate and, probably, phosphoroselenoate) (Cazenave et al., 1989; Mori et al., 1989), thus providing a catalytic mechanism for the enhancement of the oligonucleotide inhibitory action. Ribonuclease H does not hydrolyze the complexes formed by oligonucleotide methylphosphonates (Maher and Dolnick, 1988) and α-oligonucleotides (Imbach et al., 1989). This is one of the main reasons for the poor inhibitory properties of some analogs, targeted to the coding regions of mRNAs.

Oligonucleotides can be targeted to a number of different cellular RNAs (for review of the potential targets see Hélène and Toulmé 1990). However, they are not as important targets as mRNAs for achievement of specific effects.

Oligonucleotide derivatives designed to fight infectious agents can be efficiently targeted to specific sequences of their genetic programs that have no analogs in cellular programs. Thus, a common 5'-terminal mini-exon sequence present in all mRNA species of trypanosomes makes an ideal target for the arrest of translation of all proteins of the parasite (Verspieren et al., 1987). In bacteria, a sequence in the ribosomal 16 S RNA that binds to the bacterial mRNAs (Shine–Dalgarno sequence) can be targeted to in order to suppress the bacterial translation machinery without affecting eukaryotic cells (Jayaraman et al., 1981). An oligonucleotide, complementary to the intergenic sequences of vesicular stomatitis virus responsible for the viral RNA polyadenylation reinitiation of transcription on the genomic RNA, was a very efficient inhibitor of the virus multiplication (Degols et la., 1989). Effective suppression of virus multiplication was achieved using oligonucleotide derivatives targeted to the specific sequences in genomic RNAs that are responsible for specific folding of the molecules needed for their replication. Oligonucleotides targeted to the terminal reiterated sequence of the Rous sarcoma virus important for replication efficiently inhibited multiplication of the virus in cell culture (Zamecnik and Stephenson, 1978). Common terminal sequences of the influenza virus genomic RNAs proved to be good targets for oligonucleotide derivatives (Vlassov et al., 1984a,b; Zerial et al., 1987).

Several types of modifications were used for the improvement of antiviral and antimessenger potential of oligonucleotides. Coupling of intercalating groups to oligonucleotides stabilizes their complementary complexes with nucleic acids. This allows a more efficient inhibition of the target functions and using of short oligonucleotides (Hélène and Toulmé, 1989; Thuong et al., 1989). It is desirable that targeting to the coding mRNA sequences should distinguish between mRNAs with similar regulatory sequences. Therefore, attempts have been made to develop oligonucleotide derivatives capable of irreversible binding to the targets and arresting translation at any position of mRNA. Thus, it was found that a specific arrest of immunoglobulin synthesis can be achieved by an alkylating oligonucleotide derivative targeted to a sequence within the mRNA fragment coding for the variable κ chain region of the protein (Vlassov et al., 1984b, 1986b). Conjugation of photoreactive psoralen groups to oligonucleotides complementary to the coding region of rabbit globin

mRNA converted them into efficient inhibitors of mRNA translation (Kean et al., 1988). Psoralen-conjugated methylphosphonate oligonucleotide analogs complementary to RNA herpes simplex viruses considerably enhanced the ability of the compounds to suppress the development of the virus in cell culture under irradiation (Miller, 1989). Oligonucleotide derivatives equipped with intercalating and reactive groups were more active as compared to the unmodified oligonucleotides. It should be noted that the coupled reactive groups improve the inhibitory properties of oligonucleotides in two ways, by providing a more efficient blocking of the target and by protecting the oligonucleotides from degradation by cellular enzymes.

Testing of different oligonucleotide analogs on cells infected with viruses has revealed that antiviral activity of the compounds was determined essentially by their ability to enter cells and to form complexes with the viral RNA, which are attacked by RNase H (for review see Goodchild et al., 1988b; Héléne and Toulmé, 1990. Degols et al., 1991). As previously described, equipment of oligonucleotides with groups facilitating delivery into cells and protecting them from nucleases greatly increases efficiency of their action. In our experiments with HIV-1 (Abramova et al., 1990) and tick-borne encephalitis virus (Pogodina et al., 1988) it was found that antiviral properties of oligonucleotides can be greatly improved by coupling reactive $CLRCH_2NH$ groups or lipophilic groups to them. The experiments with cell cultures infected by tick-borne encephalitis virus showed that the amount of the oligonucleotide derivatives needed for efficient suppression of the virus multiplication can be reduced by a factor of about 100, if the derivatives are incorporated into membrane carriers, Sendai virus envelopes (Table 1).

The experiments on cell cultures have shown that oligonucleotide derivatives can inhibit multiplication of viruses nonspecifically. These effects deserve discussion keeping in mind the potential therapeutic applications. In many cases considerable antiviral activity was exhibited by "control" oligo-nucleotides. These were sense and mis-matched antisense oligonucleotides, homoo-ligonucleotides, and arbitrary oligonucleotides that apparently had no perfect complemen-tary binding sites in virus RNAs. One could envision several possible mechanisms by which oligonucleotides may cause these "nonspecific" effects. First, in many cases an oligonucleotide can hardly be considered as a true control. In large nucleic acids such as viral genomic and messenger RNAs, oligo-nucleotides always form some imperfect complexes with partially complementary nu-cleotide sequences. Although these com-plexes are incomparably less stable than the perfect ones, their effect may be greatly en-hanced if they are degraded by RNase H pres-ent in cells. Sometimes, the RNase H-me-diated mechanism may be helpful. Thus, in the HIV-1-infected cells, a more efficient oli-gonucleotide-catalyzed degradation of RNAs may occur due to the viral reverse transcrip-tase-associated RNase H activity. However, this enzyme may cause unwanted cleavage of nontarget RNAs in different cells. That is why the oligonucleotide derivatives, whose complexes with RNA are not sensitive to RNase H, may be considered as generally more advantageous.

Another source of the "nonspecific" effect of oligonucleotides is their ability to interact with DNA and RNA polymerases and numer-ous proteins capable of binding to nucleic acids. Being natural fragments of nucleic acids, oligonucleotides may act as classical com-petitive inhibitors of enzymes dealing with nucleic acids. The potential ability of oligo-nucleotides to interact with different targets in cells has been discussed recently (Hélène and Toulmé, 1990).

The most thorough studies of the oligo-nucleotide effect on viral multiplication were done with HIV-1 (Agrawal, 1989a; Zamecnik et al., 1986; Matsukura et al., 1987; Goodchild et al., 1988b,c; Sarin et al., 1988). It was clearly shown that the inhibition of virus multipli-cation involves two different mechanisms. A specific inhibition was apparently achieved due to the complementary interaction of the oligonucleotides with viral RNA (Agrawal et

al., 1989b). Nonspecific inhibition was ex-hibited by different analogs including those that are known to form no stable comple-mentary complexes (Agrawal et al., 1988). Homodeoxyribooligonucleotides and their derivatives appeared to be efficient inhibitors of virus production (Agrawal et al., 1988). Phosphorothioate oligonucleotides are po-tent sequence nonspecific inhibitors of HIV-1 multiplication (Stein et al., 1989; Cheng et al., 1991; Iyer et al., 1990). Their effect is dose and chain-length dependent. Among the oligonucleotide phosphorothioates tested, the oligocytidylate analog appeared to be the most efficient (Matsukura et al., 1987). On the other hand, among phosphorothioate analogs of homooligo(2'-O-methyl)ribonucleotides, the most efficient appeared to be inosine deriv-atives (Shibahara et al., 1989). These results demonstrate that the inhibition is not related to the specific complementary interaction but rather depends on some other features of the analog's structure. Therefore the nonspecific effect was attributed to the inhibition of viral polymerase (Matsukura et al., 1987, 1989), which was confirmed by the fact that the phosphorothioate analogs inhibited reverse transcriptases of lentiviruses and some other viruses (Cheng et al., 1991). The ability of oligonucleotide derivatives to inhibit viral polymerases and virus multiplication was shown in experiments with influenza virus (Vlassov et al., 1991c).

The possibility that oligonucleotides may interfere with the virus–cell interaction was suggested by results of experiments with HIV-1. Investigation of the antiviral effect of oli-gonucleotides introduced at different times prior to and postinfection showed that the nonspecific action is much more pro-nounced when the compounds are intro-duced together with the virus (Agrawal et al., 1989b).

OLIGONUCLEOTIDE DERIVATIVES IN ORGANISMS: PHARMACOLOGICAL CONSIDERATIONS

Investigations of pharmacological aspects of antisense techniques are at a very early stage. The main problems to be answered have been recently formulated by Zon (1989). Although oligonucleotides can be produced by the leading companies in amounts of hundreds of grams, their cost remains very high and seems to be keeping back detailed pharmacological studies and launching of therapeutic tests.

The design of oligonucleotide derivatives is developing in two directions. One is the attachment to natural oligonucleotides of special groups, imparting the properties needed to achieve high activity. These are the groups facilitating the uptake of oligonucleotides by cells and specific targeting groups: reactive, complex-stabilizing groups and groups protecting oligonucleotides against nucleases. Using this approach and avoiding severe damage to the oligonucleotide structure, researchers are trying to take advantage of the apparent low toxicity of natural oligonucleotides and a possibility to use natural mechanisms, such as cellular uptake and potentiation of the oligonucleotides effects by ribonuclease H. Pharmacological predictions are favorable for these compounds. Thus, an oligonucleotide−cholesterol conjugate can hardly produce any harmful effects because it consists of natural components. Another approach consists in designing artificial analogs of oligonucleotides in which only the key elements, necessary for complementary interactions, are preserved and functions are subject to modification or replaced by other groups. Indeed, in this way it is possible to develop artificial molecules that will be stable in the organism and capable of entering cells easily and affecting the target efficiently. The problem is that pharmacological properties of the compounds and the biological effects they will produce can hardly be foreseen. These compounds will be the objects of the most serious pharmacological investigation before they will find their therapeutic applications.

Fate of Oligonucleotide Derivatives in Organisms

We investigated the fate of the $5'$-^{32}P-labeled oligonucleotides bearing a benzylamine residue at the $5'$-phosphate in mice. This group is similar to the $ClRCH_2NH$ group used in reactive oligonucleotide derivatives, and it was introduced in order to protect oligonucleotides from the immediate dephosphorylation in the organism leading to the labeling of the cellular nucleic acids due to phosphate reutilization. The products of degradation retained the label at the $5'$-end and could be easily identified by electrophoresis and quantitated.

The oligonucleotides were injected into mice and their distribution among animal organs, excretion, and degradation were monitored (Bazanova et al., 1991). It was found that independently of the administration route, the oligonucleotide derivatives reached animal organs in a few minutes and rapidly distributed among them (Fig. 5). The label accumulation was more efficient in excreting organs and reticuloendothelial tissue: liver, spleen, and kidneys. The lowest concentration of the compounds was registered in brain. One and one-half hours after intravenous or intraperitoneal injections, the concentration decreased in kidneys, pancreas, spleen, blood, muscles, and somewhat increased in liver. When administered subcutaneously, the compound concentrated more in spleen and muscles. During the next 24 hr the concentration of the radiolabeled material decreased by a factor of 2 in kidneys and in liver and remained at the same level in other organs. Since, by that time, most of the oligonucleotide has been degraded, the later changes in the radioactivity distribution pattern were apparently related to the reutilization of the released inorganic ^{32}P.

A significant amount of the oligonucleotide and degradation products is excreted with urine: 30 and 50% 4 and 24 hr postinjection, respectively.

To investigate oligonucleotide degradation, tissue samples were phenol extracted and the nucleotide material of the aqueous phase was analyzed by electrophoresis. Analysis of samples of mouse tissues showed that 1 hr after injection the degradation of the oligonucleotide was considerable, although in all samples some intact compounds still remained. It should be noted that during the

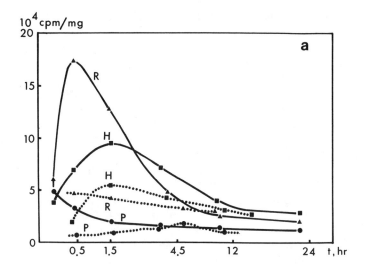

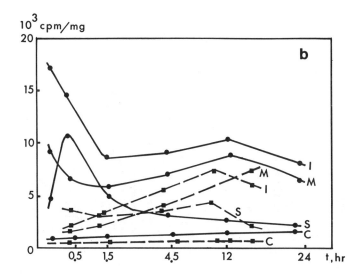

Fig. 5. Pharmacokinetics of ^{32}P-labeled oligonu-cleotide derivative in mice (a, b). Two nano-mols of 5′-end-labeled hexadecanucleotide pTGACCCTCTTCCCATC bearing a phosphatase-protective benzylamide residue at the 5′-end was injected intraperitoneally (solid lines) and sub-cutaneously (dotted and dashed lines) in mice. Specific radioactivities of tissues were deter-mined. R, Kidney; H, liver; P, pancreas; I, intestine; M, muscle; S, blood; C, brain.

first hour reutilization of the radioactive phosphate was negligible, based on separate experiments when the animals were injected with radioactive inorganic phosphate. Therefore the material labeled due to the phosphate reutilization did not interfere with the analysis performed during the first hour post-administration. Degradation was faster in the organs of excreting and reticuloendothelial systems as compared to blood. The data shown in Figure 6 demonstrate that half-lives of oligonucleotides vary in organs: from 15 min or less in liver to 30 min in blood and pancreas, apparently due to the functional differences of the cells.

One may expect that the rate of oligonucleotide degradation could depend on the doses applied. However, when the oligonucleotides were injected into 30 g mice in the amounts of 0.15, 1.5, 15, and 150 nmol (1 µg–1 mg) no noticeable difference in the degradation rate was observed. This means that a dose as high as 1 mg per mouse does not exhaust the capacity of the nucleic acid-degrading enzymatic system.

Experiments with similar oligonucleotide derivatives, containing an additional modification—an amino group at the 3'-terminus, have proved that this modification does not protect oligonucleotides from degradation.

Phosphorothioate oligonucleotide analogs were much more stable in mice. Thirty minutes postinjection, the derivative was practically undegraded in blood and pancreas; in most other organs the degradation rate was also diminished by a factor of 1.5–2 (Fig. 6). The exception was spleen, where degradation of oligonucleotides of both types oc-

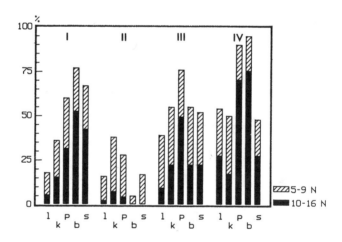

Fig. 6. Degradation of an oligonucleotide and a phosphorothioate analog in mice. Two nanomoles of 5'-end-labeled hexadecanucleotide pTGACCCTCTTCCCATC and its phosphorothioate analog bearing phosphatase-protective benzylamide residue at the 5'-end were injected intraperitoneally and subcutaneously in mice. At specified times postinjection radioactive material from different tissues was analyzed by gel electrophoresis and degradation products quantitated. Ordinate, content of oligonucleotide fractions: 10– 16 N, 10- to 16-mer fraction, and 5–9 N, 5- to 9 mer-fraction, percent of total radioactive oligonucleotide material in the gel. I and II, samples of tissues 30 and 90 min post-ip injection of BzNH$_2$(pN)$_{16}$, respectively; III, samples of tissues 90 min post-sc injection of BzNH$_2$(pN)$_{16}$; IV, samples of tissues 30 min post-ip injection of the derivative of phosphorothioate oligonucleotide. l, Liver; k, kidney; p, pancreas, b, blood; s, spleen.

curred with similar rates. The compound was still present in blood 1.5 hr postinjection, mainly in an undegraded form; in other organs 20–50% of the intact compound was found.

In another set of experiments, mice were injected with the ^{32}P-labeled reactive oligonucleotide derivative, $ClRCH_2NHp(pT)_{10}$. Distribution of the compound among the organs and the degradative process were found to be similar to those observed for the above mentioned unreactive analog. Electrophoretic analysis of the radiolabeled material extracted from the tissues has revealed a high-molecular-weight material, apparently alkylated biopolymers. The content of the labeled biopolymers increased with time postinjection. Figure 7 shows the time course of oligonucleotide degradation and alkylation. To elucidate the role of some oligonucleotide modifications on their fate in an organism, we injected the animals with ^{32}P-labeled derivatives of octathymidylate carrying a $ClRCH_2NH$ group at the 5′-phosphate (compound 1). One of the derivatives carried a 3′-end phenazinium group (compound 2), the second one a cholesterol group (compound 3), and the third one had alternating phosphodiester and methylphosphonate linkages (compound 4). (The syntheses of the compounds of this type are described in the chapter by Knorre and Zarytova, this volume). The distribution among animal organs was similar for all the derivatives; however, a 3-fold greater amount of compounds 2–4 was bound to blood cells, as compared to the parent derivative 1 (Fig. 8). Electrophoretic analysis, performed 1 hr postinjection, revealed that the modifications provided significant protection of the oligonucleotide moiety from degradation. To investigate the possibility of influencing the distribution of oligonucleotide derivatives in an organism by using membrane carriers, compound 1 was incorporated into multilamellar liposomes and into Sendai virus envelopes. The preparations were injected into mice, and it was found that this way of administration resulted in an enhanced delivery of the substance into lymphatic nodules and spleen: the concentration

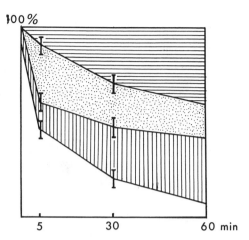

Fig. 7. Time course of degradation of a ^{32}P-labeled alkylating octathymidylate in mouse blood stream. $ClRCH_2NH(pT)_8$ (40 μg) was injected ip in mice. Oligonucleotide degradation products in blood samples were resolved by electrophoresis and the radioactivity was quantitated. Open area, undamaged compound; vertical bars, high-molecular-weight alklyated products; dotted area, degraded oligonucleotides; horizontal bars, inorganic phosphate.

of the compounds was twice as high as that observed with the free compound 1.

Methylphosphonate oligonucleotide analogs injected into the tail vein of a mouse distributed to all organs and tissues of the animal with the exception of the brain (Miller and Ts'O, 1987).

Effects Caused by Oligonucleotides in Organisms

Keeping in mind the intended therapeutic use of oligonucleotide derivatives, one has to consider their potential effects on the organism at the level of toxicity, immunogenicity, and mutagenic potential. A few studies on these subjects have been reported. Preliminary acute toxicity studies in mice and rats with natural oligonucleotides, and methylphosphonate, phosphomorpholidate, and phosphorothioate analogs (Agrawal et al., 1988; 1989a, b; Goodchild et al., 1988; Sarin et al., 1988) revealed that doses below 40 mg/kg body weight are nontoxic for animals. We tested acute toxicity of deoxyribooligo-

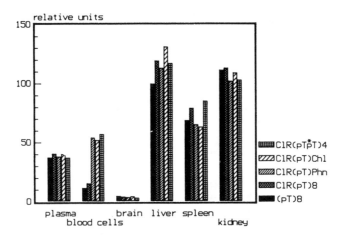

Fig. 8. Distribution of oligonucleotide derivatives in mice. Three nanomoles of ^{32}P-labeled oligonucleotide derivatives was injected ip in mice in 200 μl of physiologic salt solution. One hour postinjection, samples of tissues were analyzed for specific radioactivity. (pT)$_8$, an octathymidylate; ClR(pT)$_8$, alkylating derivative, an octathymidylate conjugated to the ClRCH$_2$HN- group;

ClR(pT)Phn, similar octathymidylate derivative bearing phenazinium group at the 3′-end; ClR(pT)Chl, the alkylating octathymidylate derivative bearning a cholesterol group at the 3′-end; ClR(pTpT)$_4$, the alkylating derivative of an octathymidylate with alternating methylphosphonate linkages.

nucleotides and the derivatives conjugated to the alkylating group ClRCH$_2$NH and observed similar relatively low toxicity for both the compounds.

One may expect mutagenicity of oligonucleotides due to direct hybridization with genomic DNA. First tests did not reveal any mutagenic effect of the normal oligodeoxyribonucleotides and the phosphorothioate analogs in a bacterial test system (Zon, 1989). In the mammalian cell culture assay some mutagenic effect was observed. It should be taken into account that in the case of the oligonucleotide derivatives two additional mechanisms for the mutagenic action of the compounds appear. Reactive oligonucleotide derivatives may introduce specific and nonspecific lesions in genomic DNA that may give rise to mutations. Products of the derivatives' metabolism, abnormal and substituted nucleosides and bases, can be incorporated into DNA or interfere with functions of cellular polymerases, which may also lead to the mutation process.

Normal nucleic acids are known to be poor

immunogens and one can hardly expect any problem with normal oligonucleotides. However, it may not be the case with the oligonucleotide derivatives. Thus, the oligonucleotides conjugated to proteins, e.g., serum albumin, can cause a specific immune response. The antibodies are formed against the junction structure of the conjugate and the oligonucleotide moiety (Brossalina et al., 1988). Oligonucleotide derivatives with reactive electrophilic groups may bind to albumin in the bloodstream and react with the protein yielding the immunogenic conjugates. The oligonucleotide analogs may produce a more intense immune response.

Animal Tests of the Antiviral Activity of Oligonucleotide Derivatives

Few in vivo studies have been reported. Miller and Ts'O (1987) tested the effect of a methylphosphonate oligonucleotide, targeted to early mRNA of the herpes simplex virus (HSV), on disease development in mice. The oligonucleotide was found to suppress viral multiplication in tissue culture. Applied repeatedly, in the form of cream, to the infected ears of mice, the oligonucleotide did not allow formation of specific lesions.

In another study (Kaji, 1987), mice infected with HSV were intracerebrally injected repeatedly (once together with the virus and then daily in the course of 3 days) with a 20-meric oligodeoxyribonucleotide complementary to the immediate early mRNA of the virus. Sixty-eight hours postinfection, the mortality of the untreated infected group was 5/19, unlike that of 0/9 and 0/10 in the groups that received 0.22 or 2.2 μg of the oligonucleotide per animal. Topical application of oligonucleotides gave positive results. Oligonucleotide solution was dropped into corneal membrane of mice at the place where the infection was introduced into injuries produced by scratching. The treatment reduced pathologic symptoms.

We investigated the effect of oligonucleotide derivatives on mice infected with tick-borne encephalitis virus (TBEV) (Pogodina et al., 1989). Oligonucleotides, pT(pCpT)$_6$ and pTGACCCTCTTCCCATC, were complementary to sequences within the ns5 gene (nucleotides 8967–8979) and within the ns3 gene (nucleotides 5317–5332). The oligonucleotides were used either unmodified or derivatized with the alkylating group ClRCH$_2$NH-. The experiment was performed according to the following scheme. The mice were infected by the intraperitoneal injection of a 200 μl suspension of the virus (strain Sofjin, 10^5 LD$_{50}$). In one group, the animals were repeatedly intraperitoneally injected with oligonucleotide derivatives (5 nmol per shot) 3, 6, 9, 18, 26, 34, 42, 50, 58, and 64 hr postinfection. All the control-infected animals who received injections of physiologic salt solution died by day 15 of the experiment (Fig. 9). Death was caused by encephalitis as revealed by histologic investigation of the central nervous systems of the animals.

TABLE 2. Inhibition of Tick-Borne Encephalitis Virus Multiplication in Cell Culture by Oligonucleotide Derivatives[a]

Experiment No.	Derivative type	Concentration of derivative (μM)	Virus titer
I	O	0	8.5
	ClRCH$_2$NHpT(pCpT)$_6$	100	3
	(OH)RCH$_2$NHpT(pCpT)$_6$	100	3.5
	ClRCH$_2$NH(pT)$_{16}$	100	8
	(OH)RCH$_2$NH(pT)$_{16}$	100	7
	ClRCH$_2$NH(pQ)$_{16}$	100	8.5
	(OH)RCH$_2$NH(pQ)$_{16}$	100	8.5
II	O	0	5.5
	ClRCH$_2$NHpT(pCpT)$_6$	0.1	3.5
	ClRCH$_2$NHpT(pCpT)$_6$	1	2
	ClRCH$_2$NHpT(pCpT)$_6$	10	1.5
	ClRCH$_2$NHpT(pCpT)$_6$	100	2
III	O	0	7
	RSVE, empty	0	5.5
	RSVE, ClRCH$_2$NHpT(pCpT)$_6$	0.1	3
	ClRCH$_2$NHpT(pCpT)$_6$	0.1	5

[a]Oligonucleotide derivatives were introduced into the cell culture wells 1 hr post infection. After 48 hr incubation culture medium was titrated by the endpoint dilution method (Pogodina et al., 1988). pT(pCpT)$_6$, tridecanucleotide complementary to the sequence within gene ns5 tick-borne encephalitis virus RNA. (pQ)$_{16}$, pTTCCTTCCCGTCTTCC-noncomplementary oligonucleotide.

Oligonucleotide injections caused a 1–2 day delay in the development of the disease, and some animals survived. It turned out that even the doses of 0.1 nmol per shot caused some protective effect. However, it was found that noncomplementary control oligonucleotides also interfered with disease development. The survivors demonstrated complete resistance to the TBEV infection (10^5 LD_{50} of the virus, intraperitoneally). Analysis of serum of the survivors 1 month after the experiment revealed the virus-neutralizing antibodies (log of the neutralizing index 3–4). It was concluded that the animals were protected by the oligonucleotides. They developed specific immunity and resistance toward repeated infection.

SUMMARY

Experimental studies demonstrate that normal phosphodiester oligonucleotides and their conjugates are taken up by animal cells. They enter the cytoplasm and nucleus and

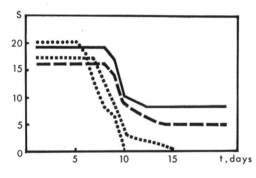

Fig. 9. Effect of the antisense oligonucleotide derivatives on the development of the disease in mice infected with the tick-borne encephalitis virus. The animals were infected with the virus and then received repeated ip injections of the oligonucleotide derivatives 5 nmol (about 30 μg) each time 3, 6, 9, 18, 26, 34, 42, 50, 58, and 66 hr after infection. Solid and broken lines: infected mice received alkylating derivative of oligonucleotides pT(pCpT)$_6$ and pTGACCCTCTTCCCATC complementary to the sequences of the genes ns5 (nucleotides 8967–8979) and ns3 (nucleotides 5217–5332) of the virus, respectively. Control groups of mice, dotted and dashed lines, received physiologic salt solution and noncomplementary oligonucleotide derivative (Pogodina et al., 1989).

affect the functions of target nucleic acids. The uptake proceeds by the endocytosis mechanism. Oligonucleotides interact with membrane proteins that seem to be specific receptors binding nucleic acids. Both the efficiency of oligonucleotide delivery into cells and the stability of oligonucleotides toward nucleolytic enzymes can be improved considerably either by using membrane carriers (liposomes, Sendai virus envelopes, erythrocyte ghosts) or by conjugation of the oligonucleotides to molecules stimulating binding to cells and endocytosis, to lipophilic groups (cholesterol) or to polycations [poly(L-lysine)]. Oligonucleotides arrest translation of target mRNAs by triggering degradation of messengers by ribonuclease H and by interfering with interaction of initiation factors and ribosomes with the mRNAs. Equipment of the oligonucleotides with reactive groups considerably improves their antimessenger activity. They interact with some complementary sequences of cellular DNA, thus suggesting the existence of some unwound DNA sites in chromatin. Oligonucleotides display low toxicity. In animals, they distribute rapidly among all organs, showing the lowest accumulation in the brain, and are excreted in the urine. Degradation halftime varies from 15 to 30 min in different organs. Oligonucleotide derivatives suppress multiplication of viruses in tissue cultures and display protective effect in virus-infected animals. Antiviral activity of some nonspecific oligonucleotides suggests that they can interfere with viral multiplication by several mechanisms.

REFERENCES

Abramova TV, Blinov VM, Vlassov VV, Gorn VV, Zarytova VF, Ivanova EM, Konevetz DA, Plyasunova OA, Pokrovsky AG, Sandakhchiev LS, Svinarchuk FP, Starostin VP, Chaplygina SR (1990): Inhibition of human immunodeficiency virus reproduction in cell culture by antisense oligonucleotide derivatives. Dokl. Akad. Nauk SSSR 312:1259–1262.

Abramova TV, Vlassov VV, Zarytova VF, Ivanova EM, Kuligina EA, Ryte AS (1991): Modification of terminal nucleotide residues in oligonucleotides increases their stability in cell culture. Mol. Biol. 25:624–632.

Agrawal S, Goodchild J, Civeira MP, Thornton AH, Sarin PS, Zamecnik PC (1988): Oligodeoxynucleoside phosphoramidates and phosphorothioates as inhibitors of human immunodeficiency virus. Proc. Natl. Acad. Sci. U.S.A. 85:7079–7083.

Agrawal S, Goodchild J, Civeira M, Sarin PS, Zamecnik PC (1989a): Phosphoramidate, phosphorothioate, and methylphosphonate analogs of oligodeoxynucleotide: Inhibitors of replication of human immunodeficiency virus. Nucleosides and Nucleotides 8:819–823.

Agrawal S, Ikeuchi T, Sun D, Sarin PS, Konopka A, Maizel J, Zamecnik PC (1989b): Inhibition of human immunodeficiency virus n early infected and chronically infected cells by antisense oligodeoxynucleotides and their phosphorothioate analogues. Proc. Natl. Acad. Sci. U.S.A. 86:7790–7794.

Babkin IV, Butorin AS, Ivanova EM, Ryte AS (1988): Study of chemical conversions of radioactive 4-(N-2-cloroethyl-N-methylamino)benzyl-5' [^{32}P]phosphoramidates of oligodeoxyribonucleotides in vivo. Biokhimiya 53:384–393.

Bazanova OM, Vlassov VV, Zarytova VF, Ivanova EM, Kuligina EA, Yakubov LA, Abdukayumov MN, Karamyshev VN, Zon G (1991): Oligonucleotide derivatives in an organism: distribution among organs, rates of release and degradation. Nucleosides and Nucleotides 10:523–526.

Belyaev ND, Vlassov VV, Kobetz ND, Ivanova EM, Yakubov LA (1986): Complementary addressed modification of DNA of metaphase chromosomes and interphase chromatin. Dokl. Akad. Nauk SSSR 291:234–236.

Bennett RM, Gabor GT, Merritt MJ (1985): DNA binding to human leukocytes: Evidence for a receptor-mediated association, internalization and degradation of DNA. J. Clin. Invest. 76:2182–2190.

Birg F, Praseuth D, Zerial A, Thuong NT, Asseline U, Le Doan T, Hélène C (1990): Inhibition of Simian virus 40 DNA replication in CV-1 cells by an oligodeoxynucleotide covalently linked to an intercalating agent. Nucl. Acids Res. 18:2901–2908.

Boutorin AS, Vlassov VV, Grachev MA, Zaychikov EF, Konevetz DA, Kutyavin IV, Tsarev IG (1987): Addressed modification of a DNA promoter region in the complex with RNA polymerase alkylating oligonucleotide derivatives. Bioorg. Khim. 13:845:848.

Boutorin AS, Guskova LV, Ivanova EM, Kobetz ND, Zarytova VP, Ryte AS, Yurchenko LV, Vlassov VV (1989): Synthesis of alkylating oligonucleotide derivatives containing cholesterol or phenazinium residues at their 3'-terminus and their interaction within mammalian cells. FEBS Lett. 254:129–132.

Brossalina EB, Vlassov VV, Ivanova EM (1988): Antibodies to specific nucleotide sequences. Investigation with a model oligonucleotide pApCpApC. Biokhimiya 53:18–25.

Caruthers MH (1989): Synthesis of oligonucleotides and oligonucleotide analogs. In Cohen JS (ed): Oligodeoxynucleotides Antisense Inhibitors of Gene Expression. Boca Raton, FL: CRC Press, pp. 7–24.

Cazenave C, Chevrier M, Thuong NT, Hélène C (1987): Rate of degradation of [α]- and [β]-oligodeoxynucleotides in Xenopus oocytes. Implications for antimessenger strategies. Nucl. Acids Res. 15:10507–10521.

Cazenave C, Stein CA, Loreau N, Thuong NT, Neckers LM, Subasinghe C, Hélène C, Cohen JS, Toulmé J-J (1989): Comparative inhibition of rabbit globin mRNA translation by modified antisense oligodeoxynucleotides. Nucl. Acids Res. 17:4255–4273.

Ceruzzi M, Draper K (1989): The intracellular and extracellular fate of oligodeoxyribonucleotides in tissue culture systems. Nucleosides and Nucleotides 8:815–818.

Cheng Y, Gao W, Han F (1991): Phosphorothioate oligonucleotides as potential antiviral compounds against human immunodeficiency virus and herpes viruses. Nucleosides and Nucleotides 10:155–166.

Cohen JS (1989): Oligonucleotides Antisense Inhibitors of Gene Expression. Boca Raton, FL: CRC Press.

Cooney M, Czernuszewicz G, Postel EH, Flint SJ, Hogan ME (1988): Site-specific oligonucleotide binding represses transcription of the human c-myc gene in vitro. Science 241:456–459.

Dash P, Lotan I, Knapp M, Kandel ER, Goelet P (1987): Selective elimination of mRNAs in vivo: Complementary oligonucleotides promote RNA degradation by an RNAse H-like activity. Proc. Natl. Acad. Sci. U.S.A. 84:7896–7900.

Degols G, Leonetti J-P, Gagnor C, Lemaitre M, Lebleu B (1989): Antiviral activity and possible mechanisms of action of oligonucleotides-poly(L-lysine) conjugates targeted to vesicular stomatitis virus mRNA and genomic RNA. Nucl. Acids Res. 17:9341–9350.

Degols G, Leonetti J-P, Lebleu B (1991): Antisense oligonucleotides as pharmacological modulators of gene expression. In Juliano RL (ed): Targeted Drug Delivery. Handbook of Experimental Pharmacology. Berlin: Springer-Verlag, in press.

Dervan PB (1989): Oligonucleotide recognition of double-helical DNA by triple-helix formation. In Cohen JS (ed): Oligonucleotides Antisense Inhibitors of Gene Expression. Boca Raton, FL: CRC Press, pp 197–210.

Diamantstein T, Blitsten-Willinger E (1978): Specific binding of poly(I)·poly(C) to the membrane of murine B lymphocyte subsets. Eur. J. Immunol. 8:896–899.

Dorsch CA (1981): Binding of single-strand DNA to human platelets. Thrombosis Res. 24:119–129.

Dudding LR, Harington A, Mizrahi V (1990): Endoribonucleolytic cleavage of RNA: Oligodeoxynucleotide hybrids by the ribonuclesase H activity of HIV-1 reverse transcriptase. Biochem. Biophys. Res. Commun. 167:244–250.

Emlen W, Rifai A, Magilavy D, Mannik M (1988): Hepatic binding of DNA is mediated by a receptor on nonparenchymal cells. Am. J. Pathol. 133: 54–60.

Furdon PJ, Dominski Z, Kole R (1989): RNAse H cleavage of RNA hybridized to oligonucleotides containing methylphosphonate, phosphorothioate and phosphodiester bonds. Nucl. Acids Res. 17:9193–9204.

Goodchild J (1989): Inhibition of gene expression by oligonucleotides. In Cohen JS (ed): Oligodeoxynucleotides Antisense Inhibitors of Gene Expression. Boca Raton, FL: CRC Press, pp 53–77.

Goodchild J (1990): Conjugates of oligonucleotides and modified oligonucleotides: A review of their synthesis and properties. Bioconjugate Chem. 1:165–187.

Goodchild J, Letsinger RL, Sarin PS, Zamecnik M, Zamecnik PC (1988a): Inhibition of replication and expression of HIV-1 in tissue culture by oligodeoxynucleotide hybridization competition. In Bolognesi D (ed): Human Retroviruses, Cancer and AIDS: Approaches to Prevention and Therapy. New York: Alan R. Liss, pp 423–438.

Goodchild J, Carroll E III, Greenberg JR (1988b): Inhibition of rabbit β-globin synthesis by complementary oligonucleotides: Identification of mRNA sites sensitive to inhibition. Arch. Biochem. Biophys. 263:401–409.

Goodchild J, Agrawal S, Civeira MP, Sarin PS, Sun D, Zamecnik PC (1988c): Inhibition of human immunodeficiency virus replication by antisense oligodeoxynucleotides. Proc. Natl. Acad. Sci. U.S.A. 85:5507–5511.

Hélène C, Toulmé J-J (1989): Control of gene expression by oligodeoxynucleotides covalently linked to intercalating agents and nucleic acid-cleaving reagents. In Cohen JS (ed): Oligodeoxynucleotides Antisense Inhibitors of Gene Expression. Boca Raton, FL: CRC Press, pp 137–172.

Hélène C, Toulmé J-J (1990): Specific regulation of gene expression by antisense, sense and antigene nucleic acids. Biochim. Biophys. Acta 1049:99–125.

Imbach J-L, Rayner B, Morvan F (1989): Sugar-modified oligonucleotides: Synthesis, physicochemical and biological properties. Nucleosides and Nucleotides 8:627–648.

Iyer RP, Uznanski B, Boal J, Storm C, Egan W, Matsukura M, Broder S, Zon G, Wilk A, Koziolkiewicz M, Stec WJ (1990): Abasic oligodeoxyribonucleoside phosphorothioates: Synthesis and evaluation as anti-HIV-1 agents. Nucl. Acids Res. 18:2855–2859.

Jayaraman K, McParland K, Miller P, Ts'o POP (1981): Selective inhibition of *Escherichia coli* protein synthesis and growth by nonionic oligonucleotides complementary to the 3' end of 16S rRNA. Proc. Natl. Acad. Sci. U.S.A. 78:1537–1541.

Kabanov AV, Vinogradov SV, Ovcharenko AV, Krivonos AV, Melik-Nubarov NS, Kiselev VI, Severin ES (1990): A new class of antivirals: Antisense oligonucleotides combined with a hydrophobic substituent effectively inhibit influenza virus reproduction and synthesis of virus-specific proteins in MDCK cells. FEBS Lett. 259:327–330.

Kaji A (1987): Method for inhibiting propagation of virus and anti-viral agent. U.S. Patent No. 4,689,320.

Karpova GG, Knorre DG, Ryte AS, Stephanovich LE (1980): Selective alkylation of poly(A) tracts of RNA inside the cell with the derivative of ethyl este of oligothymidylate bearing 2-chloroethylamino group. FEBS Lett. 122:21–24.

Kean JM, Murakami A, Blake KR, Cushman CD, Miller PS (1988): Photochemical crosslinking of psoralen-derivatized oligonucleoside methylphosphonates to rabbit globin messenger RNA. Biochemistry 27:9113–9121.

Knorre DG, Vlassov VV, Zarytova VF, Lebedev AV (1988): Reactive oligonucleotide derivatives as tools for site specific modifications of biopolymers. Sov. Sci. Rev. B Chem. 12:269–341.

Knorre DG, Vlassov VV, Zarytova VF (1989): Oligonucleotides linked to reactive groups. In Cohen JS (ed): Oligodeoxyribonucleotides: Antisense Inhibitors of Gene Expression. London: Macmillan pp 173–196.

Koshkin AA, Lebedev AV, Ryte AS, Vlassov VV (1991): Ethidium oligodeoxynucleotide derivatives. Stability against cellular nuclease hydrolysis and photochemical modification of cellular proteins in the living cells. Nucleosides and Nucleotides 10:541–542.

Lemaitre M, Bayard B, Lebleu B (1987): Specific antiviral activity of a poly(L-lysine)-conjugated oligodeoxyribonucleotide sequence complementary to vesicular stomatitis virus N protein mRNA initiation site. Proc. Natl. Acad. Sci. U.S.A. 84:648–652.

Leonetti JP, Degols G, Lebleu B (1990a): Biological activity of oligonucleotide-poly(L-lysine) conjugates: Mechanism of cell uptake. Bioconjugate Chem. 1:149–153.

Leonetti JP, Machy P, Degols G, Lebleu B, Leserman L (1990b): Antibody-targeted liposomes containing oligodeoxyribonucleotide sequences complementary to viral RNA selectively inhibit viral rep-lication. Proc. Natl. Acad. Sci. U.S.A. 87:2448–2451.

Letsinger RL, Zhang G, Sun DK, Ikeuchi T, Sarin PS (1989): Cholesteryl-conjugated oligonucleotides: Synthesis, properties, and activity as inhibitors of replication of human immunodeficiency virus in cell culture. Proc. Natl. Acad. Sci. U.S.A. 86:6553–6556.

Loke SL, Stein CA, Zhang XH, Avigan M, Cohen J, Neckers LM (1988): Delivery of c-myc antisense phosphorothioate oligodeoxynucleotides to he-matopoietic cells in culture by liposome fusion: Specific reduction in c-myc protein expression correlates with inhibition of cell growth and DNA synthesis. Curr. Top Microbiol. Immunol. 141:282–289.

Loke SL, Stein CA, Zhang XH, Mori K, Nakanishi M, Subasinghe C, Cohen JS, Neckers LM (1989): Characterisation of oligonucleotide transport into living cells. Proc. Natl. Acad. Sci. U.S.A. 86:3474–3478.

Maher LJ, Dolnick BJ (1988): Comparative hybrid arrest by tandem antisense oligodeoxyribonu-cleotides or oligodeoxyribonucleoside methyl-phosphonates in a cell free system. Nucl. Acids Res. 16:3341–3358.

Matsukura M, Shinozuka K, Zon G, Mitsuya H, Reitz M, Cohen JS, Broder S (1987): Phosphorothioate analogs of oligodeoxynucleotides: Inhibitors of replication and cytopathic effects of human im-munodeficiency virus. Proc. Natl. Acad. Sci. U.S.A. 84:7706–7710.

Matsukura M, Zon G, Shinozuka K, Robert-Guroff M, Shimada T, Stein CA, Mitsuya H, Wong-Staal F, Cohen JS, Broder S (1989): Regulation of viral expression of human immunodeficiency virus in vitro by an antisense phosphorothioate oligo-deoxynucleotide against rev (art/trs) in chroni-cally infected cells. Proc. Natl. Acad. Sci. U.S.A. 86:4244–4248.

Miller PS (1989): Non-ionic antisense oligonucleo-tides. In Cohen JS (ed): Oligodeoxynucleotides Antisense Inhibitors of Gene Expression. Boca Ra-ton, FL: CRC Press, pp 79–96.

Miller PS, Ts'o POP (1987): A new approach to chemotherapy based on molecular biology and nucleic acid chemistry: Matagen (masking tape for gene expression). Anti-Cancer Drug Design 2:117–128.

Mori K, Boiziau C, Casenave C, Matsukura M, Suba-singhe C, Cohen JS, Broder S, Toulmé JJ, Stein CA (1989): Phosphoroselenoate oligonucleotides: Synthesis, physico-chemical characterization, anti-sense inhibitory properties and anti-HIV activity. Nucl. Acids Res. 17:8207–8219.

Neckers LM (1989): Antisense oligodeoxynucleo-tides as a tool for studying cell regulation: Mech-anism of uptake and application to the study of

oncogene function. In Cohen JS (ed): Oligonu-cleotides Antisense Inhibitors of Gene Expression. Boca Raton, FL: CRC Press, pp 211–232.

Ohlbaum A, Csuzi S, Antoni F (1979): Binding of exogenous DNA by human lymphocytes and by their isolated plasma membranes. Acta Biochim. Biophis. Acad. Sci. Hung. 14:165–176.

Paoletti C (1988): Anti-sense oligonucleotides as po-tential antitumor agents: Prospective views and preliminary results. Anti-Cancer Drug Design 2:325–331.

Pogodina VV, Frolova TV, Abramova TV, Vlassov VV, Ivanova EM, Kutyavin IV, Pletnev AG, Yakubov LA (1988): Derivatives of oligonucleotides comple-mentary to viral RNA inhibit tick-borne enceph-alitis virus multiplication in cell culture. Dok. Akad. Nauk SSSR 301:1257–1260.

Pogodina VV, Frolova TV, Frolova MP, Abramova TV, Vlassov VV, Knorre DG, Pletnev AG, Yakubova LA (1989): Oligonucleotides complementary to tick-borne encephalitis virus RNA inhibit viral infec-tion in mice. Dok. Akad. Nauk SSSR 308:237–240.

Quartin RS, Brakel CL, Wetmur JG (1989): Number and distribution of methylphosphonate linkages in oligodeoxynucleotides affect exo- and endo-nuclease sensitivity and ability to form RNAse H substrates. Nucl. Acids Res. 17:7253–7262.

Sankar S, Cheah K-C, Porter AG (1989): Antisense oligonucleotide inhibition of encephalomyocar-ditis virus RNA translation. Eur. J. Biochem. 184:39–45.

Sarin PS, Agrawal S, Civeira MP, Goodchild J, Ikeuchi T, Zamecnik PC (1988): Inhibition of acquired immunodeficiency syndrome virus by oligodeox-ynucleotide methylphosphonates. Proc. Natl. Acad. Sci. U.S.A. 85:7448–7451.

Shibahara S, Mukai S, Morisawa H, Nakashima H, Ko-bayashi S, Yamamoto N (1989): Inhibition of hu-man immunodeficiency virus (HIV-1) replication by synthetic oligo-RNA derivatives. Nucl. Acids Res. 17:239–252.

Stein CA, Cohen JS (1988): Oligodeoxynucleotides as inhibitors of gene expression: A review. Cancer Res. 48:2659–2668.

Stein CA, Matsukura M, Subasinghe C, Broder S, Cohen JS (1989): Phosphorothioate oligonucleotides are potent sequence nonspecific inhibitors of de novo infection by HIV. AIDS Res. Human Retroviruses 5:639–646.

Stevenson M, Iverson PL (1989): Inhibition of human immunodeficiency virus type 1-mediated cyto-pathic effects by poly(L-lysine)-conjugated syn-thetic antisense oligodeoxyribonucleotides. J. Gen. Virol. 70:2673–2682.

Thuong NT, Asseline U, Montenay-Garestier T (1989): Oligodeoxynucleotides covalently linked to in-tercalating and reactive substances: Synthesis, characterization and physicochemical studies. In

Cohen JS (ed): Oligodeoxynucleotides Antisense Inhibitors of Gene Expression. Boca Raton, FL: CRC Press, pp 25–52.

Uhlmann E, Peyman A (1990): Antisense oligonucleotides: A new therapeutic principle. Chem. Rev. 90:543–590.

Verspieren P, Cornelissen AWCA, Thuong NT, Hélène C, Toulmé J-J (1987): An acridine-linked oligodeoxynucleotide targeted to the common 5' end of trypanosome mRNAs kills cultured parasites. Gene 61:307–315.

Vlassov·VV, Gorn VV, Kutyavin IV, Yurchenko LV, Sharova NK, Bukrinskaya AG (1984a): Possibility of blocking influenza infection by alkylating derivatives of oligonucleotides. Mol. Genet. Mikrobiol. Virusol. No11:36–41.

Vlassov VV, Godovikov AA, Zarytova VF, Ivanova VF, Knorre DG, Kutyavin IV (1984b): Arrest of immunoglobulin synthesis in MOPC 21 cells by alkylating oligonucleotide derivative complementary to mRNA coding for the light immunoglobulin chain. Dokl. Akad. Nauk SSSR 276:1263–1265.

Vlassov VV, Gorokhova OE, Ivanova EM, Kutyavin IV, Yurchenko LV, Yakubov LA, Abdukayumov MN, Skoblov YS (1986a): Interaction of alkylating oligonucleotide derivatives with mouse fibroblasts. Biopolym. Kletka 2:323–327.

Vlassov VV, Godovikov AA, Kobetz ND, Ryte AS, Yurchenko LV, Bukrinskaya AG (1986b): Nucleotide and oligonucleotide derivatives as enzyme and nucleic acid targeted irreversible inhibitors. Biochemical aspects. In Weber G (ed): Advances in Enzyme Regulation 24. London: Pergamon, pp 301–320.

Vlassov VV, Ivanova EM, Krendelev Y, Kutyavin IV, Ovander MN, Ryte AS, Svinarchuk FP, Yakubov LA (1989): Sendai virus envelopes and erythrocyte ghosts as membrane vehicles for alkylating oligonucleotide derivatives transport into mammalian cells. Biopolym. Kletka 5:52–57.

Vlassov VV, Yurchenko LV (1990): Mechanism of mRNA translation arrest by antisense oligonucleotides. Mol. Biol. 24:1157–1161.

Vlassov VV, Kobetz ND, Chernolovskaya EL, Demidova SG, Borissov RG, Ivanova EM (1990): Sequence-specific chemical modification of chromatin DNA with reactive derivatives of oligonucleotides. Mol. Biol. Rep. 14:11–15.

Vlassov VV, Deeva EA, Nechaeva MN, Rykova EN, Yakubov LA (1991a): Interaction of oligonucleotide derivatives with animal cells. Nucleosides and Nucleotides 10:581–582.

Vlassov VV, Gorn VV, Nomokonova NY, Fokina TN, Yurchenko LV (1991b): Inhibition of the influenza virus M protein mRNA translation in vitro with complementry oligonucleotides. Nucleosides and Nucleotides 10:649–650.

Vlassov VV, Frolova EI, Godovikova TS, Ivanova EM, Koshkin AA, Ledovskikh NB, Nevinsky GA (1991c): Suppression of transcription and translation of the influenza virus RNAs by oligonucleotide derivatives. Nucleosides and Nucleotides 10:645–648.

Vlassov VV, Zarytova VF, Ivanova EM, Krendelev YD, Ovander MN (1991d): Influence of terminal modification of oligonucleotides on their stability in mycoplasmas. Biopolym. Kletka (in press).

Walder RY, Walder JA (1988): Role of RNAse H in hybridarrested translation by antisense oligonucleotides. Proc. Natl. Acad. Sci. U.S.A. 85:5011–5015.

Wickstrom E (1986): Olignucleotide stability in subcellular extracts and culture media. J. Biochem. Biophys. Methods 13:97–102.

Wickstrom EL, Bacon TA, Gonzalez A, Freeman DL, Lyman GH, Wickstrom E (1988): Human promyelocytic leukemia HL-60 cell proliferation and c-myc protein expression are inhibited by an antisense pentadecadeoxynucleotide targeted against c-myc mRNA. Proc. Natl. Acad. Sci. U.S.A. 85:1028–1032.

Yakubov LA, Deeva EA, Zarytova VF, Ivanova EM, Ryte AS, Yurchenko LV, Vlassov VV (1989): Mechanism of oligonucleotide uptake by cells: Involvement of specific receptors? Proc. Natl. Acad. Sci. U.S.A. 86:6454–6458.

Yakubov LA, Savinkova IV, Karamyshev VN, Scherbakov DY, Vlassov VV (1991): Oligonucleotide-binding proteins of Drosophila: Properties and tissue specificity. Biokhimiya, in press.

Zamecnik PC, Stephenson ML (1978): Inhibition of Rous sarcoma virus replication and cell transformation by a specific oligodeoxynucleotide. Proc. Natl. Acad. Sci. U.S.A. 75:280–284.

Zamecnik PC, Goodchild J, Taguchi Y, Sarin PS (1986): Inhibition of replication and expression of human T-cell lymphotropic virus type III in cultured cells by exogeneous synthetic oligonucleotides complementary to viral RNA. Proc. Natl. Acad. Sci. U.S.A. 83:4143–4146.

Zerial A, Thuong NT, Hélène C (1987): Selective inhibition of the cytopathic effect of type A influenza viruses by oligodeoxynucleotides covalently linked to an intercalating agent. Nucl. Acids Res. 15:9909–9919.

Zon G (1988): Oligonucleotide analogues as potential chemotherapeutic agents. Pharmaceut. Res. 5:539:549.

Zon G (1989): Pharmaceutical considerations. In Cohen JS (ed): Oligonucleotides Antisense Inhibitors of Gene Expression. Boca Raton, FL: CRC Press, pp 233–248.

Index